Functional Equations and Inequalities

Solutions and Stability Results

SERIES ON CONCRETE AND APPLICABLE MATHEMATICS

ISSN: 1793-1142

Series Editor: Professor George A. Anastassiou
Department of Mathematical Sciences
University of Memphis
Memphis, TN 38152, USA

Published*

*To view the complete list of the published volumes in the series, please visit:
http://www.worldscientific/series/scaam

Series on Concrete and Applicable Mathematics – Vol. 21

Functional Equations and Inequalities

Solutions and Stability Results

John Michael Rassias

National and Kapodistrian University of Athens, Greece

E Thandapani

RIASM, University of Madras, India

K Ravi

Sacred Heart College, Tirupattur, India

B V Senthil Kumar

C Abdul Hakeem College of Engineering & Technology, India

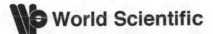

World Scientific

NEW JERSEY · LONDON · SINGAPORE · BEIJING · SHANGHAI · HONG KONG · TAIPEI · CHENNAI · TOKYO

Published by

World Scientific Publishing Co. Pte. Ltd.
5 Toh Tuck Link, Singapore 596224
USA office: 27 Warren Street, Suite 401-402, Hackensack, NJ 07601
UK office: 57 Shelton Street, Covent Garden, London WC2H 9HE

Library of Congress Cataloging-in-Publication Data
Names: Rassias, John Michael.
Title: Functional equations and inequalities : solutions and stability results / by John Michael Rassias
    (National and Kapodistrian University of Athens, Greece) [and three others].
Description: New Jersey : World Scientific, 2017. |
    Series: Series on concrete and applicable mathematics ; volume 21 |
    Includes bibliographical references and index.
Identifiers: LCCN 2016056679| ISBN 9789813147607 (hc : alk. paper) |
    ISBN 9789813149977 (pbk : alk. paper)
Subjects: LCSH: Functional equations. | Inequalities (Mathematics)
Classification: LCC QA431 .F78827 2017 | DDC 515/.72--dc23
LC record available at https://lccn.loc.gov/2016056679

British Library Cataloguing-in-Publication Data
A catalogue record for this book is available from the British Library.

Desk Editors: V. Vishnu Mohan/Kwong Lai Fun

Typeset by Stallion Press
Email: enquiries@stallionpress.com

Printed in Singapore

To all the great mathematicians from the old Greek
ARCHIMEDES (287 BC–212 BC) until the young Indian
RAMANUJAN S. (1887 AD–1920 AD)

Preface

Functional equation is a very interesting and useful area in mathematics. It involves with simple algebraic manipulations, but one can arrive at very interesting and hard solutions. It touches upon almost all parts of contemporary pure and applied mathematics.

The theory of functional equations originated very long back and one can find powerful tools useful to modern mathematics. The theory and methods contribute to the development of other branches of mathematics such as algebra, analysis, and topology, etc. The new approaches and techniques developed in the field of functional equations find a lot of applications to other fields such as Physics, Biology, Economics, Mechanics, Geometry, Statistics, Measure Theory, Algebraic Geometry, Group Theory, Astronomy, Game Theory, Fuzzy Set Theory, Information Theory, Coding Theory, Stochastic Process, etc. Many mathematical facts in different fields have become essential to the foundations of functional equations and inequalities.

In the last five decades or so, we have found that the topic functional equations has gained a lot of momentum and many research papers have been published in different journals. In particular, many interesting results on topics such as general solutions, stability results, Hyers–Ulam stability, Hyers–Ulam–Rassias stability and generalized Hyers–Ulam stability of various types of functional equations in different spaces are investigated. These scattered results that appeared in various journals are not updated regularly in the form of monographs. This motivated us to bring out this book containing the recent results on the above topics in order to fulfill the demand of the scientific community.

This book is an attempt to present the fundamentals of the topic at hand in a pedagogical manner. The book provides sufficient number of solved examples in each chapter. It will motivate the reader to solvo similar types of problems. This book will be useful to the final year undergraduate/graduate stu dents as well as for researchers who are very fond of doing research in functional equations.

The highlights of this book are as follows:

- provides a systematic way of studying functional equations,
- details basic ideas of functional equations,
- includes various methods of solving functional equations,
- exhibits general solutions of various types of functional equations and the proofs of stability results,
- presents the recent results on stability of functional equations in various spaces,
- lists sets of exercises and problems in the last chapter.

We hope that one will enjoy reading this book as the style of presentation goes from simple to complex.

The authors are very grateful to all the mathematicians who are continuously contributing to the growth of functional equations by publishing very worthwhile research papers in reputed international journals from which we extracted some of their resources for writing this book. We gratefully acknowledge each and every one of them. Ultimately, our aim is to take this vast knowledge and the beautiful results of functional equations to the younger generation, budding mathematicians, research scholars, academicians and in a way it should reach more and more people, they in turn should be instrumental in the growth of this field of mathematics.

We are indebted to Prof. George A. Anastassiou for guiding and directing us to bring out this book and Lai Fun Kwong (Ms) and World Scientific for publishing our book in the Series on Concrete and Applicable Mathematics.

J.M. Rassias
E. Thandapani
K. Ravi
B.V. Senthil Kumar

Contents

Chapter 1

Functional Equations and Applications

1.1. Introduction

In this chapter, we present definition of functional equations, classification of functional equations, solutions of functional equations, some well-known functional equations, and illustrate few applications of functional equations in various fields.

1.2. Definition of functional equation

The well-known Hungarian mathematician, Aczél [1], defines functional equation as follows:

"Functional equation is an equation in which both sides contain a finite number of functions, some are known and some are unknown."

Functional equations are classified as follows:

Ordinary functional equation

A functional equation in which all the unknown functions are of one variable is called **ordinary functional equation**.

Example 1.1. The equation $f(x+2) = f(x+1) - f(x-1)$ is an ordinary functional equation.

Partial functional equation

A functional equation in which at least one of the unknown functions is a more-place function is called **partial functional equation**.

Example 1.2. The equation $f(x+y+z) = f(x+y) + f(z)$ is a partial functional equation.

System of functional equations

If the number of functions considered in a set of functional equations is more than one, then the set of functional equations is called as **system of functional equations**.

1

Example 1.3.

(i) $f(x + y) = f(x) + f(y)$, for all $x, y \in \mathbb{R}$;

(ii) $f(\frac{1}{x}) = \frac{f(x)}{x^2}$, for all $x \neq 0$.

Mixed type functional equation

A functional equation derived from two or more combinations of additive, quadratic, cubic, quartic functions is said to be a **mixed type functional equation**.

Example 1.4. The equation

$$f(x + 2y) + f(x - 2y) + 4f(x) = 3[f(x + y) + f(x - y)] + f(2y) - 2f(y)$$

is a mixed type additive-quadratic functional equation.

1.3. Rank of functional equation

The number of independent variables occurring in a functional equation is called the **rank of functional equation**.

Example 1.5. The rank of the functional equation $f(x + y) = f(x) + f(y)$ is 2.

1.4. Solution of functional equation

A solution of a functional equation is a function which satisfies the given equation.

For example, the functions $f(x) = kx$, $f(x) = e^x$, $f(x) = x^k$, and $f(x) = k \log x$, $x > 0$ where k is a constant, are the solutions of the Cauchy functional equations

$$f(x + y) = f(x) + f(y) \quad \text{(additive functional equation)}, \tag{1.1}$$

$$f(x + y) = f(x)f(y) \quad \text{(exponential functional equation)}, \tag{1.2}$$

$$f(xy) = f(x)f(y) \quad \text{(multiplicative functional equation)}, \tag{1.3}$$

and

$$f(xy) = f(x) + f(y) \quad \text{(logarithmic functional equation)}, \tag{1.4}$$

respectively.

1.5. General solution of functional equation

The general solution of a functional equation or a system of functional equations is the totality of particular solutions.

Example 1.6. The functions $f(x) = cx + a + b$, $g(x) = cx + a$, $h(x) = cx + b$ are solutions of the functional equation $f(x + y) = g(x) + h(y)$, where a, b, c are any values.

1.6. Particular solution of functional equation

A function or a set of functions is said to be a **particular solution** of a functional equation or a system of functional equations if it satisfies the functional equation or the system of functional equations.

Example 1.7. The functional equation $f(x + y) = g(x) + h(y)$ has a particular solution $f(x) = 2x + 3$, $g(x) = 2x + 1$, $h(x) = 2x + 2$, respectively.

Note. One can observe that in a particular solution, the function occurs with particular coefficients and constant terms.

1.7. Some well-known functional equations

Usually, the functional equations are given by the name of the mathematician who discovered it or sometimes the functional equations are given names based on the property that is involved in the given functional equation. The following is a list of well-known functional equations:

(i) **Cauchy Functional Equation**

$$f(x + y) = f(x) + f(y).$$

(ii) **Jensen Functional Equation**

$$f\left(\frac{x + y}{2}\right) = \frac{1}{2}[f(x) + f(y)].$$

(iii) **D'Alembert Functional Equation**

$$f(x + y) + f(x - y) = 2f(x)f(y).$$

(iv) **Wilson Functional Equation**

$$g(x + y) + g(x - y) = 2g(x)f(y).$$

(v) **Euler–Lagrange–Rassias Quadratic Functional Equation**

$$f(ax + by) + f(bx - ay) = (a^2 + b^2)[f(x) + f(y)].$$

(vi) **Abel Functional Equation**

$$f(y(x)) = f(x) + \alpha.$$

(vii) **Pythagorean Functional Equation**

$$|f(x + iy)|^2 = |f(x)|^2 + |f(iy)|^2.$$

(viii) **Davison Functional Equation**

$$f(xy) + f(x + y) = f(xy + x) + f(y).$$

(ix) Sincov Functional Equation

$$\varphi(x, y) + \varphi(y, z) = \varphi(x, z).$$

(x) Riemann Functional Equation

$$\zeta(1 - s) = \frac{\Gamma(s)}{(2\pi)^s} 2 \cos\left(\frac{\pi s}{2}\right) \zeta(s).$$

(xi) Gauss Functional Equation

$$f(\sqrt{x^2 + y^2}) = f(x)f(y).$$

(xii) Lobachevsky Functional Equation

$$f(x + y)f(x - y) = f^2(x).$$

(xiii) Pompeiu Functional Equation

$$f(x + y + xy) = f(x) + f(y) + f(x)f(y).$$

(xiv) Drygas Functional Equation

$$f(x + y) + f(x - y) = 2f(x) + f(y) + f(-y).$$

(xv) Swiatak Functional Equation

$$f(x + y) + f(x - y) = 2f(x) + 2f(y) + g(x)g(y).$$

(xvi) Hosszu Functional Equation

$$f(x + y - \alpha xy) + g(xy) = h(x) + k(y).$$

(xvii) Baxter Functional Equation

$$f(f(x)y + f(y)x - xy) = f(x)f(y).$$

(xviii) Homogeneous Functional Equation

$$f(ax, ay) = a^\beta f(x, y).$$

(xix) Associative Functional Equation

$$f(f(x, y), z) = f(x, f(y, z)).$$

(xx) Transitivity Functional Equation

$$f(x, y) = f(f(x, z), f(y, z)).$$

(xxi) Bisymmetry Functional Equation

$$f(f(x, y), f(u, z)) = f(f(x, u), f(y, z)).$$

(xxii) Cosine Functional Equations

(1) $f(x + y) = f(x)f(y) - g(x)g(y)$,
(2) $f(x - y) = f(x)f(y) + g(x)g(y)$.

(xxiii) **Sine Functional Equations**

(1) $f(x + y) = f(x)g(y) + f(y)g(x),$

(2) $f(x - y) = f(x)g(y) - f(x)g(x).$

1.8. Significance of functional equations

Nowadays, the field of functional equations is an ever-growing branch of mathematics with far-reaching applications. It is increasingly used to investigate many problems in science and technology. The theory of functional equations is relatively new and it contributes to the development of strong tools in contemporary mathematics. Conversely, many mathematical ideas in different fields have become essential to the foundation of functional equations. Many new applied problems and theories have inspired and encouraged specialists on functional equations to develop new approaches and new methods.

1.9. Applications of functional equations

Functional equations arise in many fields of mathematics, such as mechanics, geometry, statistics, measure theory, algebraic geometry, group theory. Functional equations have many interesting applications in characterization problems of probability theory. Solutions of functional equations can be used in characterizing joint distributions from conditional distributions. Functional equations find many applications in the study of stochastic process, classical mechanics, astronomy, economics, dynamic programming, game theory, computer graphics, neural networks, digital image processing, statistics, information theory, coding theory, fuzzy set theory, decision theory, artificial intelligence, cluster analysis, multivalued logic, population ethics and many other fields.

We present some examples to illustrate how functional equations are applied to solve some interesting problems in geometry, finance, information theory, wireless sensor networks.

1.9.1. *Application of functional equation in geometry*

Area of Rectangle: In 1791, Legendre applied functional equations to obtain the area of a rectangle. Consider the rectangle whose base is b and height is a. We are interested in finding the area of the rectangle. Let us assume that the area of the rectangle is $f(a, b)$.

Now, divide the rectangle horizontally so that the rectangle is divided into two subrectangles with heights a_1 and a_2 and the same base b as in Figure 1. Then the area of subrectangles will be $f(a_1, b)$ and $f(a_2, b)$ and the area of the full rectangle is $f(a_1 + a_2, b)$. We have

$$f(a_1 + a_2, b) = f(a_1, b) + f(a_2, b). \tag{1.5}$$

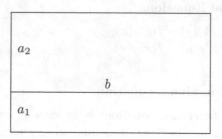

Figure 1.

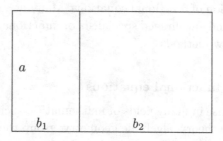

Figure 2.

In a similar manner, we divide the rectangle vertically with base heights b_1 and b_2 and the same height a as in Figure 2. Then the resulting areas are $f(a, b_1)$ and $f(a, b_2)$ and $f(a, b_1 + b_2)$. Therefore,

$$f(a, b_1 + b_2) = f(a, b_1) + f(a, b_2). \tag{1.6}$$

In equation (1.5), b is a constant and in equation (1.6), a is a constant. Both the equations are similar to Cauchy's equations $f(x + y) = f(x) + f(y)$ whose solution is $f(x) = cx$. Therefore, the solution of (1.5) and (1.6) is

$$f(a, b) = c_1(b)a = c_2(a)b. \tag{1.7}$$

From (1.7),

$$\frac{c_1(b)}{b} = \frac{c_2(a)}{a} = c. \tag{1.8}$$

From (1.8),

$$c_1(b) = cb, \quad c_2(a) = ca. \tag{1.9}$$

Substituting (1.9) in (1.7), we get

$$f(a, b) = cab,$$

where c is an arbitrary positive constant. Assume the initial conditions, that is, when $a = 1$, $b = 1$, the area of the rectangle $= 1$, which gives $c = 1$. Therefore, $f(a, b) = ab$. Hence, we arrive at the area of the rectangle.

1.9.2. *Application of functional equation in financial management*

Compound Interest: Suppose a person invests a principal of Rs. x at the rate of interest $r\%$ for the period of y years. One is interested in finding the formula for compound interest. The final amount is a function of x and y, that is, $f(x, y)$. There are two cases that arise:

(i) The final amount is the same if we invest the principal amount $x_1 + x_2$ together or principal x_1, x_2 are invested separately for y years. This can be expressed in the following functional equation:

$$f(x_1 + x_2, y) = f(x_1, y) + f(x_2, y), \tag{1.10}$$

whose solution is

$$f(x, y) = D(y)x. \tag{1.11}$$

(ii) The final amount is the same whether we invest the amount x for a period of $y_1 + y_2$ years or x invested for y_1 years and then invest the resultant amount in y_2 years. It is expressed as

$$f(x, y_1 + y_2) = f(f(x, y_1), y_2). \tag{1.12}$$

Applying (1.11) in (1.12), we get

$$D(y_1 + y_2)x = f(D(y_1)x, y_2),$$
$$D(y_1 + y_2)x = D(y_2)D(y_1)x,$$
$$D(y_1 + y_2) = D(y_2)D(y_1).$$

It is a multiplicative functional equation whose solution is

$$D(y) = a^y. \tag{1.13}$$

Substituting (1.13) in (1.11), we get

$$f(x, y) = a^y x.$$

For one year, $f(x, 1) = x + \text{interest}$. Hence,

$$ax = x + \frac{x \times 1 \times r}{100},$$
$$ax = x \left(1 + \frac{r}{100}\right),$$
$$a = \left(1 + \frac{r}{100}\right).$$

Therefore,

$$f(x,y) = x\left(1 + \frac{r}{100}\right)^y.$$

So, we arrive at the compound interest formula $= x(1 + \frac{r}{100})^y$.

1.9.3. *Application of functional equation in information theory*

Some functional equations such as

(i) $f(x) + (1-x)^\beta f(\frac{y}{1-x}) = f(y) + (1-y)^\beta f(\frac{x}{1-y})$,

(ii) $f(xy) + f((1-x)y) = f(y)\{m(x) + m(1-x)\} + m(y)\{f(x) + f(1-x)\}$,

(iii) $\sum_{i=1}^{n}\sum_{j=1}^{m} f(x_i y_j) = \sum_{i=1}^{n} f(x_i) + \sum_{j=1}^{m} f(y_j)$,

(iv) $\sum_{i=1}^{n}\sum_{j=1}^{m} f_{ij}(x_i y_j) = \sum_{i=1}^{n} g_i(x_i) + \sum_{j=1}^{m} h_j(y_j) + \sum_{i=1}^{n} k_i(x_i)\sum_{j=1}^{m} l_j(y_j)$

are applied in the information theory.

Shannon's entropy

Let $\Delta_n = \{p = (p_1, p_2, \ldots, p_n)|p_i \geq 0, \sum_i p_i = 1\}$ be the set of all finite complete discrete probability distribution on a given partition of the sure event Ω into n events $E_1, E_2, \ldots, E_n$. In 1948, Shannon, in his paper [C.E. Shannon, A mathematical theory of communication, *Bell System Tech. J.* **27** (1948), pp. 378–423 and 632–656] introduced the measure of information

$$H_n(p) = -\sum_{i=1}^{n} p_i \log p_i, \quad p \in \Delta_n$$

known as Shannon's entropy.

We have multiplicative functional equation

$$f(xy) = f(x) + f(y),$$

whose solution is $f(x) = \log x$. We note that this is a function whose value on the product of probabilities of events is equal to the sum of its values on the probabilities of the individual events. Shannon used the above functional equation in the information theory since with an intuitive notion that the information content of two independent events should be the sum of the information in each.

In particular, the functional equation is given by

$$f(x) + \alpha(1-x)g\left(\frac{y}{1-x}\right) = h(y) + \alpha(1-y)k\left(\frac{x}{1-y}\right)$$

for all $x, y \in [0,1]$ with $x + y \in [0,1]$. When $f = g = h = k$ and $\alpha =$ the identity map is known as the fundamental equation of information. It has been extensively investigated by many authors. The general solution of the above equation is dealt by P.L. Kannappan [*Can. J. Math.* **35**(5) (1983), pp. 862–872].

The effect of the information communicated in a message can be measured by the changes in the probability concerning the receiver of the message. The effect of information will depend upon the expectation of receiver before and after receiving the message. Naturally, the information received can be taken as the ratio of the logarithm of two probabilities. Thus, the information received about the event E is given by

$$I(E) = \frac{\text{Probability concerning the receiver after receiving the information}}{\text{Probability concerning the receiver before receiving the information}}.$$

(1.14)

In the case of noiseless channel, probability concerning the receiver after receiving the information equals to 1, as there will be no distortion of information during the process. The above equation becomes

$$I(E) = \log \left[\frac{1}{\text{Probability concerning the receiver before receiving the information}} \right]$$

$$= -\log[\text{Probability concerning the receiver before receiving the information}].$$

(1.15)

Shannon, with his intuitive idea, proposed a decreasing function $h(p)$, as a measure of the amount of information satisfying

$$h(p) = -\log p, \quad 0 < p \leq 1.$$

(1.16)

The function $h(p)$ is called the information function and it satisfies the additive property. Let A and B be any two events with $p(A) > 0$, $p(B) > 0$. Suppose that first we are informed that A has occurred and next we are informed that B has occurred and if A and B are independent, then

$$-\log[p(A)] - \log[p(B)] = -\log[p(AB)].$$

If $p(A) = p_1$, $p(B) = p_2$ and $p(AB) = p_1 p_2$, then

$$-\log(p_1) - \log(p_2) = -\log(p_1 p_2).$$

Hence, $h(p_1) + h(p_2) = h(p_1 p_2)$.

It shows that the information function h satisfies the Cauchy's functional equation

$$f(xy) = f(x) + f(y).$$

(1.17)

1.9.4. *Application of functional equation in wireless sensor networks*

Wireless sensor networks (WSNs) consist of small nodes with sensing, computation, and wireless communication capabilities. Many routing, power management, and data dissemination protocols have been specifically designed for WSNs where energy awareness is an essential design issue.

Due to recent technological advances, the manufacturing of small and low cost sensors became technically and economically feasible. The sensing electronics measure ambient conditions related to the environment surrounding the sensor and transforms them into an electric signal. Processing such a signal reveals some properties about objects located and/or events happening in the vicinity of the sensor. A large number of these disposable sensors can be networked in many applications that require unattended operations. A WSN contains hundreds or thousands of these sensor nodes. These sensors have the ability to communicate either among each other or directly to an external base station (BS). A greater number of sensors allows for sensing over larger geographical regions with greater accuracy. Basically, each sensor node comprises sensing, processing, transmission, mobilizer, position finding system, and power units (some of these components are optional like the mobilizer). Sensor nodes are usually scattered in a sensor field, which is an area where the sensor nodes are deployed. Sensor nodes coordinate among themselves to produce high-quality information about the physical environment. Each sensor node bases its decisions on its mission, the information it currently has, and its knowledge of its computing, communication, and energy resources. Each of these scattered sensor nodes has the capability to collect and route data either to other sensors or back to an external base station(s).

Networking unattended sensor nodes may have profound effect on the efficiency of many military and civil applications such as target field imaging, intrusion detection, weather monitoring, security and tactical surveillance, distributed computing, detecting ambient conditions such as temperature, movement, sound, light, or the presence of certain objects, inventory control, and disaster management. Deployment of a sensor network in these applications can be in random fashion (e.g., dropped from an airplane) or can be planted manually (e.g., fire alarm sensors in a facility). For example, in a disaster management application, a large number of sensors can be dropped from a helicopter. Networking these sensors can assist rescue operations by locating survivors, identifying risky areas, and making the rescue team more aware of the overall situation in the disaster area.

Routing is the process of selecting path in a network along which network traffic can be sent. Routing trees are typical structures used in WSN to deliver data to sink. To ensure robust data communication, efficient methods are required to choose routes across a network that can react quickly to communication link changes. Many algorithms have been proposed in the literature to support the routing protocols of the network. In 1958, Bellman [*Dynamic Programming*, Princeton University Press, 1957] applied the functional equation approach to devise an algorithm which converges to the solution at most $N - 1$ steps for a network with N nodes.

It is stated as follows:

"Given a set of N cities, with every two cities linked by a road. The time required to travel from i to j is not directly proportional to the distance between i and j, due to road conditions and traffic. Given the matrix $T = (t_{ij})$ not necessarily symmetric, where t_{ij} is the time required to travel from i to j. We wish to determine the path from one given city to another given city which minimizes the travel time."

The functional equation technique of dynamic programming, combined with approximation in policy space, yields an iterative algorithm which converges after at most $(N-1)$ iterations.

Let us now introduce the functional equation technique of dynamic programming. Let f_i be the time required to travel from i to N, $i = 1, 2, \ldots, N-1$, using an optimal policy with $f_n = 0$.

Employing the principle of optimality, we see that the f_i satisfy the nonlinear system of equations

$$f_i = \min[t_{ij} + f_i], \qquad i = 1, 2, \ldots, N-1,$$
$$f_N = 0. \tag{1.18}$$

Equation (1.18) is a functional equation because functions appear on both sides. We try to obtain the solution of the system (1.18) by using the method of successive approximations. Choose an initial sequence $\{f_i^{(0)}\}$, and then proceed iteratively, setting

$$f_i^{(k+1)} = \min_{i \neq j}(t_{ij} + f_i^{(k)}), \quad i = 1, 2, \ldots, N-1,$$
$$f_N^{(k+1)} = 0, \quad \text{for } k = 0, 1, 2, \ldots \tag{1.19}$$

The sequence in (1.19) converges to the solution after $(N-1)$ iterations by suitable algorithm. In this way, we can solve routing problem by functional equation.

1.9.5. *Applications of functional equations in other fields*

- The functional equation $f(x+y) + f(x-y) = 2f(x)f(y)$ has been used to derive the parallelogram law of forces. It is used to define circular functions and to show their relation to spherical trigonometry.
- Einstein used functional equations in the theory of relativity while describing the light signaling process in the space-time coordinates [*Bull. Amer. Math. Soc.* **26** (Oct. 1919 to July 1920), pp. 26–34].
- Ramanujan used functional equations to define the Bernoulli's number of negative index.
- The Cauchy additive functional $f(x + y) = f(x) + f(y)$ is used in genetics to find the combinatorial function $g_r(n)$ — the number of possible ways of picking r objects at a time from n objects allowing repetitions, since this function describes the number of possibilities from a gene pool. For further details, see [159].
- The following two-variable functional equation

$$C_1(x,y)P(x,y) = C_2(x,y)P(x,0) + C_3(x,y)P(0,y) + C_4(x,y)P(0,0),$$

where $C_i(x,y)$, $i = 1, 2, 3, 4$, are given polynomials in two complex variables x, y, arises from different communication and network systems [117].

- The system of functional equations

$$\varphi(x+y) = \frac{\varphi(x) + \{[\psi(x)]^2 - [\varphi(x)]^2\}\varphi(y)}{1 - \varphi(x)\varphi(y)}; \quad \psi(x+y) = \frac{\psi(x)\psi(y)}{1 - \varphi(x)\varphi(y)}$$

occurs in the optics and in the probability theory.

- The system of functional equations

   (i) $f(x) = \sup_{y \in D} H(x, y, f(T(x, y))), \forall x \in S$;

  (ii) $f(x) = \sup_{y \in D} \{u(x, y) + G(x, y, g(T(x, y)))\}, \forall x \in S$;

 (iii) $g(x) = \sup_{y \in D} \{u(x, y) + F(x, y, f(T(x, y)))\}, \forall x \in S$

has application in dynamic programming of multistage decision processes.

Chapter 2

Historical Development of Functional Equations

2.1. Introduction

In the previous chapter, we presented definition of functional equation, types of functional equation, some well-known functional equations, and applications of functional equations. In this chapter, we briefly present the historical development of some important functional equations developed by well-known mathematicians.

2.2. Nicole Oresme

In 1347, Nicole Oresme was the first mathematician who used the following functional equation of the form

$$\frac{f(x_1) - f(x_2)}{f(x_2) - f(x_3)} = \frac{x_1 - x_2}{x_2 - x_3}, \quad \forall \, x_1, x_2, x_3; \quad x_1 > x_2 > x_3,$$

where $x_1, x_2, x_3 \in \mathbb{R}$ describe the uniform difform motion.

2.3. Gregory of Saint-Vincent

During the years 1584–1667, the great mathematician Gregory of Saint-Vincent used the functional equation $f(xy) = f(x) + f(y)$ implicitly while describing the properties of a hyperbola. This functional equation pioneered the theory of the logarithm.

2.4. A.L. Cauchy

The functional equation associated with Cauchy (1821) is

$$f(x + y) = f(x) + f(y) \tag{2.1}$$

for all real x and y, and is called **additive Cauchy functional equation**. This functional equation is the most famous among all the functional equations because its properties are often used in the development of other functional equations arising in the field of natural and social sciences.

The other functional equations introduced by Cauchy are:

$$f(x + y) = f(x)f(y) \quad \textbf{(exponential)}, \tag{2.2}$$

$$f(xy) = f(x) + f(y) \quad \textbf{(logarithmic)}, \tag{2.3}$$

$$f(xy) = f(x)f(y) \qquad \textbf{(multiplicative)}, \tag{2.4}$$

where f is a real function of a real variable.

2.5. Jean D'Alembert

In 1769, Jean D'Alembert while studying the problem of parallelogram forces came across the following functional equation

$$f(x + y) + f(x - y) = 2f(x)f(y) \tag{2.5}$$

where f is defined on R^2, for real-valued function. The solution of this equation (3.27) is related to the composition of parallelogram forces. This functional equation (3.27) is now known as D'Alembert functional equation or cosine functional equation.

2.6. Charles Babbage

Charles Babbage wrote two major papers on functional equations in 1815 and 1816. Later in 1820, he has published a book in which while studying properties of periodic functions of various order, he has exhibited the following functional equations of the form

(i) $f(f(x)) = x$ (Periodic function of second order),
(ii) $f(f(x)) = f^{-1}(x)$ (Periodic function of third order).

He has also introduced a different type of functional equation like

$$f(x) = f(\alpha(x)),$$

and in more general form like

$$F(x, f(x), f(\alpha_1(x)), f(\alpha_2(x)), \ldots, f(\alpha_n(x))) = 0$$

where functions F and $\alpha_1, \alpha_2, \ldots, \alpha_n$ are given.

In extension to the above type of functional equations, the simultaneous functional equations, like

$$f(x) = f(\alpha(x)), \qquad f(x) = f(\beta(x))$$

for given functions $\alpha(x)$ and $\beta(x)$, were introduced by Charles Babbage.

2.7. Euler

In 1764, Euler used geometric methods to reduce the functional differential equation

$$g\left(x + g(x)g'(x)\right)^2 = g(x)^2\left(1 + g'(x)^2\right)$$

to the genuine functional equation

$$f(x + f(x)) = f(x) \qquad (2.6)$$

by defining $f(x) = g'(x)g(x)$. This is a functional equation in a single variable. Euler solved equation (2.6) using geometric methods, involving infinitesimals, which he then also applied to similar equations.

2.8. N.H. Abel

During the years 1823–1827, N.H. Abel who worked repeatedly with functional equations, published four important papers on functional equations.

(i) The first paper provided a general method for solving functional equations by differentiation method.
(ii) The second paper solved the functional equation

$$\begin{aligned}
F[x, F(y, z)] &= F[z, F(x, y)] \\
&= F[y, F(z, x)] \\
&= F(x, F(z, y)] \\
&= F[z, F(y, x)] \\
&= F[y, F(x, z)]
\end{aligned}$$

for the function $F(x, y)$.
(iii) In the third paper, the functional equation

$$g(x) + h(y) = xf(y) + yf(x)$$

is solved for the three unknown functions f, g, h. All these three papers used the method of reducing functional equations to difference equations.
(iv) Finally, in the fourth paper, N.H. Abel solved Cauchy functional equations generalized for complex variables.

2.9. Ramanujan

The theory of nested radicals is closely related to the theory of recursions. So it comes as no surprise that nested radicals can be studied using the methodology of functional equation. In 1911, Ramanujan posed the problem of evaluating

$$\sqrt{1 + 2\sqrt{1 + 3\sqrt{1 + 4\sqrt{\cdots}}}} \tag{2.7}$$

Suppose one generalizes (2.7) by defining

$$f(x) = \sqrt{1 + x\sqrt{1 + (x+1)\sqrt{\cdots}}}, \tag{2.8}$$

and squaring both sides gives us the functional equation

$$[f(x)]^2 = 1 + xf(x+1). \tag{2.9}$$

2.10. Jensen

The simplest and most elegant variation of the Cauchy additive functional equation is Jensen functional equation which is of the form:

$$f\left(\frac{x+y}{2}\right) = \frac{f(x) + f(y)}{2}. \tag{2.10}$$

Equation (2.10) is one of the equations that has been extensively explored and was solved by, among numerous authors, like Jensen (1965) [127], Aczél (1966) [1], Aczél, Chung, and Ng (1989) [2]. The functional equation (2.10) can be generalized as

$$f\left(\frac{x+y}{2}\right) = \frac{g(x) + h(y)}{2} \qquad \text{for } x, y \in \mathbb{R}$$

and it is known as Pexiderization of the Jensen Functional Equation.

2.11. Pexider

In 1903, J.V. Pexider considered the following functional equations, which are natural generalizations of Cauchy functional equations:

$$f(x + y) = g(x) + h(y), \tag{2.11}$$

$$f(x + y) = g(x)h(y), \tag{2.12}$$

$$f(xy) = g(x) + h(y), \tag{2.13}$$

and

$$f(xy) = g(x)h(y). \tag{2.14}$$

Further, the Pexiderized functional equations of the D'Alembert equation are as follows:

$$f(x+y) + f(x-y) = 2g(x)h(y),$$

$$f(x+y) + g(x-y) = 2f(x)g(y),$$

and

$$f(x+y) + g(x-y) = 2g(x)f(y).$$

2.12. Trigonometric functional equations

A functional equation is said to be a trigonometric functional equation if it has the solution in the trigonometric function form.

For example, $f(x+y) + f(x-y) = 2f(x)f(y)$ is a trigonometric functional equation, because it has the solution in the form $f(x) = \cos x$.

There are many functional equations related with trigonometric functions which arise from trigonometric identities. For example, the trigonometric identity

$$\sin(x+y)\sin(x-y) = \sin^2 x - \sin^2 y$$

leads to the trigonometric functional equation

$$f(x+y)f(x-y) = f^2(x) - f^2(y).$$

Some more trigonometric functional equations are listed below:

(1) $f(2x) = 2[f(x)]^2 - 1$,
(2) $f(2x) = 2f(x)g(x)$,
(3) $f(x+y) + f(x-y) = 2f(x)\sqrt{1 - f^2(y)}$,
(4) $s(x+y) = s(x)c(y) + c(x)s(y)$.

Many renowned mathematicians Kurepa (1960) [171], Aczél (1966) [1], Kannappan (1969) [157], Aczél and Dhombres (1989) [2], Davison (2001) [63], and Stetkaer (2002) [326] have studied various types of trigonometric functional equations.

2.13. System of functional equations

System of functional equations was used by G. Stokes in 1860 to determine the intensities of reflected and absorbed light and again by K. Weirestrass in 1886 and many successors in the characterization of determinants.

The following are examples of system of functional equations:

(a) $\phi[f(x)] = g(x)\phi(x) + F(x)$, where f, g and F are given functions and ϕ is unknown.
(b) $\phi(x) = h[x, \phi(f_1(x)), \ldots, \phi(f_n(x))]$, where ϕ is the unknown function.

(c) $\phi[\phi(x)] = g[x, \phi(x)]$, where $\phi(x)$ is an unknown function and $g(x, y)$ is a given function.

(d) The system of functional equations

 (i) $f(x) + f(\frac{1}{x}) = 1$, for all $x \in \mathbb{Q}^+$;
 (ii) $f(1 + 2x) = \frac{1}{2} f(x)$, for all $x \in \mathbb{Q}^+$.

(e) The system of functional equations

 (i) $f(x + y) = f(x) + f(y)$, for all real numbers x, y;
 (ii) $f(xy) = f(x)f(y)$, for all real numbers x, y.

(f) The system of functional equations

 (i) $f(-x) = -f(x)$, for all real x;
 (ii) $f(x + 1) = f(x) + 1$, for all real x;
 (iii) $f\left(\frac{1}{x}\right) = \frac{f(x)}{x^2}$, for all $x \neq 0$.

(g) The system of functional equations

 (i) $f(x + y) = f(x) + f(y)$, for all $x, y \in \mathbb{R}$;
 (ii) $f(\frac{1}{x}) = \frac{f(x)}{x^2}$, for all $x \neq 0$.

2.14. Contribution by other mathematicians

The great mathematicians Picard (1827), Hilbert (1832), Hardy (1840), Bellman (1855), Hille (1856) consciously or unconsciously applied the functional equations in their works. The comprehensive books by S. Pincherle's (1906, 1912), G.H. Hardy, J.E. Littlewood and G. Pólya (1934), M. Fréchet (1938) and B. Hostinsky (1939) and M. Kuczma (1968) also advanced considerably the discipline of functional equations.

In 1966, J. Aczél made an extraordinary attempt to collect all the previous work done by various mathematicians and published in a monograph namely, *Lectures on Functional Equations and their Applications*, Vol. 19, Academic Press, New York, 1966. This monograph serves as a resource book for many researchers working in the field of functional equations.

Also, the books by J. Aczél and Z. Daroczy (1975), J. Dhombres (1979, 1989), M. Kuczma (1985), J. Aczél (1987, 2006), J. Smital (1988), J. Aczél and J. Dhombres (1989), M. Kuczma, B. Choczewski and R. Ger (1990), B. Ramachandran and K.S. Lau (1991), L. Szekelyhidi (1991), E. Castillo and M.R. Ruiz-Cobo (1992), C.R. Rao and D.N. Shanbhag (1994), B.R. Ebanks, P.K. Sahoo and W. Sander (1998), P.K. Sahoo and T. Riedel (1998), D.H. Hyers, G. Isac and Th.M. Rassias (1998), S.M. Jung (2001), S. Czerwik (2002), I. Risteski and V. Covachev (2002), C.G. Small (2007) and Pl. Kannappan (2009) have contributed immensely to the further advancement of this discipline.

Most recently, the books by C. Efthiniou (2011), B.J. Venkachala (2012), Th.M. Rassias and J. Brzdek (2012), T. Andreescu, O. Mushkarov, N. Nikolov (2012), L. Szekelyhidi (2012), Th.M. Rassias (2013, 2014), Y.J. Cho, Th.M. Rassias and R. Saadati (2013), H. Stetkaer (2013) and S.M. Jung (2014) have given special attention to various advanced and applied functional equations.

Chapter 3

Methods of Solving Functional Equations

3.1. Introduction

In the previous chapter, we discussed about the historical development of functional equations. In this chapter, we present various methods of solving functional equations such as substitution method, mathematical induction method, recurrence relation method, method of fixed points, method of transformation of variables, method of undetermined coefficients, method of solving simultaneous functional equations, etc. The methods discussed to solve functional equations serve as a mode of introducing the reader to learn the beauty, simplicity of the functional equations. It is vital that these simple results and the methodology behind them are memorized as they often appear in pertinent functional equations problems.

3.2. Substitution method

We explain this method through the following examples.

Example 3.1. Find all functions $f : \mathbb{R} \to \mathbb{R}$ such that

$$x^2 f(x) + f(1 - x) = 2x - x^4 \tag{3.1}$$

for all $x \in \mathbb{R}$.

Solution. Replacing x by $1 - x$ in (3.1), we obtain

$$(1 - x)^2 f(1 - x) + f(x) = 2(1 - x) - (1 - x)^4. \tag{3.2}$$

From (3.1), we arrive at

$$f(1 - x) = 2x - x^4 - x^2 f(x). \tag{3.3}$$

Substituting (3.3) in (3.2), and then simplifying, we obtain

$$f(x)[1 - x^2 + 2x^3 - x^4] = 1 - 2x^2 + 2x^3 - 2x^5 + x^6$$
$$= (1 - x^2 + 2x^3 - x^4) - x^2(1 - x^2 + 2x^3 - x^4).$$

Therefore

$$f(x) = 1 - x^2.$$

Hence, $f(x) = 1 - x^2$ is the solution of the functional equation (3.1).

Example 3.2. Find all functions $f : \mathbb{R} \to \mathbb{R}$ satisfying the equation

$$f(x + y) + 2f(x - y) + f(x) + 2f(y) = 4x + y \tag{3.4}$$

for all $x, y \in \mathbb{R}$.

Solution. Replacing y by 0 in (3.4), and then simplifying, we obtain

$$f(x) = x - \frac{a}{2}, \quad \text{where } a = f(0). \tag{3.5}$$

Substituting (3.5) in (3.4) and again simplifying, we obtain $a = 0$. Hence, $f(x) = x$ is the solution of equation (3.4).

Example 3.3. Find all functions $f : \mathbb{R} \to \mathbb{R}$ satisfying the functional equation

$$f(x + y) - 2f(x - y) + f(x) - 2f(y) = y - 2 \tag{3.6}$$

for all $x, y \in \mathbb{R}$.

Solution. Setting $y = 0$ in (3.6) and simplifying, we obtain

$$f(0) = 1. \tag{3.7}$$

Substituting $x = 0$, $y = t$ in (3.6) and simplifying, we obtain

$$-f(t) - 2f(-t) = t - 3. \tag{3.8}$$

Replacing x by 0 and y by $-t$ in (3.6) and simplifying further, we obtain

$$-2f(t) - f(-t) = -t - 3. \tag{3.9}$$

Multiplying equation (3.9) by 2 and then subtracting from (3.8), we have

$$f(t) = t + 1.$$

Hence, the solution is $f(x) = x + 1$ of equation (3.6).

3.3. Mathematical induction method

This method depends on using the values $f(1)$, $f(n)$ for all integers n, whereas in case of rationals r, this method uses the values of $f(\frac{1}{n})$ and $f(r)$. This method is used in problems where the function is defined on set of rational numbers $\mathbb{Q}$.

Example 3.4. Find all functions $f : \mathbb{Q} \to \mathbb{Q}$ such that the Cauchy equation

$$f(x+y) = f(x) + f(y) \qquad (3.10)$$

holds for all $x, y \in \mathbb{Q}$.

Solution. Letting $x = y = 0$ in (3.10), we obtain $f(0) = 0$. Next, we will show

$$f(mx) = mf(x) \qquad (3.11)$$

for all $m \in \mathbb{N}$, $x \in \mathbb{Q}$ by mathematical induction.

Let $m = 1$; then (3.11) is obviously true. Assume that (3.11) is true for some $m \in \mathbb{Z}$. Putting $y = mx$ in (3.10), we obtain

$$f((m+1)x) = f(x+mx) = f(x) + mf(x) = (m+1)f(x).$$

Now, letting $y = -x$ in (3.10), and then simplifying, we obtain $f(-x) = -f(x)$. Therefore, $f(-mx) = -f(mx) = -mf(x)$ for $m \in \mathbb{N}, x \in \mathbb{Q}$.

Now, one can write

$$f(1) = f\left(k\left(\frac{1}{k}\right)\right) = kf\left(\frac{1}{k}\right),$$

which implies $f\left(\frac{1}{k}\right) = \frac{1}{k}f(1)$, where $k \in \mathbb{N}$.

For $m \in \mathbb{Z}, n \in \mathbb{N}$,

$$f\left(\frac{m}{n}\right) = mf\left(\frac{1}{n}\right) = \frac{m}{n}f(1).$$

Therefore, $f(x) = cx$ with $c = f(1)$, which is the general solution of (3.10).

Example 3.5. Find all functions $f : \mathbb{N} \to \mathbb{N}$ satisfying the functional equation

$$f(x+y) = f(x) + f(y) - f(x)f(y) \qquad (3.12)$$

for all $x, y \in \mathbb{N}$.

Solution. Setting $y = x$ in (3.12), we obtain

$$f(2x) = 2f(x) - [f(x)]^2 = 1 - [1 - f(x)]^2.$$

Substituting $y = 2x$ in (3.12), and simplifying, we have

$$f(3x) = 1 - [1 - f(x)]^3.$$

Similarly, setting $y = (n-1)x$ in (3.12), we arrive at

$$f(nx) = 1 - [1 - f(x)]^n.$$

Let us find the solution of equation (3.12) by the induction method. Obviously, the result is true for $n = 2$. Let us assume the result is true for n and prove the result for $n + 1$:

$$f[(n+1)x] = f(x+nx)$$
$$= f(x) + f(nx) - f(x)f(nx)$$
$$= f(x) + 1 - [1 - f(x)]^n - f(x)\{1 - [1 - f(x)]^n\}$$
$$= 1 - [1 - f(x)]^{n+1}.$$

Therefore,

$$f(nx) = 1 - [1 - f(x)]^n \tag{3.13}$$

for any positive integer $n > 1$. Substituting $x = 0$ in (3.13), we obtain

$$f(0) = 1 - [1 - f(0)]^n.$$

Substituting $x = 1$ in (3.13), we have

$$f(n) = 1 - [1 - f(1)]^n.$$

Take $a = 1 - f(1)$. Then $f(n) = 1 - a^n$ for any integer n. Therefore, the solution is $f(x) = 1 - a^x$.

3.4. Recurrence relation method

This method is usually used with the equations in which the range is bounded and in the case when we are able to find a relationship between $f(f(n))$, $f(n)$ and n. The following example illustrates the method.

Example 3.6. Solve the functional equation

$$g^{-1}[g(x) + 1] = h^{-1}[h(x) + 1]. \tag{3.14}$$

Solution. Setting $x = h^{-1}(x)$ in (3.14), and simplifying, we have

$$hg^{-1}[g(h^{-1}(x)) + 1] = h(h^{-1}(x)) + 1 \quad \text{or}$$
$$s(x) + 1 = s(x + 1) \tag{3.15}$$

which is a difference equation, where $s(x) = g(h^{-1}(x))$.

The general solution of the above difference equation (3.15) is

$$s(x) = x + c. \tag{3.16}$$

Replacing $s(x)$ in (3.16), we arrive at

$$gh^{-1}(x) = x + c$$

which gives

$$h^{-1}(x) = g^{-1}(x + c)$$

or

$$g(x) = h(x) + c,$$

which is the general solution of equation (3.14).

3.5. Method of finding fixed points

This method can be applied to get the solutions for more difficult problems. We explain this method by the following example.

Example 3.7. Find all functions $f : \mathbb{R} \to \mathbb{R}$ which satisfy

$$f(xf(y)) = yf(x) \tag{3.17}$$

for all $x, y \in \mathbb{R}$ such that $f(x) \to 0$ as $x \to \infty$.

Solution. First, let us show that $f(x)$ is surjective. For $y = \frac{x}{f(x)}$ in (3.17), we obtain

$$f\left(xf\left(\frac{x}{f(x)}\right)\right) = x.$$

Therefore, $f(x)$ is surjective. Assume $f(y) = 1$ and putting $x = 1$ in (3.17), we obtain $f(1) = yf(1)$. Therefore, $y = 1$ and $f(1) = 1$. Now, putting $x = y$ in (3.17), we obtain $f(xf(x)) = xf(x)$. Thus, $xf(x)$ is a fixed point for the function f. If a, b are fixed points of f, then

$$f(ab) = f(af(b)) = bf(a) = ab.$$

Thus, ab is also a fixed point. If a is a fixed point of f, then

$$1 = f(1) = f\left(a.\frac{1}{a}\right) = f\left(\frac{1}{a}f(a)\right) = af\left(\frac{1}{a}\right).$$

Thus, $f\left(\frac{1}{a}\right) = \frac{1}{a}$, which means that $\frac{1}{a}$ is also a fixed point. Now, if $xf(x) > 1$, then

$$f\left((xf(x))^n\right) = (xf(x))^n.$$

As $n \to \infty$, $(xf(x))^n \to \infty$ while $f\left((xf(x))^n\right) = (xf(x))^n$, which is a contradiction. If $xf(x) < 1$, then $\frac{1}{xf(x)}$ is a fixed point greater than 1, which is again a contradiction. Therefore, we must have $f(x) = \frac{1}{x}$.

Remark 3.1. In general, if we have functional equation in the form

$$f(g(x)) = h(x),$$

where $g(x)$ and $h(x)$ are known functions and if g^{-1} exists one can replace x by $g^{-1}(x)$ (by fixed point technique), we obtain the solution of the above functional equation in the form

$$f(x) = h\left(g^{-1}(x)\right).$$

3.6. Method of transformation of variables

This is one of the most common methods to solve a functional equation. In this method, we replace one variable by another (keeping the domain of the original variable unaltered) so that a new functional equation is obtained. By this method, sometimes it is easy to find the unknown function. The following examples illustrate the method.

Example 3.8. If $f(x+7) = x^2 - 5x + 2$, find $f(x)$.

Solution. Transform the variable x by $x - 7$ in the given equation, we obtain

$$f(x) = x^2 - 19x + 16.$$

Example 3.9. If

$$f\left(\frac{x+1}{x}\right) = \frac{x^2+1}{x^2} + \frac{1}{x}, \tag{3.18}$$

then find $f(x)$.

Solution. Let $u = \frac{x+1}{x}$ in (3.18), we have

$$f(u) = \frac{\left(\frac{1}{u-1}\right)^2 + 1}{\left(\frac{1}{u-1}\right)^2} + \frac{1}{\frac{1}{u-1}}$$

$$= u^2 - u + 1.$$

Therefore, $f(x) = x^2 - x + 1$.

Example 3.10. If $f(\ln x) = x^2 + x + 1$, where $x > 0$, find $f(x)$.

Solution. Let $u = \ln x$ be the given equation; we arrive at

$$f(u) = (e^u)^2 + e^u + 1.$$

Hence $f(x) = e^{2x} + e^x + 1$.

Example 3.11. If

$$f(x) + f\left(\frac{x-1}{x}\right) = 1 + x, \tag{3.19}$$

then find $f(x)$.

Solution. Transform x by $\frac{1}{1-x}$ and x by $\frac{x-1}{x}$ in (3.19), we obtain, respectively,

$$f\left(\frac{1}{1-x}\right) + f(x) = 1 + \frac{1}{1-x}, \qquad (3.20)$$

$$f\left(\frac{x-1}{x}\right) + f\left(\frac{1}{1-x}\right) = 1 + \frac{x-1}{x}. \qquad (3.21)$$

Adding (3.19)–(3.21) and then dividing by 2, we obtain

$$f(x) + f\left(\frac{x-1}{x}\right) + f\left(\frac{1}{1-x}\right) = \frac{1}{2}\left(3 + \frac{1}{1-x} + \frac{x-1}{x} + x\right). \qquad (3.22)$$

Subtracting (3.21) from (3.22), and then simplifying, we obtain

$$f(x) = \frac{-x^3 + x^2 + 1}{2x(1-x)},$$

which is the required solution.

3.7. Method of undetermined coefficients

When we know that the unknown function satisfies certain conditions, say it is a quadratic or a cubic function, we can immediately set up variables (e.g., let $f(x) = ax^2 + bx + c$ if $f(x)$ is a quadratic polynomial) and solve for them. We explain this method through the following example.

Example 3.12. If $f(x)$ is a quadratic function such that $f(x+1) - f(x) = 8x + 3$ and $f(0) = 5$, find $f(x)$.

Solution. Let $f(x) = ax^2 + bx + c$. Then $f(0) = 5$ gives $c = 5$. Now,

$$a(x+1)^2 + b(x+1) + c - ax^2 - bx - c = 8x + 3.$$

Equating coefficient of x and constant terms on both sides, we obtain $a = 4$, $b = -1$. Hence, $f(x) = 4x^2 - x + 5$.

3.8. Method of solving simultaneous functional equations

We explain this method in the following examples.

Example 3.13. Find all functions $f : \mathbb{R} \to \mathbb{R}$ such that

(a) $f(-x) = -f(x)$ for all real x;
(b) $f(x+1) = f(x) + 1$ for all real x;
(c) $f(\frac{1}{x}) = \frac{f(x)}{x^2}$ for all $x \neq 0$.

Solution.   Putting $x = 0$ in (a) and in (b), we arrive at $f(0) = 0$ and $f(1) = 1$. Using mathematical induction on (a) and (b) separately, we obtain $f(n) = n$ for all $n \in \mathbb{N}$. Another application of (a) now implies that $f(n) = n$ for all $n \in \mathbb{N}$.

Using (b) and (c), one can write

$$f\left(1 + \frac{1}{x}\right) = 1 + f\left(\frac{1}{x}\right) = 1 + \frac{f(x)}{x^2}, \tag{3.23}$$

where $x \neq 0$ and $x \neq -1$. In view of (c), we have the following expression

$$f\left(1 + \frac{1}{x}\right) = f\left(\frac{1}{x/(x+1)}\right) = \frac{f(x/(x+1))}{(x/(x+1))^2}. \tag{3.24}$$

Also from (a), (b) and (c), we have

$$f\left(\frac{x}{x+1}\right) = f\left(1 - \frac{1}{x+1}\right) = 1 - f\left(\frac{1}{x+1}\right)$$

$$= 1 - \frac{f(x+1)}{(x+1)^2} = \frac{(x+1)^2 - 1 - f(x)}{(x+1)^2}. \tag{3.25}$$

Using (3.25) in (3.24), we obtain

$$f\left(1 + \frac{1}{x}\right) = \frac{(x+1)^2 - 1 - f(x)}{x^2}. \tag{3.26}$$

From (3.23) and (3.26), we have

$$x^2 + f(x) = x^2 + 2x - f(x) \qquad \text{for all } x \neq 0, \quad x \neq -1.$$

Solving for $f(x)$, we conclude that

$$f(x) = x \qquad \text{for all } x \neq 0, \quad x \neq -1.$$

But we know that $f(0) = 0$ and $f(-1) = -f(1) = -1$. Thus, $f(x) = x$ holds good for all real numbers x.

Example 3.14. Let $f : \mathbb{R} \to \mathbb{R}$ be a function such that

(a) $f(x + y) = f(x) + f(y)$ for all $x, y \in \mathbb{R}$;
(b) $f\left(\frac{1}{x}\right) = \frac{f(x)}{x^2}$ for all $x \neq 0$.

Prove that $f(x) = cx$ for all $x \in \mathbb{R}$, for some constant c.

Solution.   It is easy to check that (a) gives $f(0) = 0$ and $f(-x) = -f(x)$ for all real x. From condition (a), one can write

$$f\left(\frac{1}{x(x-1)}\right) = f\left(\frac{1}{x-1} - \frac{1}{x}\right) f\left(\frac{1}{x-1}\right) - f\left(\frac{1}{x}\right).$$

Applying (b) and then simplifying, we obtain

$$x^2 f(x-1) - (x-1)^2 f(x) = f(x^2 - x).$$

Using (a) and $f(-y) = -f(y)$ in the last equation, we obtain

$$f(x^2) + x^2 f(1) = 2x f(x).$$

Replacing x by $x + \frac{1}{x}$ and simplifying, we obtain

$$f(x) = \left(\frac{f(2) + 2f(1)}{4} \right) x,$$

valid for all $x \neq 0$ and $x \neq -1$. Putting $x = 2$ in this relation, we see that $f(2) = 2f(1)$. Thus, we obtain $f(x) = f(1)x$, for all $x \neq 0$ and $x \neq -1$. This remains valid for $x = 0$ and $x = 1$ as may be seen by inspection. Thus, $f(x) = cx$, for all $x \in \mathbb{R}$ and for some constant c.

3.9. Method reducing functional equations to differential equations

In the following examples, we illustrate the method.

Example 3.15. Solve

$$f(x+y) = f(x) + f(y) \tag{3.27}$$

by reducing to a differential equation.

Solution. Substituting $y = 0$ in (3.27), we obtain $f(0) = 0$. Differentiating (3.27) with respect to x, we obtain

$$f'(x+y) = f'(x).$$

Therefore, $f'(x) = c$, where c is a constant.

Hence, $f(x) = cx + a$. But $f(0) = 0$ gives $a = 0$. Thus, $f(x) = cx$ is the general solution of equation (3.27).

Example 3.16. Solve

$$f(x+y) - f(x)f(y) \tag{3.28}$$

by reducing to a differential equation.

Solution. Putting $y = 0$ in (3.28), we obtain

$$f(x)[1 - f(0)] = 0.$$

Since $f(x) \neq 0$, we have $f(0) = 1$. Differentiating (3.28) with respect to x, we obtain

$$f'(x + y) = f'(x)f(y). \tag{3.29}$$

Putting x by 0 in (3.29), we obtain

$$\frac{f'(y)}{f(y)} = f'(0) = c.$$

Integrating the above equation and simplifying, we obtain

$$f(y) = e^{cy+k} = e^{cy}e^k = ae^{cy}.$$

In general, $f(x) = ae^{cx}$. But $f(0) = 1$ gives $a = 1$. Therefore, $f(x) = e^{cx}$ is the general solution of equation (3.28).

3.10. Method reducing functional equations to partial differential equations

The following example illustrates the method.

Example 3.17. Solve

$$\frac{f(x) - f(y)}{x - y} = f'\left(\frac{x+y}{2}\right) \tag{3.30}$$

by reducing to a partial differential equation.

Solution.   The given equation (3.30) can be written as

$$f(x) - f(y) = (x - y)f'(u), \qquad \text{where } u = \frac{x+y}{2}. \tag{3.31}$$

Differentiating (3.31) partially with respect to u, we obtain

$$f'(x) = f_u(u) + \frac{1}{2}(x - y)f_{uu}(u). \tag{3.32}$$

Now, differentiating (3.32) partially with respect to u, we obtain

$$0 = \frac{1}{2}f_{uu}(u) - \frac{1}{2}f_{uu}(u) + \frac{x-y}{4}f_{uuu}(u).$$

For $x \neq y$,

$$f_{uuu}(u) = 0.$$

Integrating the last equation thrice with respect to u, we obtain

$$f(u) = au^2 + bu + c.$$

Hence, $f(x) = ax^2 + bx + c$ is the general solution of equation (3.30).

3.11. Method of relating the functional equation with known functions

The following examples illustrate this method.

Example 3.18. Solve the following functional equation by relating with known functions:

$$f(x + y) = f(x) + f(y) + k \tag{3.33}$$

where k is a constant.

Solution. Consider $g(x) = f(x) + k$ in (3.33), then we obtain

$$g(x + y) = g(x) + g(y),$$

which is Cauchy's additive functional equation, whose solution is $g(x) = cx$. From this, one can easily obtain $f(x) = cx - k$, which is the solution of equation (3.33).

Example 3.19. If $f : (1, +\infty) \to \mathbb{R}$ is a continuous function such that

$$f(xy) = xf(y) + yf(x), \quad \text{for all } 1 < x, \quad y \in \mathbb{R},$$

then find $f(x)$.

Solution. The given equation can be written as

$$\frac{f(xy)}{xy} = \frac{f(x)}{x} + \frac{f(y)}{y}.$$

Take $\frac{f(x)}{x} = g(x)$. Then the above equation becomes

$$g(x + y) = g(x) + g(y).$$

Therefore, $g(x) = c \ln x$ and hence $f(x) = cx \ln x$.

3.12. Different methods of solving functional equations

Besides the above methods, we can also solve functional equations by substituting some special values, say $x = y = 0$ into the given equation. Since it is difficult to describe these techniques in words, we illustrate the methods with examples.

3.12.1. *Method of solving for special case and then for general case*

Example 3.20. Find all functions $f : \mathbb{Q} \to \mathbb{Q}$ which satisfy

$$f(xy) = f(x)f(y) - f(x + y) + 1 \tag{3.34}$$

for all $x, y \in \mathbb{Q}$ such that $f(1) = 2$.

Solution.   Set $y = 1$ in (3.34) and using $f(1) = 2$, we obtain

$$f(x + 1) = f(x) + 1.$$

By mathematical induction, we can get

$$f(x + n) = f(x) + n, \qquad \text{for } n \in \mathbb{Z}.$$

Further applying mathematical induction in condition $f(1) = 2$, we have

$$f(x) = x + 1.$$

Let $x = \frac{m}{n}$ and $y = n$, where m and n are integers, and n is not zero in (3.34); we obtain

$$f(m) = f\left(\frac{m}{n}\right) f(n) - f\left(\frac{m}{n} + n\right) + 1. \tag{3.35}$$

Since $f(x + n) = f(x) + n$, for $n \in \mathbb{Z}$, we have

$$f\left(\frac{m}{n} + n\right) = f\left(\frac{m}{n}\right) + n. \tag{3.36}$$

Substituting (3.36) in (3.35), and simplifying, we obtain

$$f\left(\frac{m}{n}\right) = \frac{m}{n} + 1.$$

Thus, we have $f(x) = x + 1$ for all $x \in \mathbb{Q}$ which is the solution of equation (3.34).

Remark 3.2. In the above example, first we solve the functional equation for a special case (we find $f(x)$ when $x \in \mathbb{Z}$) and then we solve it for the more general case (we find $f(x)$ when $x \in \mathbb{Q}$).

For $f : \mathbb{R} \to \mathbb{R}$, we can similarly apply this method. First, we find $f(x)$ when $x \in \mathbb{Z}$. Then, we find $f(x)$ when $x \in \mathbb{Q}$ by substituting $x = \frac{m}{n}$. Finally, we find $f(x)$ when $x \in \mathbb{R}$ by the density of rational numbers (for continuous functions only).

3.12.2. *Method of symmetric condition*

Definition 3.1. A functional equation is said to be symmetric if it is unaltered by interchanging the variables.

Example 3.21. If $(x - y)f(x + y) - (x + y)f(x - y) = 4xy\left(x^2 - y^2\right)$ for all x, y, find $f(x)$, where f is odd.

Solution. The given equation is equivalent to

$$\frac{f(x+y)}{x+y} - \frac{f(x-y)}{x-y} = 4xy = (x+y)^2 - (x-y)^2$$

or

$$\frac{f(x+y)}{x+y} - (x+y)^2 = \frac{f(x-y)}{x-y} - (x-y)^2 = k \text{ (say)}.$$

Due to symmetry, we obtain

$$\frac{f(x)}{x} - x^2 = k,$$

which gives

$$f(x) = x^3 + kx.$$

Remark 3.3. In the above example, we have a symmetric condition. By using the symmetry, we reduce the equation to a one-variable functional equation. This is a useful technique for symmetric functional equations.

3.12.3. *Method of replacing variables by other functions*

Example 3.22. If $f : \mathbb{R} \to \mathbb{R}$ satisfies $f(x^2 + f(y)) = y + xf(x)$ for all $x, y \in \mathbb{R}$, find $f(x)$.

Solution. Putting $x = 0$ in the given equation, we have

$$f(f(y)) = y. \tag{3.37}$$

Thus, we have

$$f(y + xf(x)) = f\left(f\left(x^2 + f(y)\right)\right) = x^2 + f(y). \tag{3.38}$$

Replacing x by $f(x)$ in (3.38) and simplifying, we obtain

$$f(y + xf(x)) = (f(x))^2 + f(y). \tag{3.39}$$

From (3.38) and (3.39), we obtain

$$(f(x))^2 = x^2. \tag{3.40}$$

Therefore, $f(x) = x$ or $f(x) = -x$.

Remark 3.4. In the above example, we replace variables by other functions (we replace x by $f(x)$ in the example). This is a very useful technique.

Some common replacements and substitutions include the following: replacing x by $f(x)$; replacing x by $f(f(x))$; substituting $x = 0$; substituting $x = y = 0$; substituting $x = 1$; etc.

3.12.4. *Various other methods of solving functional equations*

Example 3.23. Find all functions $f : \mathbb{R} \to \mathbb{R}$ such that

$$(f(x) + f(z))(f(y) + f(t)) = f(xy - zt) + f(xt + yz), \quad \text{for all } x, y, z, t \in \mathbb{R}.$$

Solution. Suppose $f(x) = c$ for all x. Then the equation implies $4c^2 = 2c$. Therefore, c can be only either 0 or $\frac{1}{2}$. It is also easy to check that $f(x) = 0$ for all x and $f(x) = \frac{1}{2}$ for all x are solutions.

Suppose the equation is satisfied by a non-constant function f. Putting $x = 0$ and $z = 0$ in the given equation, we obtain

$$2f(0) = (f(y) + f(t)) = 2f(0),$$

which implies $f(0) = 0$ or $f(y) + f(t) = 1$, for all y, t. Setting $y = t$ in $f(y) + f(t) = 1$, we obtain

$$f(y) = \frac{1}{2} \quad \text{for all } y.$$

Therefore, we may assume $f(0) = 0$. Putting $y = 1, z = 0, t = 0$, we get

$$f(x)f(1) = f(x).$$

Since $f(x) \neq 0$, $f(1) = 1$. Putting $z = 0, t = 0$, we obtain $f(x)f(y) = f(xy)$ for all x, y. In particular, $f(w) = f(w^{1/2})^2 \geq 0$, for $w > 0$. Putting $x = 0, y = 1$ and $t = 1$, we have

$$2f(1)f(z) = f(-z) + f(z),$$

which implies $f(-z) = f(z)$, for all z. Therefore, f is an even function.

Define the function $g : (0, \infty) \to \mathbb{R}$ by

$$g(w) = f(w^{1/2})^2 \geq 0.$$

Then for all $x, y > 0$,

$$g(xy) = f((xy)^{1/2}) = f(x^{1/2}y^{1/2}) = f(x^{1/2})f(y^{1/2}) = g(x)g(y).$$

Next, f is even which implies $g\left(x^2\right) = f(x)$ for all x. Setting $z = y, t = x$ in the given equation, we get

$$(g(x^2) + g(y^2))^2 = g((x^2 + y^2)^2) = g(x^2 + y^2)^2$$

for all x, y. Taking square roots and letting $a = x^2, b = y^2$, we have

$$g(a) + g(b) = g(a + b), \quad \text{for all } a, b > 0.$$

Since the function g is both additive and multiplicative, we have $g(w) = 2$, for all $w > 0$.

Since $f(0) = 0$ and f is even, we have

$$f(x) = g\left(x^2\right) = x^2, \quad \text{for all } x.$$

Example 3.24. Find all functions $f : \mathbb{R} \to \mathbb{R}$ which satisfy

$$f(x + y) + f(y + z) + f(z + x) \geq 3f(x + 2y + 3z) \tag{3.41}$$

for all $x, y, z \in \mathbb{R}$.

Solution. Put $x = a, y = z = 0$ in (3.41), we have

$$f(0) \geq f(a).$$

Put $x = y = \frac{a}{2}$ and $z = -\frac{a}{2}$ in (3.41), we obtain

$$f(a) \geq f(0).$$

Therefore, $f(a) = f(0)$, for all a. Hence, any constant function obviously satisfies the given inequality (3.41).

Example 3.25. Let $f : \mathbb{R} \to \mathbb{R}$ satisfy

$$f(x + y) = f(x)f(a - y) + f(y)f(a - x) \tag{3.42}$$

for all $x, y \in \mathbb{R}$ and a is a real constant such that $f(0) = \frac{1}{2}$. Prove that f is constant.

Solution. Put $x = y = 0$ in the given equation, we obtain

$$f(a) = \frac{1}{2}.$$

Put $y = 0$ in (3.42) and using $f(0) = \frac{1}{2}$, we have

$$f(x) = f(a - x).$$

Putting $y = a - x$ and simplifying, we obtain

$$f(x) = \frac{1}{2} \quad \text{or} \quad -\frac{1}{2}.$$

Now, for any x, we have

$$f\left(\frac{x}{2}\right) = \frac{1}{2} \quad \text{or} \quad -\frac{1}{2} \quad \text{and} \quad f\left(a - \frac{x}{2}\right) = f\left(\frac{x}{2}\right).$$

Therefore, $f(x) - f\left(\frac{x}{2} + \frac{x}{2}\right) = 2f(x)f\left(a - \frac{x}{2}\right) = \frac{1}{2} = $ a constant. This completes the proof.

Example 3.26. Find all functions $f : \mathbb{R} \to \mathbb{R}$ such that

$$f\left(x^2 + f(y)\right) = y + f(x)^2 \tag{3.43}$$

for all $x, y \in \mathbb{R}$.

Solution.   Put $x = y = 0$ and $f(0) = t$ in (3.43), we get $f(t) = t^2$. Also, $f(x^2 + t) = f(x)^2$ and $f(f(x)) = x + t^2$. Now,

$$f\left(f(1)^2 + f(t)\right) = t + f(f(1))^2$$
$$= t + \left(1 + t^2\right)^2 = 1 + t + 2t^2 + t^4.$$

Again,

$$f\left(t^2 + f(1 + t)\right) = 1 + t + f(t)^2 = 1 + t + t^4.$$

Therefore, we have $t = 0$. Hence, $f(f(x)) = x$ and $f(x^2) = f(x)^2$.
    Given any y, let $z = f(y)$. Then $y = f(z)$, and so

$$f\left(x^2 + y\right) = z + f(x)^2 = f(y) + f(x)^2.$$

Now, given any positive x, take z so that $x = z^2$. Then

$$f(x + y) = f\left(z^2 + y\right) = f(y) + f(z)^2 = f(y) + f\left(z^2\right) = f(x) + f(y).$$

Putting $y = -x$ in the above equation, we get

$$0 = f(0) = f(x + (-x)) = f(x) + f(-x).$$

Hence $f(-x) = -f(x)$. It follows that

$$f(x + y) = f(x) + f(y) \quad \text{and} \quad f(x - y) = f(x) - f(y)$$

hold for all x, y. For any x, assume $f(x) = y$. If $y > x$, then let $z = y - x$ and

$$f(z) = f(y - x) = f(y) - f(x) = x - y = -z.$$

If $y < x$, then let $z = x - y$ and

$$f(z) = f(x - y) = f(x) - f(y) = y - x = -z.$$

In either case, we get some $z > 0$ with $f(z) = -z < 0$. But now take w so that $w^2 = z$, then

$$f(z) = f\left(w^2\right) = f(w)^2 \geq 0,$$

which is a contradiction.
    Therefore, we must have $f(x) = x$, which is the solution of the given equation.

Example 3.27. Find all functions $f : \mathbb{R} \to \mathbb{R}$ which satisfy

$$f(xf(x) + f(y)) = f(x)^2 + y \tag{3.44}$$

for all $x, y \in \mathbb{R}$.

Solution. Putting $x = 0$ in (3.44), we get

$$f(f(y)) = f(0)^2 + y. \qquad (3.45)$$

Set $y = -f(0)^2$ and $k = f(y)$ in (3.45), we obtain $f(k) = 0$. Now, let $x = y = k$ in (3.44), we get $k = f(0)$. Putting $y = k, x = 0$ in (3.44), we get $k = 0$. Therefore, $f(0) = 0$. Putting $x = 0$ in (3.44), we get

$$f(f(y)) = y. \qquad (3.46)$$

Putting $y = 0$ in (3.44), we get

$$f(xf(x)) = f(x)^2. \qquad (3.47)$$

Putting $x = f(z)$ in (3.47) and then using $f(z) = x$, we get

$$f(zf(z)) = z^2.$$

Hence,

$$z^2 = f(z)^2 \qquad (3.48)$$

for all $z \in \mathbb{R}$. Hence, the solutions of equation (3.27) are

$$f(x) = \pm x, \quad \text{for all } x.$$

Chapter 4

General Solution of Euler–Lagrange Quadratic Type Functional Equations

4.1. Introduction

In the previous chapter, we discussed about various methods of solving functional equations. In this chapter, we present the general solution of several types of quadratic functional equations.

One type of the fundamental functional equations is the following **Euler–Lagrange quadratic type functional equation:**

$$f(x + y) + f(x - y) = 2f(x) + 2f(y). \tag{4.1}$$

The quadratic function $f(x) = ax^2$ is a solution of this functional equation, and so it is said that the above functional equation is quadratic. Quadratic functional equation is used to characterize inner product spaces.

Rassias [260] introduced Euler–Lagrange type quadratic functional equation of the form

$$f(ax + by) + f(bx - ay) = \left(a^2 + b^2\right)\left(f(x) + f(y)\right) \tag{4.2}$$

motivated from the following pertinent algebraic equation

$$|ax + by|^2 + |bx - ay|^2 = \left(a^2 + b^2\right)\left(|x|^2 + |y|^2\right). \tag{4.3}$$

The solution of the functional equation (4.2) is called an Euler–Lagrange quadratic type mapping. In addition, Rassias [260, 262, 264–266] generalized the standard quadratic equation to the equation

$$m_1 m_2 |a_1 x_1 + a_2 x_2|^2 + |m_2 a_2 x_1 - m_1 a_1 x_2|^2$$
$$= (m_1 |a_1|^2 + m_2 |a_2|^2)(m_2 |x_1|^2 + m_1 |x_2|^2).$$

He introduced and investigated the general pertinent Euler–Lagrange quadratic mappings. These Euler–Lagrange mappings are named Euler–Lagrange–Rassias mappings, and the corresponding Euler–Lagrange equations are called Euler–Lagrange–Rassias equations (see [118, 210, 243, 250]). These notions provide a cornerstone in analysis because of their particular interest in probability theory and stochastic analysis in merging these fields of research to functional equations via the introduction of the Euler–Lagrange–Rassias quadratic weighted means and fundamental mean equations (see [118, 264, 265]).

In Section 4.2, we present the general solution of equation (4.1) and in Section 4.3, the general solution of each other quadratic functional equation is obtained by showing equivalency of other quadratic functional equations to equation (4.1). In Sections 4.2 and 4.3, we assume that X and Y are real vector spaces.

4.2. Quadratic functional equation

A function $f : X \to Y$ is quadratic if and only if there exists a unique symmetric bi-additive function $B : X \times X \to Y$ such that $f(x) = B(x, x)$ for all $x \in X$, where the function B is given by

$$B(x,y) = \frac{1}{4}[f(x+y) - f(x-y)], \text{ for all } x, y \in X.$$

The following theorem provides the proof for the above result.

Theorem 4.1 ([2, 158]). *A function $f : X \to Y$ is quadratic if and only if there exists a symmetric bi-additive function $B : X \times X \to Y$ such that*

$$f(x) = B(x, x).$$

This B is unique.

Proof. If $f(x) = B(x, x)$, then

$$f(x+y) + f(x-y) = B(x+y, x+y) + B(x-y, x-y)$$
$$= 2B(x, x) + 2B(y, y)$$
$$= 2f(x) + 2f(y).$$

Conversely, let f be a quadratic function. Define

$$B(x,y) = \frac{1}{4}[f(x+y) - f(x-y)]. \tag{4.4}$$

By choosing $y = 0$ in (4.1), we get $f(0) = 0$ and by putting $x = y$, we obtain $f(2x) = 4f(x)$. Therefore,

$$B(x,x) = \frac{1}{4}f(2x) = f(x).$$

We now prove the symmetry of B. Interchanging x and y in (4.1), we obtain

$$f(x - y) = f(y - x),$$

which shows that f is even. This relation and (4.4) imply the following symmetry $B(x, y) = B(y, x)$, and

$$B(-x, y) = -B(x, y). \tag{4.5}$$

Thus, B is odd in each variable and it only remains to show that B is additive in the first variable. We have

$$4[B(x_1 + x_2, y) + B(x_1 - x_2, y)]$$
$$= f(x_1 + x_2 + y) + f(x_1 - x_2 + y) - (f(x_1 + x_2 - y) + f(x_1 - x_2 - y))$$
$$= 2(f(x_1 + y) - f(x_1 - y))$$
$$= 8B(x_1, y).$$

Interchange x_1 and x_2 and subtract the equation thus obtained from the above equation in order to get

$$B(x_1 - x_2, y) - B(x_2 - x_1, y) = 2B(x_1, y) - 2B(x_2, y).$$

In view of (4.5), this reduces to

$$2B(x_1 - x_2, y) = 2B(x_1, y) - 2B(x_2, y)$$

and, if we replace x_2 by $-x_2$ and take (4.5) into consideration again, we have

$$B(x_1 + x_2, y) = B(x_1, y) + B(x_2, y),$$

which is the required additivity in the first variable. This completes the proof. $\square$

4.3. Other quadratic functional equations

This section deals with the general solution of other quadratic functional equations, which are somewhat different from (4.1).

The following theorem establishes the general solution of a quadratic functional equation of the form

$$f(3x + y) + f(3x - y) = f(x + y) + f(x - y) + 16f(x). \tag{4.6}$$

Theorem 4.2 ([281]). *A function $f : X \to Y$ satisfies the functional equation (4.1) if and only if $f : X \to Y$ satisfies the functional equation (4.6). Therefore, every solution of (4.6) is a quadratic function.*

Proof. Let $f : X \to Y$ satisfy the functional equation (4.1). Putting $x = 0 = y$ in (4.1), we get $f(0) = 0$. Set $x = 0$ in (4.1) to get $f(y) = f(-y)$. Letting $y = x$ and $y = 2x$ in (4.1), respectively, we obtain that $f(2x) = 4f(x)$ and $f(3x) = 9f(x)$ for all $x \in X$. By induction, we lead to $f(kx) = k^2 f(x)$, for any positive integer k.

Now, replacing (x, y) by $(3x + y, 3x - y)$ in (4.1) and adding, we have

$$f(3x + y) + f(3x - y) = 16f(x) + f(x + y) + f(x - y) \qquad (4.7)$$

for all $x, y \in X$.

Conversely, let $f : X \to Y$ satisfy the functional equation (4.6). Putting $x = 0 = y$ in (4.6), we get $f(0) = 0$. Setting $y = 0$ and $y = x$ in (4.6), we obtain that $f(3x) = 9f(x)$ and $f(4x) = 16f(x)$ for all $x \in X$. By induction, we lead to $f(kx) = k^2 f(x)$, for all positive integers k. Setting $y = 2x$ in (4.6), we get $f(x) = f(-x)$, for all $x, y \in X$.

Replacing (x, y) by $(x + y, x - y)$ in (4.6), we have

$$f(4x + 2y) + f(2x + 4y) = f(2x) + f(2y) + 16f(x + y) \qquad (4.8)$$

for all $x, y \in X$. Again, replacing y by $x + 2y$ in (4.6), we have

$$f(4x + 2y) + f(2x - 2y) = 4f(x + y) + 4f(y) + 16f(x) \qquad (4.9)$$

for all $x, y \in X$. Interchanging x and y, we obtain

$$f(2x + 4y) + f(2x - 2y) = 4f(x + y) + 4f(x) + 16f(y) \qquad (4.10)$$

for all $x, y \in X$. Adding (4.9) and (4.10), we obtain

$$f(4x + 2y) + f(2x + 4y) + 2f(2x - 2y) = 8f(x + y) + 20f(x) + 20f(y) \qquad (4.11)$$

for all $x, y \in X$. Using (4.8) in (4.11), we obtain

$$4f(x) + 4f(y) + 16f(x + y) + 8f(x - y) = 8f(x + y) + 20f(x) + 20f(y).$$

Thus,

$$f(x + y) + f(x - y) = 2f(x) + 2f(y),$$

which completes the requested proof.   □

The following theorem provides the general solution of quadratic functional equations of the type

$$f(2x + y) + f(2x - y) = f(x + y) + f(x - y) + 6f(x), \qquad (4.12)$$

and

$$f(2x + y) + f(x + 2y) = 4f(x + y) + f(x) + f(y). \qquad (4.13)$$

Theorem 4.3 ([43]). *A function $f : X \to Y$ satisfies the functional equation (4.1) if and only if $f : X \to Y$ satisfies the functional equation (4.13) and if and only if $f : X \to Y$ satisfies the functional equation (4.12). Therefore, every solution of functional equations (4.12) and (4.13) is also a quadratic function.*

Proof. Let $f : X \to Y$ satisfy the functional equation (4.1). Putting $x = 0 = y$ in (4.1), we get $f(0) = 0$. Set $x = 0$ in (4.1) to get $f(y) = f(-y)$. Letting $y = x$ and $y = 2x$ in (4.13), respectively, we obtain that $f(2x) = 4f(x)$ and $f(3x) = 9f(x)$ for all $x \in X$. By induction, we lead to $f(kx) = k^2 f(x)$ for all positive integers k. Replacing x and y by $2x + y$ and $x + 2y$ in (4.1), respectively, and simplifying, we arrive at (4.13).

Let $f : X \to Y$ satisfy the functional equation (4.13). Putting $x = 0 = y$ in (4.13), we get $f(0) = 0$. Set $y = 0$ in (4.13) to get $f(2x) = 4f(x)$. Letting $y = x$ and $y = -2x$ in (4.13), we obtain that $f(3x) = 9f(x)$ and $f(x) = f(-x)$ for all $x \in X$. Putting x and y by $x + y$ and $x + y$ in (4.13), respectively, we obtain

$$f(2x + 3y) + f(x + 3y) = 4f(x + 2y) + f(x + y) + f(y), \qquad (4.14)$$

and

$$f(3x + y) + f(3x + 2y) = 4f(2x + y) + f(x) + f(x + y) \qquad (4.15)$$

for all $x, y \in X$. Adding (4.14) to (4.15) and using (4.13), we obtain

$$f(2x+3y)+f(3x+2y)+f(x+3y)+f(3x+y) = 18f(x+y)+5f(x)+5f(y) \quad (4.16)$$

for all $x, y \in X$. Replacing y by $2y$ and x by $2x$ in (4.13), respectively, we have

$$4f(x + y) + f(x + 4y) = 4f(x + 2y) + f(x) + 4f(y), \qquad (4.17)$$

and

$$4f(x + y) + f(4x + y) = 4f(2x + y) + 4f(x) + f(y) \qquad (4.18)$$

for all $x, y \in X$. Adding (4.17) to (4.18) and using (4.13), we get

$$f(x + 4y) + f(4x + y) = 8f(x + y) + 9f(x) + 9f(y) \qquad (4.19)$$

for all $x, y \in X$.

On the other hand, using (4.13), we obtain

$$f(x + 4y) + f(4x + y) = f(6x + 9y) + f(9x + 6y) - 4f(5x + 5y)$$
$$= 9f(2x + 3y) + 9f(3x + 2y) - 100f(x + y), \qquad (4.20)$$

which yields the relation by virtue of (4.19)

$$f(2x + 3y) + f(3x + 2y) = 12f(x + y) + f(x) + f(y) \qquad (4.21)$$

for all $x, y \in X$. Combining the last equation with (4.16), we get

$$f(x + 3y) + f(3x + y) = 6f(x + y) + 4f(x) + 4f(y). \qquad (4.22)$$

Replacing x and y by $\frac{x+y}{2}$, respectively, in (4.22), we arrive at the desired result (4.12).

Now, let $f : X \to Y$ satisfy the functional equation (4.12). Putting $x = 0 = y$ in (4.12), we get $f(0) = 0$. Letting $y = 0$ and $y = x$ in (4.12), respectively, we obtain that $f(2x) = 4f(x)$ and $f(3x) = 9f(x)$, for all $x \in X$. Putting $y = 2x$ in (4.12), we get $f(x) = f(-x)$. Replacing x and y by $x + y$ and $x - y$, respectively, in (4.12), we have

$$f(3x + y) + f(x + 3y) = 6f(x + y) + 4f(x) + 4f(y) \tag{4.23}$$

for all $x, y \in X$. Replacing y by $x + y$ in (4.12), we obtain

$$f(3x + y) + f(x - y) = 6f(x) + f(2x + y) + f(y). \tag{4.24}$$

Interchange x with y in (4.24) to get the relation

$$f(3y + x) + f(x - y) = 6f(y) + f(2y + x) + f(x). \tag{4.25}$$

Adding (4.24) to (4.25), we obtain

$$6f(x + y) + 2f(x - y) = f(2x + y) + f(x + 2y) + 3f(x) + 3f(y) \tag{4.26}$$

for all $x, y \in X$. Setting $-y$ instead of y in (4.26) and using the evenness of f, we get the relation

$$6f(x - y) + 2f(x + y) = f(2x - y) + f(2y - x) + 3f(x) + 3f(y). \tag{4.27}$$

Adding (4.26) to (4.27), we obtain equation (4.1), which completes the proof. □

The following theorem establishes the general solution of a quadratic functional equation of the form

$$f(x + 2y) + f(x - 2y) = 2f(x) + 8f(y). \tag{4.28}$$

Theorem 4.4 ([245]). *A function $f : X \to Y$ satisfies the functional equation* (4.28) *if and only if f is quadratic.*

Proof. Let $x = 0 = y$ in (4.28). Then we have $f(0) = 0$. Putting $x = 0$ in (4.28) gives

$$f(2y) + f(-2y) = 8f(y), \tag{4.29}$$

and setting $y = -y$ in (4.29), we obtain

$$f(-2y) + f(2y) = 8f(-y), \tag{4.30}$$

and so, by (4.29) and (4.30), we get $f(-y) = f(y)$, which implies f is an even function.

Hence, from (4.29) or (4.30), it follows that

$$f(2y) = 4f(y). \tag{4.31}$$

Substituting $x = 2x$ in (4.28), it follows from (4.31) that f is quadratic.

Conversely, putting $x = 0 = y$ in (4.1) yields $f(0) = 0$. Replacing y by x in (4.1), we get $f(2x) = 4f(x)$. Therefore, the substitution $y = 2y$ in (4.1) now gives equation (4.28), which completes the proof. □

So far, we have discussed about the general solution of two-dimensional quadratic functional equations. In the following theorem, we present the general solution of a three-dimensional quadratic functional equation

$$f(x + y + z) + f(x - y) + f(y - z) + f(z - x) = 3f(x) + 3f(y) + 3f(z). \tag{4.32}$$

Theorem 4.5 ([16]). *Let $f : X \to Y$ be a mapping satisfying the functional equation (4.32). Then the functional equation (4.32) is equivalent to the functional equation (4.1).*

Proof. If we replace x, y, z in (4.32) by 0, then we have $f(0) = 0$. By putting $y = z = 0$ in equation (4.32), we see that every solution of equation (4.32) is even. Putting $z = 0$ in (4.32) and using the evenness of f and $f(0) = 0$, we can transform equation (4.32) into equation (4.1).

Conversely, suppose that a function $f : X \to Y$ satisfies (4.1) for all $x, y \in X$. From (4.1), we get

$$f(x + y + z) + f(x) = 2f\left(x + \frac{y + z}{2}\right) + 2f\left(\frac{y + z}{2}\right),$$

and

$$f(y) + f(z) = 2f\left(\frac{y + z}{2}\right) + 2f\left(\frac{y - z}{2}\right).$$

According to (4.1) and the last two equalities, and using the fact $f(2x) = 4f(x)$, we obtain

$$\begin{aligned}
f(x + y + z) &+ f(x - y) + f(y - z) + f(z - x) \\
&= f(x + y + z) + f(x) + f(y) + f(z) \\
&\quad + f(x - y) + f(y - z) + f(z - x) - f(x) - f(y) - f(z) \\
&= 3f(x) + 3f(y) + 3f(z).
\end{aligned}$$

This means that equations (4.1) and (4.32) are equivalent, which completes the proof. □

The following theorem provides the general solution of a quadratic functional equation in three variables:

$$f(2x-y)+f(2y-z)+f(2z-x)+2f(x+y+z)= 7f(x)+7f(y)+7f(z). \qquad (4.33)$$

Theorem 4.6 ([208]). *A function $f : X \to Y$ satisfies the functional equation (4.33) for all $x, y, z \in X$ if and only if it satisfies the functional equation (4.1) for all $x, y \in X$.*

Proof. Suppose a function $f : X \to Y$ satisfies (4.33). Setting $(x, y, z) = (x, x, x)$ in (4.33), we simply have $f(3x) = 9f(x)$, for all $x \in X$, which in turn implies that $f(0) = 0$ by putting $x = 0$. Setting $(x, y, z) = (-x, x, x)$ in (4.33), we have

$$f(-3x) + f(x) + f(3x) + 2f(x) = 7f(-x) + 7f(x) + 7f(x).$$

Using the previously derived equation, $f(3x) = 9f(x)$, for all $x \in X$, the above equation simplifies to $f(-x) = f(x)$, for all $x \in X$. Thus, f is an even function. Setting $(x, y, z) = (x, 0, 0)$ in (4.33), we get

$$f(2x) + f(0) + f(-x) + 2f(x) = 7f(x) + 7f(0) + 7f(0),$$

which yields $f(2x) = 4f(x)$, for all $x \in X$. Letting $z = 0$ in (4.33) and simplifying, we obtain

$$f(2x - y) + 2f(x + y) = 6f(x) + 3f(y). \qquad (4.34)$$

Letting $z = -y$ in (4.33) and simplifying, we have

$$f(2x - y) + f(x + 2y) = 5f(x) + 5f(y). \qquad (4.35)$$

Eliminating $f(2x - y)$ from (4.34) and (4.35), we are left with

$$f(x + 2y) + f(x) = 2f(x + y) + 2f(y). \qquad (4.36)$$

The functional equation (4.1) follows by putting $(x, y) = (x - y, y)$ in equation (4.36).

Conversely, suppose that a function $f : X \to Y$ satisfies (4.1). Setting $(x, y) = (x + y, z)$ as well as the other two cyclic permutations of the variables x, y and z in

(4.1), we obtain a set of equations:

$$f(x+y+z) + f(x+y-z) = 2f(x+y) + 2f(z),$$
$$f(x+y+z) + f(x-y+z) = 2f(x+z) + 2f(y), \qquad (4.37)$$
$$f(x+y+z) + f(-x+y+z) = 2f(y+z) + 2f(x).$$

Setting $(x, y) = (x, y, -z)$ and all cyclic permutations of the variables (4.1), we have another set of equations:

$$f(x+y+z) + f(x-y+z) = 2f(x) + 2f(y-z),$$
$$f(-x+y+z) + f(x+y-z) = 2f(y) + 2f(z-x), \qquad (4.38)$$
$$f(x-y+z) + f(-x+y+z) = 2f(z) + 2f(x-y).$$

Subtracting half the sum of all equations in (4.38) from the sum of all equations in (4.37), we are left with

$$3f(x+y+z) = 2\left(f(x+y) + f(y+z) + f(z+x)\right)$$
$$- \left(f(x-y) + f(y-z) + f(z-x)\right) + \left(f(x) + f(y) + f(z)\right). \qquad (4.39)$$

If we rewrite (4.1) as $f(x+y) = 2f(x) + 2f(y) - f(x-y)$ and perform cyclic permutation of all variables, then (4.39) simplifies to

$$3f(x+y+z) = 9\left(f(x) + f(y) + f(z)\right) - 3\left(f(x-y) + f(y-z) + f(z-x)\right). \quad (4.40)$$

Setting $(x, y) = (x, x - y)$ and all cyclic permutations of variables in (4.1), we have

$$f(2x-y) + f(y) = 2f(x) + 2f(x-y),$$
$$f(2y-z) + f(z) = 2f(y) + 2f(y-z), \qquad (4.41)$$
$$f(2z-x) + f(x) = 2f(z) + 2f(z-x).$$

If we use (4.41) to eliminate $f(x-y)$, $f(y-z)$ and $f(z-x)$ in (4.40), then (4.33) follows, and the proof is complete. $\square$

The following theorem determines the general solution of a quadratic functional equation in three variables

$$f(x-z) + f(y-z) = \frac{1}{2}f(x-y) + 2f\left(\frac{x+y}{2} - z\right). \qquad (4.42)$$

Theorem 4.7 ([3]). *The functional equations (4.1) and (4.42) are equivalent.*

Proof. Assume that $Q : X \to Y$ is a solution of equation (4.1). Then Q is even and $Q(0) = 0$. Setting $y = x$ in (4.1), we get $Q(2x) = 4Q(x)$, Hence,

$$Q(x) = 4Q\left(\frac{x}{2}\right), \quad x \in X. \tag{4.43}$$

Replacing x and y by $x - z$ and $y - z$ in (4.1), respectively, we obtain

$$Q(x + y - 2z) + Q(x - y) = 2Q(x - z) + 2Q(y - z), \quad x, y, z \in X.$$

Therefore on account of (4.1), one can easily check that Q is a solution of (4.42).

Conversely, assume that $f : X \to Y$ is a solution of equation (4.42). Putting $x = y = z = 0$ in (4.42), we obtain $f(0) = 0$. Setting $y = z = 0$ in (4.42), we get

$$\frac{1}{2}f(x) = 2f\left(\frac{x}{2}\right), \quad x \in X.$$

Replacing x by $x + y$ in the above equality, we obtain

$$\frac{1}{2}f(x + y) = 2f\left(\frac{x + y}{2}\right), \quad x, y \in X. \tag{4.44}$$

Setting $z = 0$ in (4.42), we have

$$f(x) + f(y) = \frac{1}{2}f(x - y) + 2f\left(\frac{x + y}{2}\right), \quad x, y \in X,$$

which means that by virtue of (4.44) f satisfies (4.1). This completes the required proof. $\qquad\square$

The following theorem presents the solution of the following quadratic functional equation in three variables

$$f(2x - y) + f(2y - z) + f(2z - x) + f(x + y + z) - f(x - y + z)$$
$$- f(x + y - z) - f(x - y - z) = 3f(x) + 3f(y) + 3f(z). \tag{4.45}$$

Theorem 4.8 ([288]). *A function $f : X \to Y$ satisfies the functional equation (4.45) for all $x, y, z \in X$ if and only if it satisfies the functional equation (4.1).*

Proof. Suppose a function $f : X \to Y$ satisfies (4.45). Setting $(x, y, z) = (x, x, x)$ in (4.45), we obtain

$$f(x) + f(x) + f(x) + f(3x) - f(x) - f(x) - f(-x) = 3f(x) + 3f(x) + 3f(x)$$

which gives

$$f(3x) = 8f(x) + f(-x). \qquad (4.46)$$

Again, setting $(x, y, z) = (-x, x, x)$ in (4.45), we obtain

$$f(-3x) + f(x) + f(3x) + f(x) - f(-x) - f(-x) - f(-3x)$$
$$= 3f(-x) + 3f(x) + 3f(x)$$

which gives

$$f(3x) = 5f(-x) + 4f(x). \qquad (4.47)$$

From equations (4.46) and (4.47), we obtain $f(x) = f(-x)$. Thus, f is an even function. Putting $x = 0$ in equation (4.45), we obtain $f(0) = 0$. Setting $(x, y, z) = (x, 0, 0)$ in (4.45) and using evenness we arrive at

$$f(2x) - f(x) = 3f(x)$$

which gives

$$f(2x) = 4f(x). \qquad (4.48)$$

In equation (4.46), using the evenness, we arrive at

$$f(3x) = 9f(x) \qquad (4.49)$$

for all $x \in X$. Extending these ideas, in general, we obtain $f(nx) = n^2 f(x)$. Setting $x = 0$ in equation (4.45), using (4.48) and the evenness of f, we obtain

$$f(2y - z) - f(-y + z) - f(y - z) = 2f(y) - f(z). \qquad (4.50)$$

Again, setting $(y, z) = (y, y - z)$ in equation (4.50) and using evenness of f, we obtain

$$f(y + z) + f(y - z) = 2f(y) + 2f(z).$$

Setting $(y, z) = (x, y)$ in the above equality, we obtain (4.1).

Conversely, suppose that a function $f : X \to Y$ satisfies (4.1). Setting $(x, y) = (x + y, z)$, $(y + z, x)$, $(z + x, y)$, respectively, in equation (4.1), we obtain a set of equations:

$$f(x + y + z) + f(x + y - z) = 2f(x + y) + 2f(z),$$

$$f(y+z+x) + f(y+z-x) = 2f(y+z) + 2f(x), \qquad (4.51)$$
$$f(z+x+y) + f(z+x-y) = 2f(z+x) + 2f(y).$$

Again, setting $(x, y) = (x, y-z)$, $(y, z-x)$, $(z, x-y)$ respectively in equation (4.1), we have another set of equations:

$$f(x+y-z) + f(x-y+z) = 2f(x) + 2f(y-z),$$
$$f(y+z-x) + f(y-z+x) = 2f(y) + 2f(z-x), \qquad (4.52)$$
$$f(z+x-y) + f(z-x+y) = 2f(z) + 2f(x-y).$$

Subtracting half the sum of all equations in (4.52) from the sum of all equations in (4.51), we obtain

$$3f(x+y+z) = 2[f(x+y) + f(y+z) + f(z+x)]$$
$$- [f(x-y) + f(y-z) + f(z-x)] + [f(x) + f(y) + f(z)]. \qquad (4.53)$$

If we rewrite (4.1) as $f(x+y) = 2f(x) + 2f(y) - f(x-y)$ and perform cyclic permutation of all variables, then (4.53) simplifies to

$$3f(x+y+z) = 9[f(x) + f(y) + f(z)] - 3[f(x-y) + f(y-z) + f(z-x)]. \qquad (4.54)$$

Setting $(x, y) = (x, x-y)$ and all cyclic permutations of variables in (4.1), we have

$$f(2x-y) + f(y) = 2f(x) + 2f(x-y),$$
$$f(2y-z) + f(z) = 2f(y) + 2f(y-z), \qquad (4.55)$$
$$f(2z-x) + f(x) = 2f(z) + 2f(z-x).$$

From equation (4.55), we get

$$2[f(x-y) + f(y-z) + f(z-x)]$$
$$= f(2x-y) + f(2y-z) + f(2z-x) - [f(x) + f(y) + f(z)]. \qquad (4.56)$$

Using (4.56) in equation (4.54), we get

$$f(2x-y) + f(2y-z) + f(2z-x) + 2f(x+y+z) = 7f(x) + 7f(y) + 7f(z). \qquad (4.57)$$

Setting y by $y+z$ and $y-z$ in equation (4.1), respectively, we get

$$f(x+y+z) + f(x-y-z) = 2f(x) + 2f(y+z), \qquad (4.58)$$
$$f(x+y-z) + f(x-y+z) = 2f(x) + 2f(y-z). \qquad (4.59)$$

Adding equations (4.58) to (4.59) and using (4.1), we get

$$f(x+y+z) + f(x-y-z) + f(x+y-z) + f(x-y+z)$$
$$= 4f(x) + 4f(y) + 4f(z). \qquad (4.60)$$

Subtracting equation (4.58) from (4.57), we obtain equation (4.45). This completes the proof. $\qquad\square$

The following theorem provides the solution of a quadratic functional equation in k-variables of the form:

$$\sum_{i=2}^{k} \sum_{\epsilon_i \in \{-1,1\}} f(x_1 + \epsilon_i x_i) = 2(k-1)f(x_1) + 2\sum_{i=2}^{k} f(x_i). \qquad (4.61)$$

Theorem 4.9 ([91]). *A mapping $f : X \to Y$ satisfies the above equation (4.61) for all $x_1, x_2, \ldots, x_k \in X$ if and only if f is quadratic.*

Proof. If we replace $x_1, x_2, \ldots, x_k$ in (4.61) by 0, then we get $f(0) = 0$. Putting $x_3 = x_4 = \cdots = x_k = 0$ in equation (4.61), we see that

$$f(x_1 - x_2) + f(x_1 + x_2) + 2(k-2)f(x_1) = 2(k-1)f(x_1) + 2f(x_2).$$

Hence,

$$f(x_1 + x_2) + f(x_1 - x_2) = 2f(x_1) + 2f(x_2).$$

The converse is trivial. $\qquad\square$

The following theorem proves the general solution of a more general form of a quadratic functional equation in n-variables:

$$\sum_{1 \le k < l \le n} (f(x_k + x_l) + f(x_k - x_l)) - 2(n-1)\sum_{i=1}^{n} f(x_i) = 0. \qquad (4.62)$$

Theorem 4.10 ([142]). *A mapping $f : X \to Y$ satisfies the functional equation (4.62) if and only if f is quadratic.*

Proof. Let f be a quadratic function. Assume equation (4.62) is true for n by induction argument. By (4.1), we have

$$f(x_i + x_{n+1}) + f(x_i - x_{n+1}) - 2f(x_i) - 2f(x_{n+1}) = 0 \qquad (4.63)$$

for all $i = 1, 2, \ldots, n$. Adding up (4.62) and (4.63), we have the desired equation (4.62) for $n + 1$.

Conversely, let f satisfy equation (4.62). By letting $x_i = 0$ for all $i = 1, 2, \ldots, n$, we have $f(0) = 0$. Replacing $x_i = 0$ for all $i = 3, 4, \ldots, n$, we obtain the equation

$$f(x_1 + x_2) + f(x_1 + x_2) = 2f(x_1) + 2f(x_2),$$

which implies that f is quadratic, which completes the proof. $\qquad\square$

The following theorems contain the general solution of various other quadratic functional equations.

Theorem 4.11 ([179]). *If an even mapping $f : X \to Y$ satisfies*

$$\sum_{i=1}^{2n} f\left(x_i - \frac{1}{2n}\sum_{j=1}^{2n} x_j\right) = \sum_{i=1}^{2n} f(x_i) - 2nf\left(\frac{1}{2n}\sum_{i=1}^{2n} x_i\right) \qquad (4.64)$$

for all $x_1, x_2, \ldots, x_{2n} \in X$, then the mapping $f : X \to Y$ is quadratic.

Proof. Assume that $f : X \to Y$ satisfies (4.64). Letting $x_1 = \cdots = x_n = x, x_{n+1} = \cdots = x_{2n} = y$ in (4.64), we get

$$nf\left(x - \frac{x+y}{2}\right) + nf\left(y - \frac{x+y}{2}\right) = nf(x) + nf(y) - 2nf\left(\frac{x+y}{2}\right)$$

for all $x, y \in V$. Since $f : X \to Y$ is even, then

$$2nf\left(\frac{x-y}{2}\right) = nf(x) + nf(y) - 2nf\left(\frac{x+y}{2}\right)$$

for all $x, y \in X$. So,

$$2f\left(\frac{x+y}{2}\right) + 2f\left(\frac{x-y}{2}\right) = f(x) + f(y) \qquad (4.65)$$

for all $x, y \in V$. Letting $x = y = 0$ in (4.65), we get $f(0) = 0$. Letting $y = 0$ in (4.65), we get $f\left(\frac{x}{2}\right) = \frac{1}{4}f(x)$ for all $x \in X$. It follows from (4.65) that equation (4.1) holds. This completes the required proof. $\qquad \square$

Theorem 4.12 ([179]). *An even mapping $f : X \to Y$ satisfies*

$$f\left(x - \frac{x+y}{2}\right) + f\left(y - \frac{x+y}{2}\right) = f(x) + f(y) - 2f\left(\frac{x+y}{2}\right) \qquad (4.66)$$

for all $x, y \in X$ if and only if the mapping $f : X \to Y$ is quadratic.

Proof. By Theorem 4.11, it is enough to show that if $f : X \to Y$ is quadratic, then $f : X \to Y$ satisfies (4.66).

Assume that $f : X \to Y$ is quadratic. Since $f(2x) = 4f(x)$ for all $x \in X$, $f\left(\frac{x}{2}\right) = \frac{1}{4}f(x)$ for all $x \in V$. So,

$$2f\left(\frac{x+y}{2}\right) + 2f\left(\frac{x-y}{2}\right) = f(x) + f(y)$$

for all $x, y \in X$. Thus,

$$f\left(x - \frac{x+y}{2}\right) + f\left(y - \frac{x+y}{2}\right) = 2f\left(\frac{x-y}{2}\right) = f(x) + f(y) - 2f\left(\frac{x+y}{2}\right)$$

for all $x, y \in X$, as desired. The proof is complete. $\qquad \square$

Theorem 4.13 ([203]). *A function $f : X \to Y$ satisfies*

$$\sum_{\substack{1 \leq i < j \leq 4 \\ 1 \leq k < l \leq 4 \\ k, l \in \{1,2,3,4\}}} f\left(x_i + x_j - x_k - x_l\right) = 2 \sum_{1 \leq i < j \leq 4} f\left(x_i - x_j\right) \tag{4.67}$$

for all $x_1, x_2, x_3, x_4 \in X$, if the function f is quadratic.

Proof. Let f satisfy (4.67). Letting $x_1 = x_2 = x_3 = x_4 = 0$ in (4.67), we get that $f(0) = 0$. Setting $x_1 = x$ and $x_2 = x$ and $x_2 = x_3 = x_4 = 0$ in (4.67), we conclude that $f(-x) = f(x)$ for all $x \in X$. This means that f is an even function.

Letting $x_1 = x_2 = x$ and $x_3 = x_4 = 0$ in (4.67), and using the evenness of f, we get $f(2x) = 4f(x)$ for all $x \in X$. Letting $x_3 = x_4 = 0$ in (4.67), and using the evenness of f, we get

$$2f\left(x_1 + x_2\right) + 4f\left(x_1 - x_2\right) = 2f\left(x_1 - x_2\right) + 4f\left(x_1\right) + 4f\left(x_2\right)$$

for all $x_1, x_2 \in X$. Therefore,

$$f\left(x_1 + x_2\right) + f\left(x_1 - x_2\right) = 2f\left(x_1\right) + 2f\left(x_2\right)$$

for all $x_1, x_2 \in X$. Therefore, the function $f : X \to Y$ is quadratic. This completes the proof. $\qquad\square$

In the following theorem, we obtain the general solution of the following functional equation

$$4f\left(\frac{x + y + z}{2}\right) + f(x) + f(y) + f(z)$$
$$= 4\left[f\left(\frac{x - y}{2}\right) + f\left(\frac{y + z}{2}\right) + f\left(\frac{z - x}{2}\right)\right]. \tag{4.68}$$

To find the general solution of equation (4.68), we prove that the above functional equation (4.68) is equivalent to the following functional equation

$$4\left[f\left(\frac{x + y}{2}\right) + f\left(\frac{x - y}{2}\right)\right] = 2f(x) + 2f(y). \tag{4.69}$$

The functional equation (4.69) has the solution $f(x) = x^2$. Hence, every solution of equation (4.68) is a quadratic function.

Theorem 4.14. *A function $f : X \to Y$ satisfying the functional equation (4.68) is equivalent to the functional equation (4.69), for all $x, y \in X$.*

Proof. Putting $x = y = z = 0$ in (4.68), we get

$$f(0) = 0. \tag{4.70}$$

Putting $y = z = 0$ in equation (4.68), we obtain

$$f(x) = 4f\left(\frac{-x}{2}\right). \tag{4.71}$$

Now, replacing z by x in equation (4.68), we get

$$4f\left(\frac{-y}{2}\right) + 2f(x) + f(y) = 4\left[f\left(\frac{x-y}{2}\right) + f\left(\frac{x+y}{2}\right)\right]. \tag{4.72}$$

Applying (4.70) and (4.71) in (4.72), we obtain

$$2f(x) + 2f(y) = 4\left[f\left(\frac{x-y}{2}\right) + f\left(\frac{x+y}{2}\right)\right].$$

Therefore, equation (4.68) transforms to equation (4.69).

Conversely, assume that a function $f : X \to Y$ is a solution of equation (4.69). Putting $x = y = 0$ in (4.69), we obtain $f(0) = 0$. Substituting $y = 0$ in (4.69) and then simplifying, we have $f\left(\frac{x}{2}\right) = \frac{1}{4}f(x)$. Replacing x by 0 in (4.69) and simplifying, we arrive at $f(-y) = f(y)$. Thus, f is even. Clearly, by (4.69), we have

$$4f\left(\frac{x-y-z}{2}\right) + 4f\left(\frac{x}{2}\right) = 8f\left(\frac{x}{2} - \frac{y+z}{4}\right) + 8f\left(\frac{y+z}{4}\right) \tag{4.73}$$

and

$$4f\left(\frac{y}{2}\right) + 4f\left(\frac{z}{2}\right) = 8f\left(\frac{y+z}{4}\right) + 8f\left(\frac{y-z}{4}\right). \tag{4.74}$$

Adding (4.73) and (4.74), and then simplifying, we obtain

$$4f\left(\frac{x-y-z}{2}\right) + f(x) + f(y) + f(z)$$
$$= 8f\left(\frac{x}{2} - \frac{y+z}{4}\right) + 16f\left(\frac{y+z}{4}\right) + 8f\left(\frac{y-z}{4}\right).$$

Now, by (4.69) and using the fact that f is even, we arrive at

$$4f\left(\frac{x+y+z}{2}\right) + f(x) + f(y) + f(z)$$
$$= 4\left[f\left(\frac{x-y}{2}\right) + f\left(\frac{y+z}{2}\right) + f\left(\frac{z-x}{2}\right)\right].$$

This implies that functional equations (4.69) and (4.68) are equivalent. This completes the proof. □

Chapter 5

General Solution of Cubic Type Functional Equations

5.1. Introduction

In the previous chapter, we discussed about the general solution of quadratic functional equations involving two variables, three variables and n-variables. In this chapter, we present the general solution of various forms of cubic functional equations.

The motivation for studying cubic functional equations came from the fact that recently polynomial equations have found applications in approximate checking, self-testing and self-correcting of computer programs to compute certain polynomials.

Rassias [270] introduced the cubic functional equation as follows:

$$f(x + 2y) - 3f(x + y) + 3f(x) - f(x - y) = 6f(y).$$

This inspiring cubic functional equation was the transition from the following famous Euler–Lagrange–Rassias quadratic functional equation

$$f(x + y) - 2f(x) + f(x - y) = 2f(y)$$

to the cubic functional equations.

In Section 5.2, the general solution of the functional equation

$$f(2x + y) + f(2x - y) = 2f(x + y) + 2f(x - y) + 12f(x) \tag{5.1}$$

is presented, and in Section 5.3, the general solution of various other cubic functional equations is provided. In Sections 5.2 and 5.3, let X and Y be real vector spaces.

5.2. Cubic functional equation

This section is devoted to the general solution of the functional equation (5.1). It is easy to see that the function $f(x) = cx^3$ is a solution of the above functional equation. Thus, it is natural that equation (5.1) is called a **cubic functional equation** and every solution of the cubic functional equation (5.1) is said to be a cubic function.

The following theorem establishes the general solution of the cubic functional equation (5.1).

Theorem 5.1 ([137]). *A function $f : X \to Y$ satisfies the functional equation (5.1) if and only if there exists a function $B : X \times X \times X \to Y$ such that $f(x) = B(x,x,x)$ for all $x \in X$ and B is symmetric for each fixed one variable and is additive for two fixed variables.*

Proof. Putting $x = 0 = y$ in (5.1), we get $f(0) = 0$. Set $x = 0$ in (5.1) to get $f(y) + f(-y) = 0$. Letting $y = 0$ and $y = x$ in (5.1), we obtain that $f(2x) = 8f(x)$ and $f(3x) = 27f(x)$ for all $x \in X$, respectively. By induction, we have to $f(kx) = k^3 f(x)$ for all positive integer k. Replacing x and y by $x + y$ and $x - y$ in (5.1), respectively, we have

$$f(3x + y) + f(x + 3y) = 12f(x + y) + 16f(x) + 16f(y) \tag{5.2}$$

for all $x, y \in X$. Putting x and y by $x + y$ and $2y$ in (5.1), respectively, we obtain

$$8f(x + 2y) + 8f(x) = 12f(x + y) + 2f(x + 3y) + 2f(x - y). \tag{5.3}$$

Interchange x and y in (5.3) to get the relation

$$8f(y + 2x) + 8f(y) = 12f(x + y) + 2f(y + 3x) - 2f(x - y). \tag{5.4}$$

Adding (5.3) to (5.4) and using (5.2), we have

$$f(x + 2y) + f(2x + y) = 6f(x + y) + 3f(x) + 3f(y) \tag{5.5}$$

for all $x, y \in X$. Using (5.1), we have

$$3f(2x + z) + 3f(2x - z) + 3f(2y + z) + 3f(2y - z)$$
$$= 36f(x) + 36f(y) + 6f(x + z) + 6f(x - z) + 6f(y + z) + 6f(y - z). \tag{5.6}$$

On the other hand, using (5.5), (5.1) and simplifying, we obtain

$$3f(2x + z) + 3f(2y + z) + 3f(2x - z) + 3f(2y - z)$$
$$= 12f(2x + y) + 2f(2x + y + 3z) + 2f(2x + y - 3z)$$
$$\quad - 48f(x + y + z) + 12f(x + 2y) + 2f(x + 2y + 3z)$$
$$\quad + 2f(x + 2y - 3z) - 48f(x + y - z)$$

which yields by virtue of (5.6), the relation

$$12f(2x + y) + 2f(2x + y + 3z) + 2f(2x + y - 3z)$$
$$+ 12f(x + 2y) + 2f(x + 2y + 3z) + 2f(x + 2y - 3z)$$
$$= 36f(x) + 36f(y) + 6f(x + z) + 6f(x - z)$$
$$+ 6f(y + z) + 6f(y - z) + 48f(x + y + z) + 48f(x + y - z). \qquad (5.7)$$

Also, by virtue of (5.5) and (5.1), the left-hand side of (5.6) can be written in the form

$$3f(2x + z) + 3f(2y - z) + 3f(2x - z) + 3f(2y + z)$$
$$= 12f(2x + y) + 2f(2x + y + z) + 2f(2x + y - z)$$
$$+ 12f(x + 2y) + 2f(x + 2y + z) + 2f(x + 2y - z) - 96f(x + y). \qquad (5.8)$$

Replacing z by $3z$ in (5.8) and then using (5.7), we have

$$3f(2x + 3z) + 3f(2y - 3z) + 3f(2x - 3z) + 3f(2y + 3z)$$
$$= 36f(x) + 3yf(y) + 6f(x + z) + 6f(x - z)$$
$$+ 6f(y + z) + 6f(y - z) + 48f(x + y + z)$$
$$+ 48f(x + y - z) - 96f(x + y). \qquad (5.9)$$

Again, applying (5.5) and then (5.1) to the left-hand side of (5.9), we get

$$3f(2x + 3z) + 3f(2x - 3z) + 3f(2y + 3z) + 3f(2y - 3z)$$
$$= f(6x + 3z) + f(6x - 3z) - 6f(4x) + f(6y + 3z)$$
$$= 27(12f(x) + 2f(x + z) + 2f(x - z) + 12f(y) + 2f(y + z) + 2f(y - z))$$
$$- 384f(x) - 384f(y). \qquad (5.10)$$

Finally, we obtain from (5.9) and (5.10) that

$$f(x + y + z) + f(x + y - z) + 2f(x) + 2f(y)$$
$$= 2f(x + y) + f(x + z) + f(x - z) + f(y + z) + f(y - z) \qquad (5.11)$$

for all $x, y \in X$. Hereafter, the last relation plays an important role in proving our statement.

Define $B : X \times X \times X \to Y$ by

$$B(x, y, z) = \frac{1}{24}[f(x + y + z) + f(x - y - z) - f(x + y - z) - f(x - y + z)] \qquad (5.12)$$

for all $x, y, z \in X$. Then B is symmetric for any one fixed variable since f is odd function and

$$B(x, x, x) = \frac{1}{24}[f(3x) + f(-x) - f(x) - f(x)] = f(x)$$

holds for all $x \in X$.

Finally, we claim that B is additive for any two fixed variables. Since B is symmetric, it is sufficient to show that

$$B(u+v,y,z) = B(u,y,z) + B(v,y,z) \tag{5.13}$$

for all $u,v,y,z \in X$. Now, using equation (5.11) and the definition of B, we obtain

$$24B(u+v,y,z)$$
$$= f(u+v+y+z) + f(u+v-y-z) - f(u+v+y-z) - f(u+v-y+z)$$
$$= 24B(u,y,z) + 24B(v,y,z). \tag{5.14}$$

Conversely, if there exists a function $B : X \times X \times X \to Y$ such that $f(x) = B(x,x,x)$ for all $x \in X$ and B is symmetric for fixed one variable and B is additive for fixed two variables, it is obvious that f satisfies equation (5.1). This completes the proof of theorem. $\qquad\square$

5.3. Other cubic functional equations

The following theorems comprise the general solution of other cubic functional equations.

Theorem 5.2 ([204]). *A mapping $f : X \to Y$ satisfies the functional equation* (5.1) *if and only if $f : X \to Y$ satisfies the functional equation*

$$2f(x+2y) + f(2x-y) = 5f(x+y) + 5f(x-y) + 15f(y) \tag{5.15}$$

for all $x,y \in X$.

Proof. Let $f : X \to Y$ satisfy the functional equation (5.1). Putting $x = y = 0$ in (5.1), we get $f(0) = 0$. Set $x = 0$ in (5.1) to get $f(-y) = -f(y)$ for all $y \in X$, so the mapping f is odd. Letting $y = 0$ and $y = x$ in (5.1), respectively, we obtain that $f(2x) = 8f(x)$ and $f(3x) = 27f(x)$ for all $x \in X$. Replacing y with $2y$ in (5.1), we have

$$4f(x+y) + 4f(x-y) = f(x+2y) + f(x-2y) + 6f(x) \tag{5.16}$$

for all $x,y \in X$. Replacing x and y by $-y$ and x in (5.1), respectively, and using the oddness of f, we get that

$$f(x-2y) = f(x+2y) - 2f(x+y) + 2f(x-y) - 12f(y) \tag{5.17}$$

for all $x,y \in X$. Hence, we obtain from (5.16) and (5.17) that

$$f(x+2y) = 3f(x+y) + f(x-y) - 3f(x) + 6f(y) \tag{5.18}$$

for all $x,y \in X$. Replacing x by $2x$ in (5.18), we get

$$f(2x+2y) = 3f(2x+y) + f(2x-y) - 3f(2x) + 6f(y)$$

for all $x, y \in X$. Therefore, we have

$$8f(x+y) = 2f(2x+y) + [f(2x+y) + f(2x-y)] - 24f(x) + 6f(y)$$

for all $x, y \in X$. Then by using (5.1) in the last equation, we obtain

$$f(2x+y) = 3f(x+y) - f(x-y) + 6f(x) - 3f(y)$$

for all $x, y \in X$. Replacing y by $-y$ in the above equation, we have

$$f(2x-y) = 3f(x-y) - f(x+y) + 6f(x) + 3f(y) \qquad (5.19)$$

for all $x, y \in X$. Therefore, it follows from (5.18) and (5.19) that

$$2f(x+2y) + f(2x-y) = 5f(x+y) + 5f(x-y) + 15f(y)$$

for all $x, y \in X$.

Conversely, let $f : X \to Y$ satisfy the functional equation (5.15). Putting $x = y = 0$ in (5.15), we get $f(0) = 0$. Letting $y = 0$ and $y = x$ in (5.15), respectively, we obtain that $f(2x) = 8f(x)$ and $f(3x) = 27f(x)$, for all $x \in X$. By setting $x = 2y$ in (5.15), we infer $f(4y) = 64f(y)$, for all $y \in X$. Set $x = 0$ in (5.15) to get $f(-y) = -f(y)$, for all $y \in X$, so the mapping f is odd. Replacing x by $2x$ in (5.15), we have

$$16f(x+y) + f(4x-y) = 5f(2x+y) + 5f(2x-y) + 15f(y) \qquad (5.20)$$

for all $x, y \in X$. Replacing y by $4y$ in (5.20), we get with a simple computation that

$$2f(x+4y) + 8f(x-y) = 5f(x+2y) + 5f(x-2y) + 120f(y) \qquad (5.21)$$

for all $x, y \in X$. Replacing x and y by $-y$ and x in (5.21), respectively, and using the oddness of f, we get

$$2f(4x-y) - 8f(x+y) = 5f(2x-y) - 5f(2x+y) + 120f(x) \qquad (5.22)$$

for all $x, y \in X$. Hence, we obtain from (5.20) and (5.22) that

$$10[f(2x+y) + f(2x-y)] = 32f(x+y) + 8f(x+y) + 5f(2x-y) \\ - 5f(2x+y) + 120f(x) - 30f(y)$$

for all $x, y \in X$. Therefore, we have

$$3f(2x+y) + f(2x-y) = 8f(x+y) + 24f(x) - 6f(y) \qquad (5.23)$$

for all $x, y \in X$. Replacing y by $-y$ in (5.23) and using the oddness of f, we get

$$3f(2x-y) + f(2x+y) = 8f(x-y) + 24f(x) + 6f(y) \qquad (5.24)$$

for all $x, y \in X$. Adding (5.23) and (5.24), we have

$$4[f(2x-y) + f(2x+y)] = 8f(x+y) + 8f(x-y) + 48f(x)$$

for all $x, y \in X$. Hence, we obtain (5.1). This completes the proof. $\qquad \square$

Theorem 5.3 ([293]). *A function $f : X \to Y$ satisfies the functional equation*

$$3f(x + 3y) - f(3x + y) = 12[f(x + y) + f(x - y)] + 80f(y) - 48f(x) \qquad (5.25)$$

if and only if $f : X \to Y$ satisfies the functional equation (5.15), for all $x, y \in X$.

Proof. Putting $x = y = 0$ in (5.25), we get $f(0) = 0$. Let $y = 0$ in (5.25), we obtain

$$f(3x) = 27f(x) \qquad (5.26)$$

for all $x \in X$. Setting $x = 0$ in equation (5.25) and using (5.26), we get

$$f(-y) = -f(y) \qquad (5.27)$$

for all $y \in X$. Replacing y by x in (5.25), we obtain

$$f(4x) = 6f(2x) + 16f(x) \qquad (5.28)$$

for all $x \in X$. Again replacing x by $3y$ in (5.25), and using equation (5.28), we get

$$f(10y) = -72f(2y) - 192f(y) - 12f(2) - 80f(y) + 1296f(y) + 81f(2y) \qquad (5.29)$$

for all $y \in X$. Substituting y by $3x$ in equation (5.25), and using (5.28) and (5.29), we obtain

$$f(2x) = 8f(x) \qquad (5.30)$$

for all $x \in X$. Replacing y by $y - x$ in (5.25) and using oddness of f, we arrive at

$$3f(-2x + 3y) - f(2x + y) = 12[f(y) + f(2x - y)] + 80f(y - x) - 48f(x) \qquad (5.31)$$

for all $x, y \in X$. Interchanging x and y in (5.31), we get

$$3f(3x - 2y) - f(2y + x) = 12[f(x) + f(2y - x)] + 80f(x - y) - 48f(y) \qquad (5.32)$$

for all $x, y \in X$. Replacing x by $x - y$ in (5.25), we obtain

$$3f(x + 2y) - f(3x - 2y) = 12[f(x) + f(x - 2y)] + 80f(y) - 48f(x - y) \qquad (5.33)$$

for all $x, y \in X$. Multiplying the above equation by 3 and using (5.32), we get

$$f(x + 2y) - 3f(x - 2y) = 6f(x) + 24f(y) - 8f(x - y) \qquad (5.34)$$

for all $x, y \in X$. Replacing y by $-y$ in (5.34), we obtain

$$f(x - 2y) - 3f(x + 2y) = 6f(x) - 24f(y) - 8f(x + y) \qquad (5.35)$$

for all $x, y \in X$. Using (5.34) in (5.35), we arrive at

$$f(x + 2y) = 3f(x + y) + f(x - y) - 3f(x) + 6f(y) \qquad (5.36)$$

for all $x, y \in X$. Replacing (x, y) by $(-y, x)$ in (5.36), we have

$$f(2x - y) = 3f(x - y) - f(x + y) + 3f(y) + 6f(x) \qquad (5.37)$$

for all $x, y \in X$. Multiplying equation (5.37) by 2 and adding with (5.36), we arrive at (5.15), for all $x, y \in X$.

Conversely, assume f satisfies the functional equation (5.15). Letting (x, y) by $(0, 0)$ in (5.15), we get $f(0) = 0$. Set $y = 0$ in (5.15), we obtain $f(2x) = 8f(x)$, for all $x \in X$. Putting $x = 0$ in (5.15), we obtain $f(-y) = -f(y)$, for all $y \in X$. Thus, f is an odd function. Replacing x by $x + y$ in (5.15), we get

$$2f(x + 3y) + f(2x + y) = 5f(x + 2y) + 5f(x) + 15f(y) \tag{5.38}$$

for all $x, y \in X$. Again replacing y by $y - x$ in (5.15), we get

$$-2f(x - 2y) + f(3x - y) = 5f(y) + 5f(2x - y) + 15f(y - x) \tag{5.39}$$

for all $x, y \in X$. Substituting (x, y) by $(y, -x)$ in (5.39), we arrive at

$$-2f(2x + y) + f(x + 3y) = -5f(x) + 5f(x + 2y) - 15f(x + y) \tag{5.40}$$

for all $x, y \in X$. Adding (5.38) and (5.40), we get

$$3f(x + 3y) - f(2x + y) = 10f(x + 2y) - 15f(x + y) + 15f(y) \tag{5.41}$$

for all $x, y \in X$. Again adding (5.41) to (5.39) and using (5.15), we arrive at

$$3f(x + 3y) + f(3x - y) = 15f(x + y) + 15f(x - y) + 80f(y) \tag{5.42}$$

for all $x, y \in X$. Replacing y by $-y$ in (5.39), we arrive at

$$f(3x + y) = 2f(x + 2y) - 5f(y) + 5f(2x + y) - 15f(x + y) \tag{5.43}$$

for all $x, y \in X$. Adding (5.42) to (5.43) and using (5.41) and (5.15), we get

$$f(3x + y) + f(3x - y) = -5f(x + y) - 5f(x - y) + 4f(2x + y) + 4f(2x - y) \tag{5.44}$$

for all $x, y \in X$. Setting (x, y) by $(x - y, y)$ in (5.15), we obtain

$$2f(x + y) + (2x - 3y) = 5f(x) + 5f(x - 2y) + 15f(y) \tag{5.45}$$

for all $x, y \in X$. Setting (x, y) by $(x, x - y)$ in (5.15), we obtain

$$2f(3x - 2y) + f(x + y) = 5f(2x - y) + 5f(y) + 15f(x - y) \tag{5.46}$$

for all $x, y \in X$. Replacing x by y in (5.46) and using oddness of f, we arrive at

$$-2f(2x - 3y) + f(x + y) = -5f(x - 2y) + 5f(x) - 15f(x - y) \tag{5.47}$$

for all $x, y \in X$. Multiplying equation (5.45) by 2, we get

$$4f(x + y) + 2f(2x - 3y) = 10f(x) + 10f(x - 2y) + 30f(y) \tag{5.48}$$

for all $x, y \in X$. Adding (5.47) and (5.48), we obtain

$$5f(x - 2y) = 5f(x + y) - 15f(x) - 30f(y) + 15f(x - y) \tag{5.49}$$

for all $x, y \in X$. Replacing x by y and using oddness of f in (5.49) and then dividing by 5, we arrive at

$$f(2x - y) = -f(x + y) + 3f(y) + 6f(x) + 3f(x - y) \qquad (5.50)$$

for all $x, y \in X$. Replacing y by $-y$ in equation (5.50), we obtain

$$f(2x + y) = -f(x - y) - 3f(y) + 6f(x) + 3f(x + y) \qquad (5.51)$$

for all $x, y \in X$. Using equations (5.50) and (5.51) in equation (5.44), we obtain

$$f(3x + y) + f(3x - y) = 3f(x + y) + 3f(x - y) + 48f(x) \qquad (5.52)$$

for all $x, y \in X$. Subtracting (5.52) from (5.42), we obtain (5.25). This completes the proof. □

Theorem 5.4 ([201]). *A mapping $f : X \to Y$ satisfies the functional equation (5.15) if and only if $f : X \to Y$ satisfies the functional equation*

$$3f(x + 3y) + f(3x - y) = 15f(x + y) + 15f(x - y) + 80f(y) \qquad (5.53)$$

for all $x, y \in X$.

Proof. Putting $x = y = 0$ in (5.15), we get $f(0) = 0$. Set $y = 0$ in (5.15) to get $f(2x) = 8f(x)$, for all $x \in X$. Letting $x = 0$ in (5.15), we obtain that $f(-y) = -f(y)$, for all $y \in X$. Replacing x by $x + y$ in (5.15), we have

$$2f(x + 3y) + f(2x + y) = 5f(x + 2y) + 5f(x) + 15f(y) \qquad (5.54)$$

for all $x, y \in X$. Since f is odd, replacing y by $y - x$ in (5.15), we get

$$f(3x - y) - 2f(x - 2y) = 5f(2x - y) - 15f(x - y) + 5f(y) \qquad (5.55)$$

for all $x, y \in X$. Replacing x and y by y and $-x$ in (5.55), respectively, we obtain

$$f(3y + x) - 2f(2x + y) = 5f(x + 2y) - 15f(x + y) - 5f(x) \qquad (5.56)$$

for all $x, y \in X$. Adding (5.54) and (5.56), we get

$$3f(x + 3y) - f(2x + y) = 10f(x + 2y) - 15f(x + y) + 15f(y) \qquad (5.57)$$

for all $x, y \in X$. Once again adding (5.57) and (5.55), we obtain

$$\begin{aligned} 3f(x + 3y) + f(3x - y) &= [2f(x - 2y) + f(2x + y)] \\ &\quad + 5[2f(x + 2y) + f(2x - y)] \\ &\quad - 15f(x - y) - 15f(x + y) + 20f(y) \end{aligned} \qquad (5.58)$$

for all $x, y \in X$. Therefore, (5.53) follows from (5.15) and (5.58).

Conversely, suppose that $f : X \to Y$ satisfy (5.53). Putting $x = y = 0$ in (5.53), we get $f(0) = 0$. Set $y = 0$ in (5.53) to get $f(3x) = 27f(x)$, for all $x \in X$. Letting

$x = 0$ in (5.53), we obtain that $f(-y) = -f(y)$, for all $y \in X$. Replacing x by $x - y$ in (5.53), we have

$$3f(x + 2y) + f(3x - 4y) = 15f(x - 2y) + 15f(x) + 80f(y) \qquad (5.59)$$

for all $x, y \in X$. Since f is odd, replacing y by $x + y$ in (5.53), we get

$$3f(4x + 3y) + f(2x - y) = 15f(2x + y) + 80f(x + y) - 15f(y) \qquad (5.60)$$

for all $x, y \in X$. Replacing x and y by y and $-x$ in (5.60), respectively, and multiplying both sides of (5.60) to (-1), we obtain

$$3f(3x - 4y) - f(x + 2y) = 15f(x - 2y) + 80f(x - y) - 15f(x) \qquad (5.61)$$

for all $x, y \in X$. Adding (5.59) and (5.61), we have

$$2f(3x - 4y) + f(x + 2y) = 15f(x - 2y) + 40f(x - y) + 40f(y) \qquad (5.62)$$

for all $x, y \in X$. Therefore, we infer from (5.59) and (5.62) that

$$f(x + 2y) - 3f(x - 2y) = 6f(x) + 24f(y) - 8f(x - y) \qquad (5.63)$$

for all $x, y \in X$. Replacing y by $-y$ in (5.63), we obtain

$$f(x - 2y) - 3f(x + 2y) = 6f(x) - 24f(y) - 8f(x + y) \qquad (5.64)$$

for all $x, y \in X$. It follows from (5.63) and (5.64) that

$$f(x + 2y) = 3f(x + y) + f(x - y) + 6f(y) - 3f(x) \qquad (5.65)$$

for all $x, y \in X$. Replacing x and y by $-y$ and x in (5.65), respectively, we get

$$f(2x - y) = 3f(x - y) - f(x + y) + 6f(x) + 3f(y) \qquad (5.66)$$

for all $x, y \in X$. Hence, (5.15) follows from (5.65) and (5.66).   □

Theorem 5.5 ([202]). *A function* $f : X \to Y$ *satisfies the functional equation* (5.1) *if and only if* $f : X \to Y$ *satisfies the functional equation*

$$f(mx + y) + f(mx - y) = mf(x + y) + mf(x - y) + 2\left(m^3 - m\right) f(x) \qquad (5.67)$$

for all $x, y \in X$. *Therefore, every solution of functional equation* (5.67) *is also a cubic function.*

Proof. Let $f : X \to Y$ satisfy the functional equation (5.1). Putting $x = y = 0$ in (5.1), we get $f(0) = 0$. Set $x = 0$ in (5.1) to get $f(-y) = -f(y)$. Letting $y = x$ and $y = 2x$ in (5.1), respectively, we obtain that $f(2x) = 8f(x)$ and $f(3x) = 27f(x)$ for all $x \in X$. By induction, we obtain to $f(kx) = k^3 f(x)$, for all positive integer k. Replacing y by $x + y$ in (5.1), we have

$$f(3x + y) + f(x - y) = 2f(2x + y) - 2f(y) + 12f(x) \tag{5.68}$$

for all $x, y \in X$. Once again, replacing y by $y - x$ in (5.1), we have

$$f(x + y) + f(3x - y) = 2f(y) + 2f(2x - y) + 12f(x) \tag{5.69}$$

for all $x, y \in X$. Adding (5.68) to (5.69) and using (5.1), we obtain

$$f(3x + y) + f(3x - y) = 3f(x + y) + 3f(x - y) + 48f(x) \tag{5.70}$$

for all $x, y \in X$. By using the above method, by induction, we infer that

$$f(mx + y) + f(mx - y) = mf(x + y) + mf)x - y) + 2\left(m^3 - m\right) f(x) \tag{5.71}$$

for all $x, y \in X$ and each positive integer $m \geq 3$.

Let $f : X \to Y$ satisfy the functional equation (5.67) with the positive integer $m \geq 3$. Putting $x = y = 0$ in (5.67), we get $f(0) = 0$. Set $x = 0$ in (5.67) to get $f(-y) = -f(y)$.

Let k be a positive integer. Replacing y by $kx + y$ in (5.67), we have

$$f((m + k)x + y) + f((m - k)x - y)$$
$$= mf((k + 1)x + y) - mf((k - 1)x + y) + 2\left(m^3 - m\right) f(x) \tag{5.72}$$

for all $x, y \in X$. Replacing y by $y - kx$ in (5.67), we have

$$f((m - k)x + y) + f((m + k)x - y)$$
$$= mf((k + 1)x - y) - mf((k - 1)x - y) + 2\left(m^3 - m\right) f(x) \tag{5.73}$$

for all $x, y \in X$. Adding (5.72) and (5.73), we obtain

$$f((m + k)x + y) + f((m + k)x - y) + f((m - k)x + y) + f((m - k)x - y)$$
$$= m[f((k + 1)x + y) + f((k + 1)x - y)]$$
$$- m[f((k - 1)x + y) + f((k - 1)x - y)] + 4\left(m^3 - m\right) f(x) \tag{5.74}$$

for all $x, y \in X$ and for all integer $k \geq 1$. Let $\varphi_n(x, y) = f(nx + y) + f(nx - y)$ for each integer $n \geq 0$. Then (5.74) means that

$$\varphi_{m+k}(x, y) + \varphi_{m-k}(x, y) = m\varphi_{k+1}(x, y) - m\varphi_{k-1}(x, y) + 4\left(m^3 - m\right) f(x) \tag{5.75}$$

for all $x, y \in X$ and for all integer $k \geq 1$. For $k = 1$ and $k = m$ in (5.74), we obtain

$$\varphi_{m+1} + \varphi_{m-1} = m\varphi_2 + 4\left(m^3 - m\right) f(x) \tag{5.76}$$

and

$$\varphi_{2m} = m\varphi_{m+1} - m_{m-1} + 4\left(m^3 - m\right) f(x) \tag{5.77}$$

for all $x, y \in X$. By the proof of the first part, since $f : X \to Y$ satisfies the functional equation (5.67) with the positive integer $m \geq 3$, then f satisfies the functional equation (5.67) with the positive integer $k \geq m$. It follows from (5.76) and (5.77) that f satisfies the functional equation (5.1) and

$$
\begin{aligned}
f((m-1)x + y) &+ f((m-1)x - y) \\
&= (m-1)f(x+y) + (m-1)f(x-y) + 2\left((m-1)^3 - (m-1)\right)f(x) \quad (5.78)
\end{aligned}
$$

for all $x, y \in X$, which completes the proof. $\qquad\square$

Chapter 6

General Solution of Quartic Type Functional Equations

6.1. Introduction

In the previous chapter, we have discussed about the general solutions of various forms of cubic type functional equations. In this chapter, we focus on the general solution of different types of quartic form of functional equations.

Rassias [267] introduced the following functional equation:

$$f(x + 2y) + f(x - 2y) + 6f(x) = 4f(x + y) + 4f(x - y) + 24f(y). \tag{6.1}$$

It is easy to see that $f(x) = x^4$ is a solution of equation (6.1). For this reason, equation (6.1) is called a **quartic functional equation**. The general solution of (6.1) is determined without assuming any regularity conditions on the unknown function (see [57]). The function $f : \mathbb{R} \to \mathbb{R}$ is a solution of equation (6.1) if and only if $f(x) = A(x, x, x, x)$, where the function $A : \mathbb{R}^4 \to \mathbb{R}$ is symmetric and additive in each variable. Since the solution of equation (6.1) is even, we can rewrite (6.1) as

$$f(2x + y) + f(2x - y) = 4f(x + y) + 4f(x - y) + 24f(x) - 6f(y). \tag{6.2}$$

Every solution of the quartic functional equation is said to be quartic mapping. In Section 6.2, the general solution of equation (6.2) is presented, and in Section 6.3, the general solution of other recent quartic functional equations is studied by a simple approach. Throughout this chapter, let X and Y be vector spaces.

6.2. Quartic functional equation

The following theorem presents the general solution of the quartic functional equation (6.2).

Theorem 6.1 ([178]). *A function $f : X \to Y$ satisfies the functional equation (6.2) if and only if there exists a symmetric bi-quadratic function $F : X \times X \to Y$ such that $f(x) = F(x, x)$, for all $x \in X$.*

Proof. Assume that f satisfies the functional equation (6.2). Putting $x = y = 0$ in (6.2), we have $f(0) = 0$. Putting $x = 0$ in (6.2), we have $f(y) = f(-y)$, for all $y \in X$. Putting $y = 0$ and $y = x$ in (6.2), we obtain that $f(2x) = 16f(x)$ and $f(3x) = 81f(x)$, for all $x \in X$, respectively. Actually, we can lead to $f(nx) = n^4 f(x)$, for all $x \in X$ and all $n \in \mathbb{N}$. Replacing x and y by $x + y$ and $x - y$ in (6.2), respectively, we get

$$f(3x + y) + f(x + 3y) = 64f(x) + 64f(y) + 24f(x + y) - 6f(x - y) \qquad (6.3)$$

for all $x, y \in X$. Replacing x and y by $x + y$ and $2y$ in (6.2), respectively, we obtain

$$4f(x + 2y) + 4f(x) = f(x + 3y) + f(x - y) + 6f(x + y) - 24f(y) \qquad (6.4)$$

for all $x, y \in X$. Interchange x and y in (6.4) to get the relation

$$4f(y + 2x) + 4f(y) = f(y + 3x) + f(y - x) + 6f(y + x) - 24f(x) \qquad (6.5)$$

for all $x, y \in X$. Adding (6.4) to (6.5) and using (6.3), we are led to

$$f(x + 2y) + f(2x + y) = 9f(x) + 9f(y) + 9f(x + y) - f(x - y) \qquad (6.6)$$

for all $x, y \in X$. Using (6.2), we have

$$9f(2x + z) + 9f(2x - z) + 9f(2y + z) + 9f(2y - z)$$
$$= 36f(x + z) + 36f(x - z) + 216f(x) - 54f(z)$$
$$+ 36f(y + z) + 36f(y - z) + 216f(y) - 54f(z) \qquad (6.7)$$

for all $x, y, z \in X$.

On the other hand, replacing x and y by $2x + z$ and $2y + z$ in (6.6), respectively, we obtain

$$9f(2x + z) + 9f(2y + z)$$
$$= f(2x + 4y + 3z) + f(4x + 2y + 3z) - 9f(2x + 2y + 2x) + f(2x - 2y) \quad (6.8)$$

for all $x, y, z \in X$. Replacing x and y by $2x - z$ and $2y - z$ in (6.6), respectively, we obtain

$$9f(2x - z) + 9f(2y - z)$$
$$= f(2x + 4y - 3z) + f(4x + 2y - 3z) - 9f(2x + 2y - 2z) + f(2x - 2y) \quad (6.9)$$

for all $x, y, z \in X$. Adding (6.8) to (6.9) and using (6.2), we obtain

$$9f(2x + z) + 9f(2y + z) + 9f(2x - z) + 9f(2y - z)$$
$$= 4f(x + 2y + 3z) + 4f(x + 2y - 3z) + 24f(x + 2y) - 6f(3z)$$
$$+ 4f(2x + y + 3z) + 4f(2x + y - 3z) + 24f(2x + y) - 6f(3z)$$
$$- 144f(x + y + z) - 144f(x + y - z) + 32f(x - y) \qquad (6.10)$$

for all $x, y, z \in X$. By (6.7) and (6.10), we have

$$36f(x+z) + 36f(x-z) + 216f(x) - 54f(z)$$
$$+ 36f(y+z) + 36f(y-z) + 216f(y) - 54f(z)$$
$$= 4f(x+2y+3z) + 4f(x+2y-3z) + 24f(x+2y) - 6f(3z)$$
$$+ 4f(2x+y+3z) + 4f(2x+y-3z) + 24f(2x+y) - 6f(3z)$$
$$- 144f(x+y+z) - 144f(x+y-z) + 32f(x-y) \tag{6.11}$$

for all $x, y, z \in X$. Referring to the process of (6.8)–(6.10), when $x = 2x + z$, $y = 2y - z$ in (6.6) and $x = 2x - z$, $y = 2y + z$ in (6.6), we obtain

$$9f(2x+z) + 9f(2y-z) + 9f(2x-z) + 9f(2y+z)$$
$$= 4f(x+2y+z) + 4f(x+2y-z) + 24f(x+2y) - 6f(z)$$
$$+ 4f(2x+y+z) + 4f(2x+y-z) + 24f(2x+y) - 6f(z)$$
$$- 288f(x+y) + 16f(x-y+z) + 16f(x-y-z) \tag{6.12}$$

for all $x, y, z \in X$. Replacing z by $3z$ in (6.12) and then using (6.11), we have

$$9f(2x+3z) + 9f(2y-3z) + 9f(2x-3z) + 9f(2y+3z)$$
$$= 36f(x+z) + 36f(x-z) + 216f(x) - 54f(z)$$
$$+ 36f(y+z) + 36f(y-z) + 216f(y) - 54f(z)$$
$$+ 36f(y+z) + 36f(y-z) + 216f(y) - 54f(z)$$
$$+ 144f(x+y+z) + 144f(x+y-z) - 32f(x-y)$$
$$- 288f(x+y) + 16f(x-y+3z) + 16f(x-y-3z) \tag{6.13}$$

for all $x, y, z \in X$. On the other hand, putting $x = x - y + 3z$ and $y = x - y - 3z$ in (6.6), we have

$$9f(x-y+3z) + 9f(x-y-3z)$$
$$= 81f(x-y+z) + 81f(x-y-z) - 144f(x-y) + 1296f(z) \tag{6.14}$$

for all $x, y, z \in X$. Multiplying both sides of (6.14) by $\frac{16}{9}$, we obtain

$$16f(x-y+3z) + 16f(x-y-3z)$$
$$= 144f(x-y+z) + 144f(x-y-z) - 256f(x-y) + 2304f(z) \tag{6.15}$$

for all $x, y, z \in X$. Applying (6.15) and (6.13), we have

$$9f(2x+3z) + 9f(2y-3z) + 9f(2x-3z) + 9f(2y+3z)$$
$$= 36f(x+z) + 36f(x-z) + 216f(x) - 54f(z)$$
$$+ 36f(y+z) + 36f(y-z) + 216f(y) - 54f(z)$$
$$+ 144f(x+y+z) + 144f(x+y-z) - 32f(x-y) - 288f(x+y)$$
$$+ 144f(x-y+z) + 144f(x-y-z) - 256f(x-y) + 2304f(z) \tag{6.16}$$

for all $x, y, z \in X$.

Referring to the process of (6.8)–(6.10), when $x = 2x + 3z$, $y = 2x - 3z$ in (6.6) and $x = 2y - 3z$, $y = 2y + 3z$ in (6.6), and using (6.2), we obtain

$$
\begin{aligned}
9f(2x + 3z) &+ 9f(2x - 3z) + 9f(2y - 3z) + 9f(2y + 3z) \\
&= 324f(x + z) + 324f(x - z) + 1944f(x) - 486f(z) \\
&\quad + 324f(y + z) + 324f(y - z) + 1944f(y) - 486f(z) \\
&\quad - 2304f(x) - 2304f(y) + 2592f(z)
\end{aligned} \tag{6.17}
$$

for all $x, y, z \in X$. By (6.16) and (6.17), we obtain

$$
\begin{aligned}
f(x + y + z) &+ f(x - y + z) + f(x + y - z) + f(-x + y + z) \\
&= 2[f(x + y) + f(x - y) + f(y + z) + f(y - z) + f(z + x) + f(z - x)] \\
&\quad - 4[f(x) + f(y) + f(z)]
\end{aligned} \tag{6.18}
$$

for all $x, y, z, \in X$.

Define a function $F : X \times X \to Y$ by

$$
F(x, y) = \frac{1}{12}[f(x + y) + f(x - y) - 2f(x) - 2f(y)]
$$

for all $x, y \in X$. Then we have $F(x, x) = f(x)$, for all $x \in X$ and F is symmetric since f is even.

Now, we claim that $Q = F(\cdot, y) : X \to Y$ defined by $Q(x) = F(x, y)$ is quadratic for each fixed $y \in X$. Using (6.18) and evenness of f, we obtain

$$
\begin{aligned}
12[f(x + z, y) &+ f(x - z, y) - 2F(x, y) - 2F(z, y)] \\
&= f(x + z + y) + f(x + z - y) - 2f(x + z) - 2f(y) \\
&\quad + f(x - z + y) + f(x - z - y) - 2f(x - z) - 2f(y) \\
&\quad - 2f(x + y) - 2f(x - y) + 4f(x) + 4f(y) \\
&\quad - 2f(z + y) - 2f(z - y) + 4f(z) + 4f(y) = 0
\end{aligned}
$$

for all $x, y, z \in X$. This shows that $Q = F(\cdot, y)$ is quadratic. Since F is symmetric, $Q' = F(x, \cdot) : X \to Y$ defined by $Q' = F(x, y)$ is quadratic for fixed $x \in X$.

Conversely, assume that a function $F : X \times X \to Y$ is symmetric bi-quadratic such that $f(x) = F(x, x)$ for all $x \in X$. Then

$$
\begin{aligned}
f(2x + y) &+ f(2x - y) - 4f(x + y) - 4f(x - y) - 24f(x) + 6f(y) \\
&= F(2x + y, 2x + y) + F(2x - y, 2x - y) \\
&\quad - 4F(x + y, x + y) - 4F(x - y, x - y) - 24F(x, x) + 6F(y, y) \\
&= 2F(x, 2x + y) + 2F(x + y, 2x + y) - F(y, 2x + y)
\end{aligned}
$$

$$+ 2F(x, 2x - y) + 2F(x - y, 2x - y) - F(y, 2x - y)$$
$$- 4F(x + y, x + y) - 4F(x - y, x - y) - 24F(x, x) + 6F(y, y)$$
$$= 8[2F(x, x) + 2F(x, y)] - 4[2F(y, x) + 2F(y, y)]$$
$$- 16F(x, x) - 8F(x, y) + 8F(y, y)$$
$$= 0$$

for all $x, y \in X$. Hence, f satisfies the functional equation (6.2). This completes the proof. $\square$

6.3. Other quartic functional equations

6.3.1. *Quartic functional equation* (6.19)

Using a different approach, the general solution of the following quartic functional equation

$$f(3x + y) + f(x + 3y) = 64f(x) + 64f(y) + 24f(x + y) - 6f(x - y) \qquad (6.19)$$

is obtained in the following theorem.

Theorem 6.2 ([249]). *A function $f : X \to Y$ satisfies the functional equation (6.2) if and only if there exists a 4-additive symmetric function $A_4 : X^4 \to Y$ such that $f(x) = A_4(x, x, x, x)$ for all $x \in X$.*

Proof. Assume that f satisfies the functional equation (6.19). Putting $x = y = 0$ in (6.3), we have $f(0) = 0$. Replacing x and y by $x + y$ and $x - y$, respectively, in (6.19), we obtain

$$f(4x + 2y) + f(4x - 2y) = 64f(x + y) + 64f(x - y) + 24f(2x) - 6f(2y). \qquad (6.20)$$

Replacing y by $-y$ in (6.20), we can see that

$$f(y) = f(-y)$$

for all $y \in X$. That is, f is an even function. Replacing y by $-x$ in (6.19) and using the evenness of f, we obtain

$$f(2x) = 16f(x) \qquad (6.21)$$

for all $x \in X$. Applying (6.21) to (6.20), we obtain

$$f(2x + y) + f(2x - y) = 4f(x + y) + 4f(x - y) + 24f(x) - 6f(y). \qquad (6.22)$$

Replacing y by $y + 3x$ and $y + 2x$, respectively, in (6.22), then taking the difference of the two newly obtained equations, we have

$$f(5x + y) - 5f(4x + y) + 10f(3x + y) - 10f(2x + y) + 5f(x + y) - f(y) = 0.$$

Hence, f satisfies the difference functional equation $\Delta_x^5 f(y) = 0$. Consequently, f is a polynomial function of order 4. Then there exist n-additive symmetric functions $A_n : X^n \to Y$, $n = 0, \ldots, 4$, such that

$$f(x) = A^0 + A^1(x) + A^2(x) + A^3(x) + A^4(x), \qquad (6.23)$$

where $A^n : X \to Y$ is the diagonalization of A_n, for each $n = 0, \ldots, 4$. Since f is an even function, then $A^1(x)$ and $A^3(x)$ must vanish. Moreover, since $f(0) = 0$, we have $A^0 = 0$. Then (6.7) is reduced to

$$f(x) = A^2(x) + A^4(x). \qquad (6.24)$$

By using the symmetry and the additivity of $A_2(x, y)$, one can verify that

$$A^2(x + y) + A^2(x - y) = 2A^2(x) + 2A^2(y). \qquad (6.25)$$

Substituting (6.24) into (6.19) and using the property (6.25), we obtain $A^2(x) = 0$. Hence, we conclude that $f(x) = A^4(x)$, for all $x \in X$.

Conversely, assume that there exists a 4-additive symmetric function $A_4 : X^4 \to Y$ such that $f(x) = A^4(x)$, for all $x \in X$. Note that $\Delta_x^4 A^4(y) = 4! A^4(x)$. Thus, we obtain

$$A^4(4x + y) - 4A^4(3x + y) + 6A^4(2x + y) - 4A^4(x + y) + A^4(y) = 24A^4(x). \qquad (6.26)$$

Replacing y by $y - x$ in (6.26), we obtain

$$A^4(3x + y) - 4A^4(2x + y) + 6A^4(x + y) - 4A^4(y) + A^4(y - x) = 24A^4(x). \qquad (6.27)$$

Replacing x and y by $x + y$ and $-2y$, respectively, in (6.27), we obtain

$$A^4(3x + y) - 4A^4(2x) + 6A^4(x - y) - 4A^4(-2y) + A^4(-3y - x) = 24A^4(x + y). \qquad (6.28)$$

On account of the additivity of $A_4(x_1, x_2, x_3, x_4)$, we have $A^4(nx) = n^4 A^4(x)$, for all $n \in \mathbb{Z}$. Then we have

$$A^4(3x + y) + A^4(3y + x) = 64A^4(x) + 64A^4(y) + 24A^4(x + y) - 6A^4(x - y). \qquad (6.29)$$

By the assumption, we arrive at the functional equation (6.19), and the proof is completed. $\square$

6.3.2. *Quartic functional equation* (6.30)

In this section, we find the general solution of the quartic functional equation

$$f(x+ny)+f(x-ny)+2(n^2-1)f(x) = n^2[f(x+y)+f(x-y)+2(n^2-1)f(y)] \quad (6.30)$$

for all $x, y \in \mathbb{R}$ without assuming any regularity conditions on the unknown function f.

We recall some preliminary results on m-additive function, and symmetric function which will be used for further developments.

A function $B : \mathbb{R} \to \mathbb{R}$ is said to be additive if $B(x + y) = B(x) + B(y)$, for all $x, y \in \mathbb{R}$.

A function $B_m : \mathbb{R}^m \to \mathbb{R}$ is called m-additive (for $m \in \mathbb{N}$) if it is additive in each of its variable.

A function B_m is called symmetric if

$$B_m(x_1, x_2, \ldots, x_m) = B_m\left(x_{\pi(1)}, x_{\pi(2)}, \ldots, x_{\pi(m)}\right),$$

for every permutation $\{\pi(1), \pi(2), \ldots, \pi(m)\}$ of $\{1, 2, \ldots, m\}$.

We denote the diagonal $B_m(x, x, \ldots, x)$ by $B^m(x)$, if $B_m(x_1, x_2, \ldots, x_m)$ is an m-additive symmetric map. If we substitute $x_1 = x_2 = \cdots = x_l = x$, and $x_{l+1} = x_{l+2} = \cdots = x_m = y$ in $B_m(x_1, x_2, \ldots, x_m)$, then we denote the resulting expression as $B^{l,m-l}(x, y)$.

The difference operator Δ_h for the function $f : \mathbb{R} \to \mathbb{R}$ is defined as follows:

$$\Delta_h f(x) = f(x + h) - f(x) \quad \text{for } h \in \mathbb{R},$$

and

$$\Delta_h^0 f(x) = f(x), \quad \Delta_h^1 f(x) = \Delta_h f(x) \quad \text{and} \quad \Delta_h \circ \Delta_h^m f(x) = \Delta_h^{m+1} f(x)$$

for all $m \in \mathbb{N}$ and $h \in \mathbb{R}$, $\Delta_h \circ \Delta_h^m$ denotes the composition of the operators Δ_h, and Δ_h^m. Using the above definition, consider the functional equation

$$\Delta_h^{m+1} f(x) = 0$$

which can be written in explicit form as

$$\sum_{k=0}^{m+1} (-1)^{m+1-k} \binom{m+1}{k} f(x + kh) = 0,$$

and is equivalent to the Fréchet functional equation

$$\Delta_{h_1, h_2, \ldots, h_{m+1}} f(x) = 0,$$

where $x, h_1, h_2, \ldots, h_{m+1} \in \mathbb{R}$ and $\Delta_{h_1, \ldots, h_k} = \Delta_{h_k} \circ \cdots \circ \Delta_{h_1}$ for $k = 2, 3, \ldots, m+1$.

Lemma 6.1. *If any function f satisfies the quartic functional equation (6.30) for all $x, y \in \mathbb{R}$, then it also satisfies the functional equation*

$$
\begin{aligned}
f(x + ny) &+ f(x - ny) \\
&= f(nx + y) + f(nx - y) + 2(n^4 - 1)f(y) + 2(1 - n^4)f(x) \quad (6.31)
\end{aligned}
$$

for all $x, y \in \mathbb{R}$.

Proof. Taking $x = y = 0$ in (6.30), we obtain $f(0) = 0$. Again, setting $x = 0$ and $y = -x$ in (6.30), we have

$$
f(-nx) + f(nx) - n^2 f(x) + n^2(1 - 2n^2)f(-x) = 0 \quad (6.32)
$$

for all $x \in \mathbb{R}$. Similarly, setting $x = 0$ and $y = x$ in (6.30), and then using $f(0) = 0$, we obtain

$$
f(nx) + f(-nx) - n^2 f(-x) + n^2(1 - 2n^2)f(x) = 0. \quad (6.33)
$$

From equations (6.32) and (6.33), we obtain $f(x) = f(-x)$ for all $x \in \mathbb{R}$, that is, f is an even function. Next, interchanging x with y in (6.30) and using the fact that f is an even function, we have

$$
\begin{aligned}
f(nx + y) &+ f(nx - y) + 2(n^2 - 1)f(y) \\
&= n^2[f(x + y) + f(x - y) + 2(n^2 - 1)f(x)]. \quad (6.34)
\end{aligned}
$$

Equating (6.30) and (6.34), and simplifying, we obtain the result (6.31). This proves the lemma. $\qquad\square$

Lemma 6.2. *If any function $G : \mathbb{R} \to \mathbb{R}$ satisfies the functional equation*

$$
\begin{aligned}
G(x + ny) &+ G(x - ny) \\
&= G(nx + y) + G(nx - y) + 2(n^4 - 1)G(y) + 2(1 - n^4)G(x) \quad (6.35)
\end{aligned}
$$

for all $x, y \in \mathbb{R}$, then G also satisfies the Fréchet functional equation

$$
\Delta_{x_1, \ldots, x_5} G(x_0) = 0 \quad (6.36)
$$

for all $x_0, x_1, \ldots, x_5 \in \mathbb{R}$.

Proof. Let $G_1(x) = 2(1 - n^4)G(x)$ and $G_2(y) = 2(n^4 - 1)G(y)$; then equation (6.35) can be written as

$$
G(x + ny) + G(x - ny) = G(nx + y) + G(nx - y) + G_1(x) + G_2(y). \quad (6.37)
$$

Now, we substitute $x_0 = x + ny$ and $y_1 = x - ny$ in (6.37), we get

$$G(x_0) + G(y_1) = G\left(\frac{n^2+1}{2n}x_0 + \frac{n^2-1}{2n}y_1\right) + G\left(\frac{n^2-1}{2n}x_0 + \frac{n^2+1}{2n}y_1\right)$$
$$+ G_1\left(\frac{x_0+y_1}{2}\right) + G_2\left(\frac{x_0-y_1}{2n}\right). \tag{6.38}$$

Now, replacing x_0 by x_0+x_1 in (6.38), and then subtracting (6.38) from the resulting expression, we have

$$G(x_0 + x_1) - G(x_0)$$
$$= G\left(\frac{n^2+1}{2n}(x_0 + x_1) + \frac{n^2-1}{2n}y_1\right)$$
$$+ G\left(\frac{n^2-1}{2n}(x_0 + x_1) + \frac{n^2+1}{2n}y_1\right) - G\left(\frac{n^2+1}{2n}x_0 + \frac{n^2-1}{2n}y_1\right)$$
$$- G\left(\frac{n^2-1}{2n}x_0 + \frac{n^2+1}{2n}y_1\right) + G_1\left(\frac{x_0+x_1+y_1}{2}\right)$$
$$+ G_2\left(\frac{x_0+x_1-y_1}{2n}\right) - G_1\left(\frac{x_0+y_1}{2}\right) - G_2\left(\frac{x_0-y_1}{2n}\right). \tag{6.39}$$

Setting $y_2 = \frac{n^2+1}{2n}x_0 + \frac{n^2-1}{2n}y_1$ in (6.39), we obtain

$$G(x_0 + x_1) - G(x_0)$$
$$= G\left(\frac{n^2+1}{2n}x_1 + y_2\right) + G\left(\frac{n^2-1}{2n}x_1 + \frac{n^2+1}{n^2-1}y_2 - \frac{2n}{n^2-1}x_0\right)$$
$$+ G_1\left(\frac{x_1}{2} + \frac{n}{n^2-1}y_2 - \frac{1}{n^2-1}x_0\right) + G_2\left(\frac{x_1}{2n} - \frac{1}{n^2-1}y_2 + \frac{n}{n^2-1}x_0\right)$$
$$- G_1\left(\frac{n}{n^2-1}y_2 - \frac{1}{n^2-1}x_0\right) - G_2\left(-\frac{1}{n^2-1}y_2 + \frac{n}{n^2-1}x_0\right)$$
$$- G(y_2) - G\left(\frac{n^2+1}{n^2-1}y_2 - \frac{n}{n^2-1}x_0\right). \tag{6.40}$$

Now, replacing x_0 by x_0+x_2 in (6.40) and then subtracting (6.40) from the resulting expression, we have

$$G(x_0 + x_1 + x_2) - G(x_0 + x_1) - G(x_0 + x_2) + G(x_0)$$
$$= G\left(\frac{n^2+1}{2n}x_1 + \frac{n^2+1}{n^2-1}y_2 - \frac{2n}{n^2-1}(x_0 + x_2)\right)$$
$$+ G_1\left(\frac{x_1}{2} + \frac{n}{n^2-1}y_2 - \frac{1}{n^2-1}(x_0 + x_2)\right)$$

$$+ G_2\left(\frac{x_1}{2n} - \frac{1}{n^2-1}y_2 + \frac{n}{n^2-1}(x_0+x_2)\right)$$

$$- G\left(\frac{n^2+1}{n^2-1}y_2 - \frac{2n}{n^2-1}(x_0+x_2)\right)$$

$$- G_1\left(\frac{n}{n^2-1}y_2 - \frac{1}{n^2-1}(x_0+x_2)\right)$$

$$- G_2\left(-\frac{1}{n^2-1}y_2 + \frac{n}{n^2-1}(x_0+x_2)\right)$$

$$- G\left(\frac{n^2+1}{2n}x_1 + \frac{n^2+1}{n^2-1}y_2 - \frac{2n}{n^2-1}x_0\right)$$

$$- G_1\left(\frac{x_1}{2} + \frac{n}{n^2-1}y_2 - \frac{1}{n^2-1}x_0\right)$$

$$- G_2\left(\frac{x_1}{2n} - \frac{1}{n^2-1}y_2 + \frac{n}{n^2-1}x_0\right) + G\left(\frac{n^2+1}{n^2-1}y_2 - \frac{2n}{n^2-1}x_0\right)$$

$$+ G_1\left(\frac{n}{n^2-1}y_2 - \frac{1}{n^2-1}x_0\right) + G_2\left(-\frac{1}{n^2-1}y_2 + \frac{n}{n^2-1}x_0\right). \quad (6.41)$$

Similar to the process of obtaining equations (6.40) and (6.41), substitute $y_3 = \frac{(n^2+1)y_2 - 2nx_0}{n^2-1}$ in the last equation, replace x_0 by $x_0 + x_3$ in the resulting equation and then subtract the former resulting equation from the later resulting equation.

In the same way, proceeding further to the resultant expression obtained above, set $y_4 = \frac{ny_3 + x_0}{n^2+1}$, replace x_0 by $x_0 + x_4$ in the resulting equation.

Likewise, set $y_5 = \frac{y_4}{n} + \frac{x_0}{n}$, replace x_0 by $x_0 + x_5$ and carry out the procedure as done before to get

$$G(x_0 + x_1 + x_2 + x_3 + x_4 + x_5) - G(x_0 + x_1 + x_2 + x_3 + x_4)$$
$$- G(x_0 + x_1 + x_2 + x_3 + x_5) - G(x_0 + x_1 + x_2 + x_4 + x_5)$$
$$- G(x_0 + x_1 + x_3 + x_4 + x_5) - G(x_0 + x_2 + x_3 + x_4 + x_5)$$
$$+ G(x_0 + x_1 + x_2 + x_3) + G(x_0 + x_1 + x_2 + x_4)$$
$$+ G(x_0 + x_1 + x_2 + x_5) + G(x_0 + x_1 + x_3 + x_4)$$
$$+ G(x_0 + x_1 + x_3 + x_5) + G(x_0 + x_1 + x_4 + x_5)$$
$$+ G(x_0 + x_2 + x_3 + x_4) + G(x_0 + x_2 + x_3 + x_5)$$
$$+ G(x_0 + x_2 + x_4 + x_5) + G(x_0 + x_3 + x_4 + x_5)$$
$$- G(x_0 + x_1 + x_2) - G(x_0 + x_1 + x_3) - G(x_0 + x_1 + x_4)$$
$$- G(x_0 + x_1 + x_5) - G(x_0 + x_2 + x_3)$$
$$- G(x_0 + x_2 + x_4) - G(x_0 + x_2 + x_5)$$
$$- G(x_0 + x_3 + x_4) - G(x_0 + x_3 + x_5)$$

$$- G(x_0 + x_4 + x_5) + G(x_0 + x_1) + G(x_0 + x_2) + G(x_0 + x_3)$$
$$+ G(x_0 + x_4) + G(x_0 + x_5) - G(x_0) = 0 \tag{6.42}$$

which is (6.31). This completes the proof. $\square$

Hosszu [121] has obtained the solution of Fréchet functional equation in the following manner.

Consider the functional equation

$$\Delta^n_{x_1,\ldots,x_n} f = \sum_{\text{comb.}} (-1)^k f(x_{i_1} + x_{i_2} + \cdots + x_{i_n}) = 0 \tag{6.43}$$

due to Fréchet. Here the sum consists of such terms in which $(i_1, i_2, \ldots, i_n)$ is a k-tuple combination of the indices $1, 2, \ldots, n$. The variables and the values of the unknown function f are taken from any abelian groups A and A', respectively. We shall reduce (6.43) to the Cauchy's equation.

Let us define

$$a(x_1, x_2, \ldots, x_{n-1}) = \Delta^{n-1}_{x_1,\ldots,x_{n-1}} f$$

which is, clearly, a symmetric function of its variables; moreover, it is additive (e.g., in the first variable). In fact, if f satisfies (6.43) then

$$a(t + s, x_2, \ldots, x_{n-1}) = a(t, x_2, \ldots, x_{n-1}) + a(s, x_2, \ldots, x_{n-1})$$

holds as we have

$$\Delta^{n-1}_{t+s,x_2,\ldots,x_{n-1}} f - \Delta^{n-1}_{t,x_2,\ldots,x_{n-1}} f - \Delta^{n-1}_{s,x_2,\ldots,x_{n-1}} f = -\Delta^{n-1}_{t,s,x_2,\ldots,x_{n-1}} f = 0.$$

If the multilinear (additive in each variable) functions $a(x_1, \ldots, x_k) : A \times A \times \cdots \times A \to A'$ can easily be determined, then the solution of (6.43) just reduces to a simpler equation. In fact, every solution f of (6.43) satisfies an equation of the form

$$\Delta^{n-1}_{x_1,\ldots,x_{n-1}} f = a(x_1, \ldots, x_{n-1}), \tag{6.44}$$

where $a(x_1, \ldots, x_{n-1})$ is a multilinear symmetric function. However, for a given function $a(x_1, \ldots, x_{n-1})$, (6.44) can be solved as a linear inhomogeneous functional equation. We can look for the solution of (6.44) in the form

$$f = g + p_{n-1},$$

where p_{n-1} is a particular solution. Then, clearly, g must satisfy

$$\Delta^{n-1}_{x_1,\ldots,x_{n-1}} g = 0.$$

By the repeated application of this method, we arrive at

$$f = p_{n-1} + \cdots + p_2 + p_1, \tag{6.45}$$

where p_i is a particular solution of

$$\Delta^i_{x_1,\ldots,x_i} p = a_i(x_1,\ldots,x_i), \qquad \text{for } i = 1, 2, \ldots, n-1. \tag{6.46}$$

Here, a_i is an additive symmetric function.

For example, if every equation $mx = a$ $(m = 2, 3, \ldots)$ has a solution $x = a/m$ in A', then $p_i(x) = a_i(x, x, \ldots, x)/m$, with a suitable integer m, is a particular solution of (6.46). This can be verified by an easy calculation.

On the other hand, if p satisfies (6.46) for a given additive symmetric function a_i, $(i < n)$, then $\Delta^{i+1} p = 0$ is true and, *a fortiori*, p satisfies (6.43).

Lemma 6.3. *The function* $G : \mathbb{R} \to \mathbb{R}$ *satisfies the functional equation* (6.36) *for all* $x_0, x_1, x_2, x_3, x_4, x_5 \in \mathbb{R}$ *if and only if* G *is of the form*

$$G(x) = F^4(x) + F^3(x) + F^2(x) + F^1(x) + F^0(x),$$

where $F^m(x)$ *is the diagonal of a* m-*additive symmetric function* $F_m : \mathbb{R}^m \to \mathbb{R}$ *for* $m = 1, 2, 3, 4$ *and* $F^0(x) = F^0$ *is an arbitrary constant.*

Proof. The proof is obtained similar to that of Hosszu's method of solving Fréchet's functional equation given above, and the details are omitted. $\square$

Theorem 6.3. *The function* $f : \mathbb{R} \to \mathbb{R}$ *satisfies the quartic functional equation* (6.30) *for all* $x, y \in \mathbb{R}$ *if and only if* f *is of the form*

$$f(x) = F^4(x),$$

where $F^4(x)$ *is the diagonal of the 4-additive map* $F_4 : \mathbb{R}^4 \to \mathbb{R}$.

Proof. From Lemma 6.1, we see that the functional equation (6.30) implies the functional equation (6.31). Again by Lemma 6.2, we see that f satisfies the Fréchet functional equation

$$\Delta_{x_1,\ldots,x_5} f(x_0) = 0 \tag{6.47}$$

for all $x_0, x_1, x_2, x_3, x_4, x_5 \in \mathbb{R}$. Now, using Lemma 6.3, we see that the general solution of equation (6.47) can be obtained

$$f(x) = F^4(x) + F^3(x) + F^2(x) + F^1(x) + F^0(x), \tag{6.48}$$

where $F^m(x)$ is the diagonal of a m-additive symmetric function $F_m : \mathbb{R}^m \to \mathbb{R}$ for $m = 1, 2, 3, 4$ and $F^0(x) = F^0$ is an arbitrary constant. But f is even function (see Lemma 6.1), therefore, we obtain $F^3(x) = 0$ and $F'(x) = 0$. Hence, equation (6.48) becomes

$$f(x) = F^4(x) + F^2(x) + F^0(x). \tag{6.49}$$

Letting (6.49) into (6.30), and noting that

$$F^4(ny) = n^4 F^4(y), F^2(ny) = n^2 F^2(y), F^{2,2}(x, ny) = n^2 F^{2,2}(x, y),$$

$$F^4(x+y) + F^4(x-y) = 2F^4(x) + 2F^4(y) + 12F^{2,2}(x, y),$$

and

$$F^2(x+y) + F^2(x-y) = 2F^2(x) + 2f^2(y)$$

for all $x, y \in \mathbb{R}$, we get

$$2n^2(n^2-1)F^2(y) + 2(n^4-n^2)F^0 = 0 \tag{6.50}$$

for all $y \in \mathbb{R}$. This gives $F^2(y) \equiv 0$, and $F^0 \equiv 0$. Substituting in equation (6.49) we obtain $f(x) = F^4(x)$, and this completes the proof of the theorem.

6.3.3. *Quartic functional equation* (6.51)

The following lemmas and theorem present the general solution of the general quartic functional equation

$$f(ax+by) + f(ax-by) = (ab)^2[f(x+y) + f(x-y)]$$
$$+ 2(b^2-a^2)[b^2 f(y) - a^2 f(x)] \tag{6.51}$$

where $a \neq b$; $a, b \neq 0, \pm 1$ using an elementary technique, and the Fréchet functional equation

$$\Delta_{h_1, h_2, \ldots, h_{m+1}} f(x) = 0, \tag{6.52}$$

where $x, h_1, \ldots, h_{m+1} \in \mathbb{R}$ and $\Delta_{h_1, \ldots, h_k} = \Delta_{h_k} \circ \cdots \circ \Delta_{h_1}$ for $k = 2, 3, \ldots, m+1$.

□

Lemma 6.4. *If $f : \mathbb{R} \to \mathbb{R}$ satisfies the quartic functional equation* (6.51) *for all $x, y \in \mathbb{R}$, then it also satisfies the functional equation*

$$f(ax+by) + f(ax-by) - f(bx+ay) - f(ax-by)$$
$$= 2(b^4-a^4)[f(y) - f(x)] \tag{6.53}$$

for all

$$a \neq b; a, b \neq 0, \pm 1 \quad and \quad x, y \in \mathbb{R}. \tag{6.54}$$

Proof. Substituting $x = y = 0$ in (6.51), we obtain

$$2[(b^2 - a^2) + a^2 b^2 - 1]f(0) = 0,$$

which gives $f(0) = 0$, because of (6.54). Now, substitute $x = 0$, $y = -x$ in (6.51), then we obtain

$$f(bx) + f(-bx) = (ab)^2[f(x) + f(-x)] + 2(b^2 - a^2)[b^2 f(x)]. \tag{6.55}$$

Again substituting $x = 0$, $y = x$ in (6.51), we arrive at

$$f(-bx) + f(bx) = (ab)^2[f(-x) + f(x)] + 2(b^2 - a^2)[b^2 f(-x)]. \tag{6.56}$$

From (6.55) and (6.56), we obtain

$$2(b^2 - a^2)[b^2 f(x)] = 2(b^2 - a^2)[b^2 f(-x)]$$

which proves $f(x) = f(-x)$, that is, f is even. Equation (6.51) can be written as

$$(ab)^2[f(x + y) + f(x - y)] = f(ax + by) + f(ax - by)$$
$$- 2(b^2 - a^2)[b^2 f(y) - a^2 f(x)]. \tag{6.57}$$

Interchanging x and y in (6.57) and using the evenness of f, we obtain

$$(ab)^2[f(x + y) + f(x - y)] = f(bx + ay) + f(bx - ay)$$
$$- 2(b^2 - a^2)[b^2 f(x) - a^2 f(y)]. \tag{6.58}$$

Equating (6.57) and (6.58), we obtain (6.53). This completes the proof of the lemma. $\square$

Lemma 6.5. *If the function $H : \mathbb{R} \to \mathbb{R}$ satisfies the functional equation*

$$H(ax + by) + H(ax - by) - H(bx + ay) - H(ax - by)$$
$$= 2(b^4 - a^4)[H(y) - H(x)] \tag{6.59}$$

for all $x, y \in \mathbb{R}$, then H also satisfies the Fréchet functional equation

$$\Delta_{y_1, y_2, \ldots, y_5} H(y_0) = 0 \tag{6.60}$$

for all $y_0, y_1, y_2, y_3, y_4, y_5 \in \mathbb{R}$.

Proof. Assume $H_1(y) = 2(b^4 - a^4)H(y)$ and $H_2(x) = -2(b^4 - a^4)H(x)$, then (6.59) is transformed into

$$H(ax + by) + H(ax - by) - H(bx + ay) - H(bx - ay) = H_1(y) + H_2(x). \tag{6.61}$$

Substituting $y_0 = ax + by$ and $z_1 = ax - by$ in (6.61), we obtain

$$H(y_0) + H(z_1) - H\left(\frac{a^2 + b^2}{2ab}y_0 + \frac{b^2 - a^2}{2ab}z_1\right) - H\left(\frac{b^2 - a^2}{2ab}y_0 + \frac{a^2 + b^2}{2ab}z_1\right)$$

$$= H_1\left(\frac{y_0 - z_1}{2b}\right) + H_2\left(\frac{y_0 + z_1}{2a}\right). \tag{6.62}$$

Replacing y_0 by $y_0 + y_1$ in (6.62) and then subtracting (6.62) from the resulting equation, we obtain

$$H(y_0 + y_1) - (y_0) - H\left(\frac{a^2 + b^2}{2ab}(y_0 + y_1) + \frac{b^2 - a^2}{2ab}z_1\right)$$

$$+ H\left(\frac{a^2 + b^2}{2ab}y_0 + \frac{b^2 - a^2}{2ab}z_1\right) - H\left(\frac{b^2 - a^2}{2ab}(y_0 + y_1)\frac{a^2 + b^2}{2ab}z_1\right)$$

$$+ H\left(\frac{b^2 - a^2}{2ab}y_0 + \frac{a^2 + b^2}{2ab}z_1\right)$$

$$= H_1\left(\frac{y_0 + y_1 - z_1}{2b}\right) - H_1\left(\frac{y_0 - z_1}{2b}\right)$$

$$+ H_2\left(\frac{y_0 + y_1 + z_1}{2a}\right) - H_2\left(\frac{y_0 + z_1}{2a}\right). \tag{6.63}$$

Substituting $z_1 = \frac{2abz_2 - (b^2 - a^2)y_0}{(a^2 + b^2)}$ in (6.63), $z_2 = \frac{b^2 - a^2}{2ab}y_0 + \frac{a^2 + b^2}{2ab}z_1$. Equation (6.63) becomes

$$H(y_0 + y_1) - H(y_0) - H\left(\frac{a^2 + b^2}{2ab}y_1 + \frac{b^2 - a^2}{a^2 + b^2}z_2 + \frac{2ab}{a^2 + b^2}y_0\right)$$

$$+ H\left(\frac{b^2 - a^2}{a^2 + b^2} + \frac{2ab}{a^2 + b^2}y_0\right) - H\left(\frac{b^2 - a^2}{2ab}y_1 + z_2\right) + H(z_2)$$

$$= H_1\left(\frac{y_1}{2b} - \frac{a}{a^2 + b^2}z_2 + \frac{b}{a^2 + b^2}y_0\right) - H_1\left(-\frac{a}{a62 + b^2}z_2 + \frac{b}{a^2 + b^2}y_0\right)$$

$$+ H_2\left(\frac{y_1}{2a} + \frac{b}{a^2 + b^2}z_2 + \frac{a}{a^2 + b^2}y_0\right) - H_2\left(\frac{b}{a^2 + b^2}z_2 + \frac{a}{a^2 + b^2}y_0\right). \tag{6.64}$$

Now, replacing y_0 by $y_0 + y_2$ in (6.64) and then subtracting (6.64) from the resulting equation, we obtain

$$H(y_0 + y_1 + y_2) - H(y_0 + y_1) - H(y_0 + y_2) + H(y_0)$$

$$- H\left(\frac{a^2 + b^2}{2ab}y_1 + \frac{b^2 - a^2}{a^2 + b^2}z_2 + \frac{2ab}{a^2 + b^2}(y_0 + y_2)\right)$$

$$+ H\left(\frac{a^2 + b^2}{2ab}y_1 + \frac{b^2 - a^2}{a^2 + b^2}z_2 + \frac{2ab}{a^2 + b^2}y_0\right)$$

$$+ H\left(\frac{b^2 - a^2}{a^2 + b^2}z_2 + \frac{2ab}{a^2 + b^2}(y_0 + y_2)\right) - H\left(\frac{b^2 - a^2}{a^2 + b^2}z_2 + \frac{2ab}{a^2 + b^2}y_0\right)$$

$$= H_1\left(\frac{y_1}{2b} - \frac{a}{a^2 + b^2}z_2 + \frac{b}{a^2 + b^2}(y_0 + y_2)\right)$$

$$- H_1\left(\frac{y_1}{2b} - \frac{a}{a^2 + b^2}z_2 + \frac{b}{a^2 + b^2}y_0\right)$$

$$- H_1\left(-\frac{a}{a^2 + b^2}z_2 + \frac{b}{a^2 + b^2}(y_0 + y_2)\right) + H_1\left(-\frac{a}{a^2 + b^2}z_2 + \frac{b}{a^2 + b^2}y_0\right)$$

$$+ H_2\left(\frac{y_1}{2a} + \frac{b}{a^2 + b^2}z_2 + \frac{a}{a^2 + b^2}(y_0 + y_2)\right)$$

$$- H_2\left(\frac{y_1}{2a}\frac{b}{a^2 + b^2}z_2 + \frac{a}{a^2 + b^2}y_0\right)$$

$$- H_2\left(\frac{b}{a^2 + b^2}z_2 + \frac{a}{a^2 + b^2}(y_0 + y_2)\right) + H_2\left(\frac{b}{a^2 + b^2}z_2 + \frac{a}{a^2 + b^2}y_0\right).$$

$$(6.65)$$

Similar to the process of obtaining equations (6.64) and (6.65), substitute $z_2 = \frac{(a^2 + b^2)z - 3 - 2aby_0}{b^2 - a^2}$ in (6.65), replace y_0 by $y_0 + y_3$ in the resulting equation and then subtract the former resulting equation from the later resulting equation.

In the same way, proceeding further to the resultant expression above, substitute $z_3 = \frac{(b^2 - a^2)z_4 ay_0}{b}$, $z_4 = \frac{bz_3 - ay_0}{b^2 - a^2}$ and then replace y_0 by $y_0 + y_4$ in the resulting equation.

Likewise, set $z_5 = -\frac{a}{b}z_4 + \frac{y_0}{b}$, replace y_0 by $y_0 + y_5$ and carry out the procedure as done before, we obtain

$$H(y_0 + y_1 + y_2 + y_3 + y_4 + y_5) - H(y_0 + y_1 + y_2 + y_3 + y_4)$$
$$- H(y_0 + y_1 + y_2 + y_3 + y_5) - H(y_0 + y_1 + y_2 + y_4 + y_5)$$
$$- H(y_0 + y_1 + y_3 + y_4 + y_5) - H(y_0 + y_2 + y_3 + y_4 + y_5)$$
$$+ H(y_0 + y_1 + y_2 + y_3) + H(y_0 + y_1 + y_2 + y_4) + H(y_0 + y_1 + y_2 + y_5)$$
$$+ H(y_0 + y_1 + y_3 + y_4) + H(y_0 + y_1 + y_3 + y_5) + H(y_0 + y_1 + y_4 + y_5)$$
$$+ H(y_0 + y_2 + y_3 + y_4) + H(y_0 + y_2 + y_3 + y_5) + H(y_0 + y_2 + y_4 + y_5)$$
$$+ H(y_0 + y_3 + y_4 + y_5) - H(y_0 + y_1 + y_2) - H(y_0 + y_1 + y - 3)$$
$$- H(y_0 + y_1 + y_4) - H(y_0 + y_1 + y_5) - H(y_0 + y_2 + y_3)$$
$$- H(y_0 + y_2 + y_4) - H(y_0 + y_2 + y_5) - H(y_0 + y_3 + y_4)$$
$$- H(y_0 + y_3 + y_5) - H(y_0 + y_4 + y_5) + H(y_0 + y_1) + H(y_0 + y_2)$$
$$+ H(y_0 + y_3) + H(y_0 + y_4) + H(y_0 + y_5) - H(y_0) = 0, \qquad (6.66)$$

which gives

$$\Delta_{y_1,y_2,\ldots,y_5} H(y_0) = 0.$$

Hence, the lemma is proved. □

Lemma 6.6. *The function $H : \mathbb{R} \to \mathbb{R}$ satisfies the functional equation (6.60) for all $y_0, y_1, y_2, y_3, y_4, y_5 \in \mathbb{R}$ if and only if H is expressed as*

$$H(y) = B^4(y) + B^3(y) + B^2(y) + B^1(y) + B^0(y),$$

where $B^m(y)$ is the diagonal of the m-additive symmetric function $B_m : \mathbb{R}^m \to \mathbb{R}$ for $m = 1, 2, 3, 4$ and $B^0(y) = B^0$ is an arbitrary constant.

Proof. The proof is similar to that of Lemma 6.3. □

Lemma 6.7. *The m-additive symmetric function $B^m(y)$, for $m = 2, 4$, satisfies*

$$B^4(ax + by) + B^4(ax - by) = 2B^4(ax) + 2B^4(by) + 12B^{2,2}(ax, by), \quad (6.67)$$

$$B^4(x + y) + B^4(x - y) = 2B^4(x) + 2B^4(y) + 12B^{2,2}(x, y), \quad (6.68)$$

$$B^2(ax + by) + B^2(ax - by) = 2B^2(ax) + 2B^2(by), \quad (6.69)$$

$$B^2(x + y) + B^2(x - y) = 2B^2(x) + 2B^2(y). \quad (6.70)$$

Proof. Using the definition of m-additive symmetric function and expanding, we get

$$\begin{aligned}
B^4(ax + by) &= B_4(ax + by, ax + by, ax + by, ax + by) \\
&= B^4(ax) + B^4(by) + 4B^{1,3}(ax, by) \\
&\quad + 4B^{3,1}(ax, by) + 6B^{2,2}(ax, by).
\end{aligned} \quad (6.71)$$

Also,

$$\begin{aligned}
B^4(ax - by) &= B_4(ax - by, ax - by, ax - by, ax - by) \\
&= B^4(ax) + B^4(by) - 4B^{1,3}(ax, by) \\
&\quad - 4B^{3,1}(ax, by) + 6B^{2,2}(ax, by).
\end{aligned} \quad (6.72)$$

Adding (6.71) and (6.72), we obtain the result (6.67). Again

$$\begin{aligned}
B^2(ax &+ by) + B^2(ax - by) \\
&= B_2(ax + by, ax + by) + B_2(ax - by, ax - by) \\
&= B_2(ax, ax) + B_2(ax, by) + B_2(by, ax) + B_2(by, by) \\
&\quad + B_2(ax, ax) - B_2(ax, by) - B_2(by, ax) + +B_2(by, by) \\
&= 2B^2(ax) + 2B^2(by).
\end{aligned}$$

Using similar arguments, we obtain (6.68) and (6.70). □

Theorem 6.4. *If $B^4(x)$ is the diagonal of the 4-additive symmetric function $B_4 : \mathbb{R}^4 \to \mathbb{R}$, then the function $f : \mathbb{R} \to \mathbb{R}$ satisfies the quartic functional equation (6.51) for all $x, y \in \mathbb{R}$ if and only if f is of the form $f(x) = B^4(x)$.*

Proof. By Lemma 6.3, we find that the functional equation (6.51) implies the functional equation (6.53). Again by Lemma 6.4, we see that f which satisfies the Fréchet functional equation (6.60) is given by

$$f(x) = B^4(x) + B^3(x) + B^2(x) + B^1(x) + B^0(x), \tag{6.73}$$

where $B^m(x)$ is the diagonal of the m-additive symmetric function $B_m : \mathbb{R}^m \to \mathbb{R}$ for $m = 1, 2, 3, 4$. But f is even function. Hence, $B^3(x) \equiv 0$ and $B^1(x) \equiv 0$. Hence, equation (6.73) reduces to

$$f(x) = B^4(x) + B^2(x) + B^0(x). \tag{6.74}$$

Substituting (6.74) in (6.51), we get

$$2B^4(ax) + 2B^4(by) + 12B^{2,2}(ax, by) + 2B^2(ax) + 2B^2(by) + 2B^0$$
$$= (ab)^2 \left[2B^4(x) + 2B^4(y) + 12B^{2,2}(x, y) + 2B^2(x) + 2B^2(y) + 2B^0\right]$$
$$+ 2(b^2 - a^2) \left[b^2 \left\{B^4(y) + B^2(y) + B^0\right\} - a^2 \left\{B^4(x) + B^2(x) + B^0\right\}\right]. \tag{6.75}$$

In equation (6.75), comparing the corresponding terms on both sides, and observing that

$$B^4(ax) = a^4 B^4(x), B^4(by) = b^4 B^4(y),$$

$$B^{2,2}(ax, by) = a^2 b^2 B^{2,2}(x, y), B^2(ax) = a^2 B^2(x), B^2(by) = b^2 B^2(y),$$

we obtain

$$2(b^4 - b^2)B^2(y) + 2(a^4 - a^2)B^2(x) + 2(b^4 - a^2 b^2 + a^4 - 1)B^0 = 0 \tag{6.76}$$

which gives $B^2(y) \equiv 0$, $B^2(x) \equiv 0$, $B^0 \equiv 0$. Thus, (6.73) reduces to

$$f(x) = B^4(x),$$

which completes the proof of the theorem. $\qquad\square$

Chapter 7

General Solution of Quintic and Sextic Functional Equations

7.1. Introduction

In the previous chapter, the general solution of several types of quartic functional equations are discussed. In this chapter, we present the general solution of quintic functional equation, and sextic functional equation. In Sections 7.2 and 7.3, the general solution of quintic functional equation and the general solution of sextic functional equation are presented in a lucid manner, respectively. Throughout this chapter, let X and Y be vector spaces.

7.2. Quintic functional equation

Consider the following functional equation

$$f(x + 3y) - 5f(x + 2y) + 10f(x + y) - 10f(x) + 5f(x - y) - f(x - 2y) = 120f(y),$$
$$(7.1)$$

which is called a **quintic functional equation**, and every solution of the quintic functional equation is said to be quintic mapping. It is easy to show that the function $f(x) = kx^5$ is the solution of the quintic functional equation (7.1).

The following theorem presents the solution of the quintic functional equation (7.1).

Theorem 7.1 ([343]). *A function $f : X \to Y$ is a solution of the functional equation (7.1) if and only if f is of the form $f(x) = A^5(x)$, for all $x \in X$, where $A^5(x)$ is the diagonal of the 5-additive symmetric map $A_5 : X^5 \to Y$.*

Proof. Assume that f satisfies the functional equation (7.1). Replacing $x = y = 0$ in equation (7.1), one gets $f(0) = 0$. Replacing (x, y) with $(0, x)$ and $(x, -x)$ in (7.1), respectively, and adding the two resulting equations, we obtain $f(-x) = -f(x)$. Replacing (x, y) with $(3x, x)$ and $(0, 2x)$ in (7.1), respectively, and subtracting the two resulting equations, we get

$$5f(5x) - 14f(4x) + 10f(3x) - 120f(2x) + 121f(x) = 0. \tag{7.2}$$

Replacing (x, y) by $(2x, x)$ in (7.1), we have

$$5f(5x) - 25f(4x) + 50f(3x) - 50f(2x) - 575f(x) = 0 \tag{7.3}$$

for all $x \in X$. Subtracting (7.3) from (7.2), we find

$$11f(4x) - 40f(3x) - 70f(2x) + 696f(x) = 0 \tag{7.4}$$

for all $x \in X$. Replacing (x, y) with (x, x) in (7.1) and multiplying the resulting equation by 11, we obtain

$$11f(4x) - 55f(3x) + 110f(2x) - 1419f(x) = 0 \tag{7.5}$$

for all $x \in X$. Subtracting (7.5) from (7.4), one gets

$$15f(3x) - 180f(2x) + 2115f(x) = 0 \tag{7.6}$$

for all $x \in X$. Replacing (x, y) with $(0, x)$ in (7.1) and multiplying the resulting equation by 15, one finds

$$15f(3x) - 60f(2x) - 1725f(x) = 0 \tag{7.7}$$

for all $x \in X$. Subtracting (7.7) from (7.6), we arrive at

$$f(2x) = 2^5 f(x) \tag{7.8}$$

for all $x \in X$.

On the other hand, one can rewrite the functional equation (7.1) in the form

$$f(x) - \frac{1}{10}f(x + 3y) + \frac{1}{2}f(x + 2y) - f(x + y)$$

$$- \frac{1}{2}f(x - y) + \frac{1}{10}f(x - 2y) + 12f(y) = 0 \tag{7.9}$$

for all $x \in X$. Hence f is a generalized polynomial function of degree at most 6, that is, f is of the form

$$f(x) = A^5(x) + A^4(x) + A^3(x) + A^2(x) + A^1(x) + A^0(x) \tag{7.10}$$

for all $x \in X$, where $A^0(x) = A^0$ is an arbitrary element of Y, and $A^i(x)$ is the diagonal of the i-additive symmetric map $A_i : X^i \to Y$ for $i = 1, 2, 3, 4, 5$. By $f(0) = 0$ and $f(-x) = -f(x)$ for all $x \in X$, we get $A^0(x) = A^0 = 0$, and the function f is odd. Thus, we have $A^4(x) = A^2(x) = 0$. It follows that $f(x) = A^5(x) + A^3(x) + A^1(x)$. By (7.8) and $A^n(rx) = r^n A^n(x)$ whenever $x \in X$ and $r \in \mathbb{Q}$, we obtain $2^5 A^3(x) + 2^5 A^1(x) = 2^3 A^3(x) + 2A^1(x)$. Hence $A^1(x) = -4A^3(x)/5$, and so $A^3(x) = A^1(x) = 0$ for all $x \in X$. Therefore, $f(x) = A^5(x)$.

Conversely, assume that $f(x) = A^5(x)$ for all $x \in X$, where $A^5(x)$ is the diagonal of the 5-additive symmetric map $A_5 : X^5 \to Y$. From $A^5(x + y) = A^5(x) + A^5(y) + 5A^{4,1}(x,y) + 10A^{3,2}(x,y) + 10A^{2,3}(x,y) + 5A^{1,4}(x,y)$, $A^5(rx) = r^5A^r(x)$, $A^{4,1}(x,ry) = rA^{4,1}(x,y)$, $A^{3,2}(x,ry) = r^2A^{3,2}(x,y)$, $A^{2,3}(x,ry) = r^3A^{2,3}(x,y)$, $A^{2,3}(x,ry) = r^3A^{2,3}(x,y)$ and $A^{1,4}(x,ry) = r^4A^{1,4}(x,y)$ $(x,y \in X, r \in \mathbb{Q})$, we see that f satisfies (7.1), which completes the proof. $\qquad\square$

7.3. Sextic functional equation

Consider the following functional equation

$$f(x + 3y) - 6f(x + 2y) + 15f(x + y) - 20f(x)$$
$$+ 15f(x - y) - 6f(x - 2y) + f(x - 3y) = 720f(y), \qquad (7.11)$$

which is called a **sextic functional equation**, and every solution of the sextic functional equation is said to be sextic mapping. It is easy to show that the function $f(x) = kx^6$ is the solution of the sextic functional equation (7.11).

The following theorem provides the general solution of the sextic functional equation (7.11).

Theorem 7.2 ([343]). *A function $f : X \to Y$ is a solution of the functional equation (7.11) if and only if f is of the form $f(x) = A^6(x)$, for all $x \in X$, where $A^6(x)$ is the diagonal of the 6-additive symmetric map $A_6 : X^6 \to Y$.*

Proof. Assume that f satisfies the functional equation (7.11). Replacing $x = y = 0$ in (7.11), one gets $f(0) = 0$. Substituting y by $-y$ in (7.11) and subtracting the resulting equation from (7.11) and then y by x, we obtain $f(-x) = f(x)$. Replacing (x, y) with $(0, 2x)$ and $(3x, x)$ in (7.11), respectively, we obtain

$$f(6x) - 6f(4x) - 345f(2x) = 0, \qquad (7.12)$$

$$f(6x) - 6f(5x) + 15f(4x) - 20f(3x) + 15f(2x) - 726f(x) = 0 \qquad (7.13)$$

for all $x \in X$. Subtracting (7.13) from (7.12), we find

$$6f(5x) - 21f(4x) + 20f(3x) - 360f(2x) + 726f(x) = 0 \qquad (7.14)$$

for all $x \in X$. Replacing (x, y) with $(2x, x)$ in (7.11), and from $f(0) = 0$ and $f(-x) = f(x)$, and then multiplying the resulting equation by 6, we obtain

$$6f(5x) - 36f(4x) + 90f(3x) - 120f(2x) - 4224f(x) = 0 \qquad (7.15)$$

for all $x \in X$. Subtracting (7.15) from (7.14), one gets

$$15f(4x) - 70f(3x) - 240f(2x) + 4950f(x) = 0 \qquad (7.16)$$

for all $x \in X$. Replacing (x, y) with $(0, x)$ (and then multiplying by 10) and (x, y) with (x, x) (and then multiplying by 15) in (7.11), respectively, then using $f(0) = 0$

and $f(-x) = f(x)$, we arrive at

$$20f(3x) - 120f(2x) - 6900f(x) = 0 \tag{7.17}$$

and

$$15f(4x) - 90f(3x) + 240f(2x) - 11190f(x) = 0 \tag{7.18}$$

for all $x \in X$. Subtracting (7.18) from (7.16), one gets

$$20f(3x) - 480f(2x) + 16140f(x) = 0 \tag{7.19}$$

for all $x \in X$. Subtracting (7.19) from (7.17), we have

$$360f(2x) = 23040f(x) \tag{7.20}$$

for all $x \in X$. Hence,

$$f(2x) = 2^6 f(x) \tag{7.21}$$

for all $x \in X$.

On the other hand, one can rewrite the functional equation (7.11) in the form

$$f(x) - \frac{1}{20}f(x + 3y) + \frac{3}{10}f(x + 2y) - \frac{3}{4}f(x + y) - \frac{3}{4}f(x - y)$$

$$+ \frac{3}{10}f(x - 2y) - \frac{1}{20}f(x - 3y) + 36f(y) = 0 \tag{7.22}$$

for all $x \in X$. Hence, f is a generalized polynomial function of degree at most 6, that is, f is of the form,

$$f(x) = A^6(x) + A^5(x) + A^4(x) + A^3(x) + A^2(x) + A^1(x) + A^0(x) \tag{7.23}$$

for all $x \in X$, where $A^0(x) = A^0$ is an arbitrary element of Y, and $A^i(x)$ is the diagonal of the i-additive symmetric map $A_i : X^i \to Y$ for $i = 1, 2, 3, 4, 5, 6$. By $f(0) = 0$ and $f(-x) = f(x)$ for all $x \in X$, we get $A^0(x) = A^0 = 0$, and the function f is even. Thus, we have $A^5(x) = 0$, $A^3(x) = 0$, and $A^1(x) = 0$. It follows that $f(x) = A^6(x) + A^4(x) + A^2(x)$. By (7.21) and $A^n(rx) = r^n A^n(x)$ whenever $x \in X$ and $r \in \mathbb{Q}$, we obtain $2^6 A^4(x) + 2^6 A^2(x) = 2^4 A^4(x) + 2^2 A^2(x)$. Hence $A^2(x) = -4A^4(x)/5$, and so $A^2(x) = A^4(x) = 0$ for all $x \in X$. Therefore, $f(x) = A^6(x)$.

Conversely, assume that $f(x) = A^6(x)$ for all $x \in X$, where $A^6(x)$ is the diagonal of the 6-additive symmetric map $A_6 : X^6 \to Y$. From $A^6(x+y) = A^6(x) + A^6(y) + 6$ $A^{5,1}(x,y) + 15A^{4,2}(x,y) + 20A^{3,3}(x,y) + 15A^{2,4}(x,y) + 6A^{1,5}(x,y)$, $A^6(rx) = r^6 A^6(x)$, $A^{5,1}(x,ry) = rA^{5,1}(x,y)$, $A^{4,2}(x,ry) = r^2 A^{4,2}(x,y)$, $A^{3,3}(x,ry) = r^3 A^{3,3}(x,y)$, $A^{2,4}$ $(x,ry) = r^4 A^{2,4}(x,y)$ and $A^{1,5}(x,ry) = r^5 A^{1,5}(x,y)$ $(x,y \in X, r \in \mathbb{Q})$, we see that f satisfies (7.11), which completes the proof of the theorem. $\square$

Chapter 8

Mixed Type Functional Equations

8.1. Introduction

In previous chapters, we have discussed about the general solution of quadratic, cubic, quartic, quintic, and sextic type of functional equations. In this chapter, we elucidate the method of obtaining the general solution of a variety of mixed type functional equations such as additive-quadratic, additive-cubic, additive-quartic, quadratic-cubic, quadratic-quartic, and cubic-quartic functional equations. Throughout this chapter, we assume, X and Y as vector spaces.

8.2. Mixed type additive-quadratic functional equations

Consider the following mixed type functional equation derived from quadratic and additive functions

$$f(x + 2y) + f(x - 2y) + 4f(x) = 3[f(x + y) + f(x - y)] + f(2y) - 2f(y).$$

$$(8.1)$$

It is easy to see that the function $f(x) = ax^2 + bx$ is a solution of the functional equation (8.1), and hence it is called mixed type additive-quadratic functional equation.

8.2.1. *Mixed type functional equation* (8.1)

The following theorem establishes the general solution of the functional equation (8.1).

Lemma 8.1 ([197]). *If an even function $f : X \to Y$ satisfies (8.1) for all $x, y \in X$, then f is quadratic.*

Proof. In view of the evenness of f, we have $f(-x) = f(x)$, for all $x \in X$. Putting $x = y = 0$ in (8.1), we get $f(0) = 0$. Setting $x = 0$ in (8.1), we obtain that

$f(2y) = 4f(y)$, for all $y \in X$, and equation (8.1) implies the following equation

$$f(x + 2y) + f(x - 2y) + 4f(x) = 3f(x + y) + 3f(x - y) + 2f(y) \qquad (8.2)$$

for all $x, y \in X$. Replacing x by $2x$ in (8.2), we obtain

$$4f(x + y) + 4f(x - y) + 16f(x) = 3f(2x + y) + 3f(2x - y) + 2f(y) \qquad (8.3)$$

for all $x, y \in X$. Replacing x and y by y and x in (8.2), respectively, we get by evenness of f that

$$f(2x + y) + f(2x - y) + 4f(y) = 3f(x + y) + 3f(x - y) + 2f(x) \qquad (8.4)$$

for all $x, y \in X$. So we obtain from (8.3) and (8.4) that

$$f(x + y) + f(x - y) = 2f(x) + 2f(y)$$

for all $x, y \in X$. Therefore, the function $f : X \to Y$ is quadratic. $\qquad \square$

Lemma 8.2 ([197]). *If an odd function $f : X \to Y$ satisfies equation (8.1) for all $x, y \in X$, then f is additive.*

Proof. In view of the oddness of f, we have $f(-x) = -f(x)$ for all $x \in X$. Therefore $f(0) = 0$, and equation (8.1) implies the following equation:

$$f(x + 2y) + f(x - 2y) + 4f(x) = 3f(x + y) + 3f(x - y) \qquad (8.5)$$

for all $x, y \in X$. Letting $x = 0$ in (8.5), we obtain

$$f(2y) = 2f(y) \qquad (8.6)$$

for all $x \in X$. Replacing x by $2x$ in (8.5) and using (8.6), we obtain

$$2f(x + y) + 2f(x - y) + 8f(x) = 3f(2x + y) + 3f(2x - y) \qquad (8.7)$$

for all $x, y \in X$. Replacing x and y by y and x in (8.5), we have

$$f(2x + y) - f(2x - y) + 4f(y) = 3f(x + y) - 3f(x - y) \qquad (8.8)$$

for all $x, y \in X$. It follows from (8.7) and (8.8) that

$$6f(2x - y) = -7f(x + y) + 11f(x - y) + 8f(x) + 12f(y) \qquad (8.9)$$

for all $x, y \in X$. Replacing y by $x - y$ in (8.9), we obtain

$$7f(2x - y) = -6f(x + y) + 12f(x - y) + 8f(x) + 11f(y) \qquad (8.10)$$

for all $x, y \in X$. Hence it follows from (8.9) and (8.10) that

$$f(2x - y) = f(x + y) + f(x - y) - f(y) \qquad (8.11)$$

for all $x, y \in X$. Replacing y by $-y$ in (8.11) and using oddness of f, we obtain

$$f(2x + y) = f(x - y) + f(x + y) + f(y) \qquad (8.12)$$

for all $x, y \in X$. It follows from (8.7), (8.11) and (8.12) that

$$3f(x+y) - 3f(x-y) = 6f(y) \tag{8.13}$$

for all $x, y \in X$. Replacing x and y by y and x in (8.13), respectively, and using oddness of f, we obtain

$$3f(x+y) + 3f(x-y) = 6f(x) \tag{8.14}$$

for all $x, y \in X$. Adding (8.13) and (8.14), we conclude that

$$f(x+y) = f(x) + f(y)$$

for all $x, y \in X$. So the function $f : X \to Y$ is additive, and the proof is completed. $\square$

Theorem 8.1 ([197]). *A function $f : X \to Y$ satisfies (8.1) for all $x, y \in X$ if and only if there exist a symmetric bi-additive function $B : X \times X \to Y$ and an additive function $A : X \to Y$ such that $f(x) = B(x,x) + A(x)$ for all $x \in X$.*

Proof. If there exist a symmetric bi-additive function $B : X \times X \to Y$ and an additive function $A : X \to Y$ such that $f(x) = B(x,x) + A(x)$ for all $x \in X$, it is easy to show that

$$f(x+2y) + f(x-2y) + 4f(x) = 6B(x,x) + 8B(y,y) + 6A(x)$$
$$= 3f(x+y) + 3f(x-y) + f(2y) - 2f(y)$$

for all $x, y \in X$. Therefore, the function $f : X \to Y$ satisfies equation (8.1).

Conversely, we decompose f into the even part and odd part by setting

$$f_e(x) = \frac{f(x) + f(-x)}{2} \quad \text{and} \quad f_o(x) = \frac{f(x) - f(-x)}{2},$$

respectively, for all $x \in X$. It is clear that $f(x) = f_e(x) + f_o(x)$ for all $x \in X$. It is easy to show that the functions f_e, and f_o satisfy equation (8.1). Hence by Lemmas 8.1 and 8.2, we achieve that the functions f_e and f_o are quadratic and additive, respectively. Therefore, there exists a symmetric bi-additive function $B : X \times X \to Y$ such that $f_e(x) = B(x,x)$ for all $x \in X$. So

$$f(x) = B(x,x) + A(x),$$

where $A(x) = f_o(x)$, for all $x \in X$. This completes the proof. $\square$

8.2.2. *Mixed type functional equation (8.15)*

The general solution of the mixed type additive-quadratic functional equation

$$f(x+2y) + 2f(x-y) = f(x-2y) + 2f(x+y) \tag{8.15}$$

is discussed in the following theorem.

Theorem 8.2 ([139]). *A function $f : X \to Y$ satisfies the functional equation (8.15) if and only if there exist functions $B : X \times X \to Y$, $A : X \to Y$ and a constant c in Y such that $f(x) = B(x, x) + A(x) + c$ for all $x \in X$, where B is symmetric bi-additive, and A is additive.*

Proof. Let $f : X \to Y$ satisfy the functional equation (8.15). If we put $g(x) = f(x) - f(0)$, we obtain that g is also a solution of (8.15) and $g(0) = 0$. So we may assume, without loss of generality, that f is a solution of (8.15) and $f(0) = 0$. Let $f_e(x) = \frac{f(x) + f(-x)}{2}$, $f_o(x) = \frac{f(x) - f(-x)}{2}$ for all $x \in X$. Then $f_e(0) = 0 = f_o(0)$ and $f(x) = f_e(x) + f_o(x)$, f_e is even and f_o is odd. Since f is a solution of (8.15), f_e and f_o also satisfy equation (8.15).

Thus, we may assume that f is a solution of the functional equation (8.15) and f is odd, $f(0) = 0$. Putting $x = 0$ and $y = x$ in (8.15) separately, we get

$$f(2y) = 2f(y) \quad \text{and} \quad f(3y) = 3f(y)$$

for all $x \in X$. Thus equation (8.15) can be written as

$$f(2x + y) + f(2x - y) = 2f(x + y) + 2f(x - y), \tag{8.16}$$

$$f(x + 2y) + f(-x + 2y) = f(2x + 2y) - f(2x - 2y), \tag{8.17}$$

which yield

$$f(u) + f(v) = f\left(\frac{3u - v}{2}\right) - f\left(\frac{u - 3v}{2}\right) \tag{8.18}$$

for all $u, v \in X$. Replacing x by $x - y$ in (8.15) and using the oddness of f, one arrives at

$$f(x + y) + 2f(x - 2y) = f(x - 3y) + 2f(x). \tag{8.19}$$

Setting y by $x - y$, $x + y$ in (8.16), separately, we obtain the following two equations

$$f(3x - y) + f(x + y) = 2f(2x - y) + 2f(y), \tag{8.20}$$

and

$$f(3x + y) + f(x - y) = 2f(2x + y) - 2f(y). \tag{8.21}$$

Using (8.20) and (8.21), one obtains that

$$f(3x - y) + f(x - 3y) = 2f(2x - y) + 2f(x - 2y) + 2f(y) - 2f(x). \tag{8.22}$$

On the other hand, utilizing equation (8.18), we obtain

$$f(3x - y) + f(x - 3y) = f(4x) - f(4y) = 4f(x) - 4f(y), \tag{8.23}$$

which induces by (8.22)

$$f(2x - y) + f(x - 2y) = 3f(x) - 3f(y). \tag{8.24}$$

Putting $-x + y$ instead of y in (8.24), one has by the oddness of f

$$f(3x - y) + f(3x - 2y) = 3f(x) + 3f(x - y). \tag{8.25}$$

Replacing y in (8.25) by $-y$ and then adding the resulting relation to (8.25), we have that

$$f(3x+y)+f(3x-y)+f(3x+2y)+f(3x-2y) = 6f(x)+3f(x+y)+3f(x-y). \tag{8.26}$$

In turn, it follows from (8.20), (8.21), and (8.16) that

$$f(3x + y) + f(3x - y) = 3f(x + y) + 3f(x - y), \tag{8.27}$$

from which we deduce the following relation together with (8.26)

$$f(3x + 2y) + f(3x - 2y) = 6f(x) = 2f(3x). \tag{8.28}$$

Here equation (8.28) is equivalent to $f(X + Y) + f(X - Y) = 2f(X)$, which is in fact the Cauchy–Jensen equation. Hence in this case $f(x) = A(x)$ for some additive mapping A.

Second, we assume that f is a solution of the functional equation (8.15) and f is even, $f(0) = 0$. Thus equation (8.15) is written by

$$f(x + 2y) + 2f(x - y) = f(x - 2y) + 2f(x + y) \tag{8.29}$$

for all $x, y \in X$. Putting $y = x$ and $y = \frac{x}{2}$ in (8.29) separately, we get

$$f(3x) = f(x) + 2f(2x),$$

$$f(2x) + 2f\left(\frac{x}{2}\right) = 2f\left(\frac{3x}{2}\right) = 2f\left(\frac{x}{2}\right) + 4f(x),$$

which imply that $f(2x) = 4f(x)$, $f(3x) = 9f(x)$ for all $x \in X$. Interchanging x and y in (8.29) yields the relation

$$f(2x + y) + 2f(x - y) = f(2x - y) + 2f(x + y) \tag{8.30}$$

for all $x, y \in X$. Setting x by $x + y$ in (8.29), one obtains that

$$f(x + 3y) + 2f(x) = f(x - y) + 2f(x + 2y) \tag{8.31}$$

for all $x, y \in X$. Replacing y by $-y$ in (8.31), we obtain that

$$f(x - 3y) + 2f(x) = f(x + y) + 2f(x - 2y) \tag{8.32}$$

for all $x, y \in X$. Setting y by $x - y$ in (8.29) and then putting y by $\frac{3y}{2}$ in the resulting relation, we have

$$9f(x - y) + \frac{9}{2}f(y) = f(x - 3y) + \frac{1}{2}f(4x - 3y). \tag{8.33}$$

Adding equation (8.32) to (8.33) side by side, one obtains that

$$9f(x-y) + \frac{9}{2}f(y) + 2f(x) = f(x+y) + 2f(x-2y) + \frac{1}{2}f(4x-3y) \qquad (8.34)$$

for all $x, y \in X$. Exchanging x with y in (8.34) and then subtracting the resulting relation from (8.34), we have

$$\frac{5}{2}f(y) - \frac{5}{2}f(x) = 2f(x-2y) - 2f(2x-y) + \frac{1}{2}f(4x-3y) - \frac{1}{2}f(3x-4y) \qquad (8.35)$$

for all $x, y \in X$. Replacing x by $4x$ in (8.32), we get

$$f(4x-3y) + 32f(x) = f(4x+y) + 8f(2x-y) \qquad (8.36)$$

for all $x, y \in X$. Interchanging x and y in (8.36), we have by the evenness of f

$$f(3x-4y) + 32f(y) = f(x+4y) + 8f(x-2y) \qquad (8.37)$$

for all $x, y \in X$. Subtracting (8.37) from (8.36) and dividing it by 2, we arrive at the equation

$$\frac{1}{2}f(4x-3y) - \frac{1}{2}f(3x-4y) + 16f(x) - 16f(y)$$

$$= \frac{1}{2}f(4x+y) - \frac{1}{2}f(x+4y) + 4f(2x-y) - 4f(x-2y). \qquad (8.38)$$

Combining (8.35) and (8.38), we easily see that

$$\frac{27}{2}f(x) - \frac{27}{2}f(y)$$

$$= \frac{1}{2}f(4x+y) - \frac{1}{2}f(x+4y) + 2f(2x-y) - 2f(x-2y) \qquad (8.39)$$

for all $x, y \in X$. Putting $x+y$ instead of x in (8.31), we obtain that

$$f(x+4y) + 2f(x+y) = f(x) + 2f(x+3y) \qquad (8.40)$$

for all $x, y \in X$. Replacing y by $2x+y$ in (8.30), one obtains that

$$f(4x+y) + 2f(x+y) = f(y) + 2f(3x+y) \qquad (8.41)$$

for all $x, y \in X$. Subtracting equation (8.40) from (8.41), we yield the relation

$$f(4x+y) - f(x+4y) = 2f(3x+y) - 2f(x+3y) + f(y) - f(x) \qquad (8.42)$$

for all $x, y \in X$. Multiplying equation (8.42) by $\frac{1}{2}$, and then adding the resulting relation to (8.39), we have

$$14f(x) - 14f(y) = 2f(2x - y) - 2f(x - 2y) + f(3x + y) - f(x + 3y) \qquad (8.43)$$

for all $x, y \in X$.

In turn, interchanging x and y in (8.31), and then subtracting (8.31) from the resulting relation, one obtains that

$$f(3x + y) - f(x + 3y) + 2f(y) - 2f(x) = 2f(2x + y) - 2f(x + 2y) \qquad (8.44)$$

for all $x, y \in X$. Adding the relation (8.44) to (8.43) side by side and dividing it by 2, we arrive at the equation

$$6f(x) - 6f(y) = f(2x + y) - f(x + 2y) + f(2x - y) - f(x - 2y). \qquad (8.45)$$

Applying the relations (8.29) and (8.30) to (8.45), we have the following equation

$$3f(x) - 3f(y) = f(2x + y) - f(x + 2y) \qquad (8.46)$$

for all $x, y \in X$.

Now utilizing (8.46) one obtains the following two relations:

$$f(x + y) - f\left(x - \frac{y}{2}\right) = \frac{1}{3}f\left(3\left(x + \frac{y}{2}\right)\right) - \frac{1}{3}f(3x),$$

and

$$f(x - y) - f\left(x + \frac{y}{2}\right) = \frac{1}{3}f\left(3\left(x - \frac{y}{2}\right)\right) - \frac{1}{3}f(3x).$$

Since $f(2x) = 4f(x)$, $f(3x) = 9f(x)$ for all $x \in X$, adding the above two relations we get the equation

$$f(x + y) + f(x - y) + 6f(x) = f(2x + y) + f(2x - y), \qquad (8.47)$$

which is equivalent to the quadratic functional equation $f(x+y)+f(x-y) = 2f(x)+ 2f(y)$. Therefore, $f(x) = Q(x, x)$, where Q is a symmetric bi-additive function.

That is, if $f : X \to Y$ satisfies the functional equation (8.15), then $f(x) = f_e(x) + f_o(x) = B(x, x) + A(x)$ for all $x \in X$, where B, A are mappings stated in the theorem. Since we regard $f(x)$ as $f(x)-f(0)$, we get $f(x) = B(x,x)+A(x)+f(0)$ for all $x \in X$ and we obtain the desired results.

Conversely, if there exist functions $B : X \times X \to Y$, $A : X \to Y$ and a constant c in Y such that $f(x) = B(x, x) + A(x) + c$ for all $x \in X$, where A is additive and B is symmetric bi-additive, then it is obvious that f satisfies equation (8.15). This completes the proof. $\qquad \square$

8.3. Mixed type additive-cubic functional equations

8.3.1. *Mixed type functional equation* (8.48)

The general solution of mixed type additive-cubic functional equation

$$3f(x + y + z) + f(-x + y + z)$$
$$+ f(x - y + z) + f(x + y - z) + 4[f(x) + f(y) + f(z)]$$
$$= 4[f(x + y) + f(x + z) + f(y + z)] \tag{8.48}$$

is obtained in the following lemma.

Lemma 8.3 ([277]). *Let* $f : X \to Y$ *be a function satisfying the functional equation* (8.48). *Then* f *is an odd function.*

Proof. Letting (x, y, z) be $(0, 0, 0)$ in (8.48), we obtain

$$f(0) = 0. \tag{8.49}$$

Replacing (x, y, z) by $(x, 0, 0)$ in (8.48), we obtain

$$f(-x) = -f(x) \tag{8.50}$$

for all $x \in X$. Hence f is an odd function, which completes the proof. $\qquad\square$

Lemma 8.4 ([277]). *Let* $f : X \to Y$ *be a function satisfying the functional equation* (8.48). *Then* f *is a cubic function.*

Proof. Replacing (x, y, z) by $(x - y, y - z, z - x)$ and using (8.49), (8.50) in (8.48), we obtain

$$f(2(x - y)) + f(2(y - z)) + f(2(z - x)) = 8[f(x - y) + f(y - z) + f(z - x)] \tag{8.51}$$

for all $x, y, z \in X$. Setting $(x - y, y - z, z - x)$ by (u, v, w) in (8.51), we arrive at

$$f(2u) + f(2v) + f(2w) = 8[f(u) + f(v) + f(w)] \tag{8.52}$$

for all $u, v, w \in X$. Replacing (u, v, w) by (u, u, u) in (8.52), we obtain

$$f(2u) = 8f(u) \tag{8.53}$$

for all $u \in X$. Again replacing (x, y, z) by (x, x, x) in (8.48), we obtain

$$f(3x) = 4f(2x) - 5f(x) \tag{8.54}$$

for all $x \in X$. Now with the help of (8.53), we arrive at $f(3x) = 27f(x)$ for all $x \in X$. In general for any positive integer n, we obtain $f(nx) = n^3 f(x)$ for all $x \in X$. Therefore f is a cubic function. This completes the proof. $\qquad\square$

Lemma 8.5 ([277]). *Let* $f : X \to Y$ *be a function satisfying the functional equation* (8.48). *Then* f *is an additive function.*

Proof. Replacing z by $x + y$ in (8.48), we get

$$3f(2(x+y)) + f(2y) + f(2x) + 4f(x) + 4f(y) = 4f(2x+y) + 4f(x+2y)$$
$$(8.55)$$

for all $x, y \in X$. Again replacing z by $-x - y$ in (8.48), we obtain

$$f(2x) + f(2y) + 8f(x+y) = 8f(x) + 8f(y) + f(2(x+y)) \qquad (8.56)$$

for all $x, y \in X$. Multiplying (8.56) by 3 and using (8.55), we obtain

$$f(2x) + f(2y) + 6f(x+y) = 5f(x) + 5f(y) + f(2x+y) + f(x+2y) \qquad (8.57)$$

for all $x, y \in X$. Subtracting (8.57) from (8.56), we have

$$f(2x+y) + f(x+2y) = 3f(x) + 3f(y) + f(2(x+y)) - 2f(x+y) \qquad (8.58)$$

for all $x, y \in X$. Substituting y by $-y$ in (8.58), we arrive at

$$f(2x-y) + f(x-2y) = 3f(x) - 3f(y) + f(2(x-y)) - 2f(x-y) \qquad (8.59)$$

for all $x, y \in X$. Adding (8.58) and (8.59), we obtain

$$f(2x+y) + f(x+2y) + f(2x-y) + f(x-2y)$$
$$= 6f(x) + f(2(x+y)) - 2f(x+y) + f(2(x-y)) - 2f(x-y) \qquad (8.60)$$

for all $x, y \in X$. Setting $(x+y, x-y)$ by (u, v) in (8.60), we arrive at

$$f(u+x) + f(u+y) + f(x+v) + f(x-v)$$
$$= 6\left(\frac{u+v}{2}\right) + f(2u) + f(2v) - 2f(u) - 2f(v) \qquad (8.61)$$

for all $x, y, u, v \in X$. Replacing (x, y, u, v) by (z, z, z, z) in (8.61), we obtain

$$f(2z) = 2f(z) \qquad (8.62)$$

for all $z \in X$. Again replacing (x, y, z) by (x, x, x) in (8.48), we obtain

$$f(3x) = 4f(2x) - 5f(x) \qquad (8.63)$$

for all $x \in X$. With the help of (8.62), we arrive at $f(3x) = 3f(x)$ for all $x \in X$. In general for any positive integer n, we obtain $f(nx) = nf(x)$ for all $x \in X$. Therefore f is an additive function. This completes the proof. $\qquad \square$

Theorem 8.3 ([277]). *A mapping $f : X \to Y$ is a function satisfying the functional equation (8.48) for all $x, y \in X$, if and only if there exists two mappings $T : X \times X \times X \to Y$ and $R : X \times X \times X \to Y$ such that $f(x) = T(x, x, x) + R(x)$ for all $x \in X$, where T is symmetric for each fixed one variable and is additive for fixed two variables and R is additive.*

Proof. Replacing z by $-z$ in (8.48), we get

$$3f(x+y-z) + f(-x+y-z) + f(x-y-z) + f(x+y+z)$$
$$+ 4[f(x) + f(y) - f(z)] = 4[f(x+y) + f(x-z) + f(y-z)] \qquad (8.64)$$

for all $x, y, z \in X$. Adding (8.48) and (8.64), and dividing by 4, we arrive at

$$f(x+y+z) + f(x+y-z) + 2f(x) + 2f(y)$$
$$= 2f(x+y) + f(x+z) + f(y+z) + f(x-z) + f(y-z) \qquad (8.65)$$

for all $x, y, z \in X$, which is a cubic functional equation. Hence the proof is complete.

$\square$

8.4. Mixed type additive-quartic functional equations

8.4.1. *Mixed type functional equation* (8.66)

The general solution of a mixed type additive-quartic functional equation

$$f(2x+y) + f(2x-y)$$

$$= 4f(f(x+y) + f(x-y)) - \frac{3}{7}(f(2y) - 2f(y)) + 2f(2x) - 8f(x) \qquad (8.66)$$

is achieved in the following theorems.

Theorem 8.4 ([103]). *Let* $f : X \to Y$ *be a function satisfies* (8.66). *Then the following assertions hold:*

(a) *If* f *is even function, then* f *is quartic.*
(b) *If* f *is odd function, then* f *is additive.*

Proof. (a) Putting $x = y = 0$, we get $f(0) = 0$. Setting $x = 0$ in (8.66), then by evenness of f, we obtain

$$f(2y) = 16f(y) \qquad (8.67)$$

for all $y \in X$. Hence (8.66) can be written as

$$f(2x+y) + f(2x-y) = 4(f(x+y) + f(x-y)) + 24f(x) - 6f(y) \qquad (8.68)$$

for all $x, y \in X$. This means that f is a quartic function.

(b) Setting $x = y = 0$ in (8.66) to obtain $f(0) = 0$. Putting $x = 0$ in (8.66), then by oddness of f, we have

$$f(2y) = 2f(y) \qquad (8.69)$$

for all $y \in X$. We obtain from (8.66) and (8.69) that

$$f(2x+y) + f(2x-y) = 4(f(x+y) + f(x-y)) - 4f(x) \qquad (8.70)$$

for all $x, y \in X$. Replacing y by $-2y$ in (8.70), it follows that

$$f(2x - 2y) + f(2x + 2y) = 4(f(x - 2y) + f(x + 2y)) - 4f(x). \tag{8.71}$$

Combining (8.69) and (8.71) yields

$$f(x - y) + f(x + y) = 2(f(x - 2y) + f(x + 2y)) - 2f(x). \tag{8.72}$$

Interchanging x and y in (8.72) yields the relation

$$f(x + y) - f(x - y) = 2(f(y - 2x) + f(y + 2x)) - 2f(y). \tag{8.73}$$

Replacing y by $-y$ in (8.73), and using the oddness of f, we get

$$f(x - y) - f(x + y) = 2(f(2x - y) - f(2x + y)) + 2f(y). \tag{8.74}$$

From (8.70) and (8.74), in which replacing y by $-y$, we obtain

$$4f(2x + y) = 9f(x + y) + 7f(x - y) - 8f(x) + 2f(y). \tag{8.75}$$

Replacing $x + y$ by y in (8.75), one obtains that

$$7f(2x - y) = 4f(x + y) + 2f(x - y) - 9f(y) + 8f(x). \tag{8.76}$$

By using (8.75) and (8.71), we have

$$f(2x + y) + f(2x - y) = \frac{79}{28}f(x + y) + \frac{57}{28}f(x - y) - \frac{6}{7}f(x) - \frac{11}{14}f(y). \tag{8.77}$$

From (8.70) and (8.77) we get that

$$3f(x + y) + 5f(x - y) = 8f(x) - 28f(y). \tag{8.78}$$

Replacing x by $2x$ in (8.70), one obtains that

$$f(4x + y) + f(4x - y) = 16(f(x + y) + f(x - y)) - 24f(x). \tag{8.79}$$

Setting $2x + y$ instead of y in (8.70), we arrive at

$$f(4x + y) - f(y) = 4(f(3x - y) + f(x - y)) - 4f(x). \tag{8.80}$$

Replacing y by $-y$ in (8.80), and using oddness of f, we get

$$f(4x - y) + f(y) = 4(f(3x + y) + f(x + y)) - 4f(x). \tag{8.81}$$

Adding (8.80) and (8.81) yields the relation

$$f(4x + y) + f(4x - y) = 4(f(3x + y) + f(3x - y)) - 4(f(x + y) + f(x - y)) - 8f(x). \tag{8.82}$$

Replacing y by $x + y$ in (8.70), we obtain

$$f(3x + y) + f(x - y) = 4(f(2x + y) - f(y)) - 4f(x). \tag{8.83}$$

Replacing y by $-y$ in (8.83), and using the oddness of f, we obtain

$$f(3x - y) + f(x + y) = 4(f(2x - y) + f(y)) - 4f(x). \qquad (8.84)$$

Combining (8.83) and (8.84) yields

$$f(3x + y) + f(3x - y) = 15(f(x + y) + f(x - y)) - 24f(x). \qquad (8.85)$$

Using (8.82) and (8.85), we get

$$f(4x + y) + f(4x - y) = 56(f(x + y) + f(x - y)) - 104f(x). \qquad (8.86)$$

Combining (8.79) and (8.86), we arrive at

$$f(x + y) + f(x - y) = 2f(x). \qquad (8.87)$$

Hence, by using (8.78) and (8.87), it is easy to see that f is additive, which completes the proof of theorem. $\qquad \square$

Theorem 8.5 ([103]). *Let $f : X \to Y$ be a function. Then f satisfies (8.66) if and only if there exist a unique symmetric multiadditive function $B : X \times X \times X \times X \to Y$ and a unique additive function $A : X \to Y$ such that $f(x) = B(x, x, x, x) + A(x)$ for all $x \in X$.*

Proof. Suppose f satisfies equation (8.66). We decompose f into the even part and odd part by setting

$$f_e(x) = \frac{f(x) + f(-x)}{2}, \quad f_o(x) = \frac{f(x) - f(-x)}{2},$$

respectively, for all $x \in X$. By (8.66), we have

$$f_e(2x + y) + f_e(2x - y)$$

$$= \frac{1}{2}[f(2x + y) + f(-2x - y) + f(2x - y) + f(-2x + y)]$$

$$= \frac{1}{2}\left[4(f(x + y) + f(x - y)) - \frac{3}{7}(f(2y) - 2f(y)) + 2f(2x) - 8f(x)\right]$$

$$+ \frac{1}{2}\left[4(f(-x - y) + f(-x - (-y))) - \frac{3}{7}(f(-2y) - 2f(-y))\right.$$

$$\left. + 2f(-2x) - 8f(-x)\right]$$

$$= 4\left[\frac{1}{2}(f(x+y) + f(-x-y)) + \frac{1}{2}(f(-x+y) + f(x-y))\right]$$

$$- \frac{3}{7}\left[\frac{1}{2}(f(2y) + f(-2y)) - (f(y) - f(-y))\right]$$

$$+ 2\left[\frac{1}{2}(f(2x) + f(-2x))\right] - 8\left[\frac{1}{2}(f(x) + f(-x))\right]$$

$$= 4\left(f_e(x+y) + f_e(x-y)\right) - \frac{3}{7}\left(f_e(2y) - 2f_e(y)\right) + 2f_e(2x) - 8f_e(x)$$

for all $x, y \in X$. It means that f_e holds in (8.66). Similarly, we can show that f_o satisfies (8.66). By the above theorem, f_e and f_o are quartic and additive, respectively. Thus, there exists a unique symmetric multiadditive function $B : X \times X \times X \times X \to Y$ such that $f_e(x) = B(x, x, x, x)$ for all $x \in X$. Put $A(x) = f_o(x)$ for all $x \in X$. It follows that $f(x) = B(x) + A(x)$ for all $x \in X$. The proof of the converse is trivial. This completes the proof. $\square$

8.5. Mixed type quadratic-cubic functional equations

8.5.1. *Preliminaries*

In this section, we introduce generalized polynomial functions. For further details, one can refer to [61].

Let X and Y be linear spaces over the field $\mathbb{Q}$ of rational numbers, and let $s = 0, 1, 2, \ldots$. A function $f : X \to Y$ is called a *polynomial function of order s* if f satisfies the functional equation

$$\sum_{i=0}^{s+1} (-1)^{s+1-i}\binom{s+1}{i} f(x+iy) = 0 \tag{8.88}$$

for all $x, y \in X$. For instance, when $s = 1$, a function f fulfilling the functional equation

$$f(x+2y) - 2f(x+y) + f(x) = 0 \tag{8.89}$$

is a polynomial function of order 1. The following theorem gives a formula of the general solution of the polynomial functions.

Theorem 8.6 ([333]). *Let $n = 0, 1, 2, \ldots$. A function $f : X \to Y$ is a polynomial function of order n if and only if there exist k-additive symmetric functions $A_k : X^k \to Y$, $k = 0, 1, 2, \ldots, n$, such that*

$$f(x) = A^0(x) + A^1(x) + A^2(x) + \cdots + A^n(x)$$

for all $x \in X$ where $A^k : X \to Y$, $k = 0, 1, 2, \ldots, n$, is the diagonalization of A_k and is defined by

$$A^k(x) = A_k(x, \ldots, x), \quad \text{for all } x \in X.$$

By the above theorem, a function f satisfying (8.89) takes the form of $f(x) = A^0(x) + A^1(x)$. Let us consider a k-additive symmetric function $A_k(x_1, \ldots, x_k)$ for $x_1, x_2, \ldots, x_k \in X$ and its diagonalization, $A^k(x)$. It can be proven that the additive of A_k in the ith variable leads to

$$A_k(x_1, \ldots, x_{i-1}, r x_i, x_{i+1}, \ldots, x_k) = r A_k(x_1, \ldots, x_k) \quad \text{for each } r \in \mathbb{Q}.$$

Thus $A^k(rx) = r^k A^k(x)$. In particular, $A^k(-x) = (-1)^k A^k(x)$. Since the function A^1 satisfies the additive functional equation

$$A^1(x + y) = A^1(x) + A^1(y)$$

for all $x \in X$, $A^1(x)$ will also be called an *additive function*. In addition, the functional equation $A^2(x)$ and $A^3(x)$ will be referred as a *quadratic function* and a *cubic function*, respectively.

In the following, we will call a function $f : X \to Y$ given by

$$f(x) = A^0(x) + A^2(x) + A^3(x)$$

for all $x \in X$, a *mixed type quadratic-cubic function*.

8.5.2. *Mixed type functional equation* (8.90)

This section deals with the general solution of mixed type quadratic-cubic functional equation

$$f(x + 3y) - 3f(x + 2y) + 3f(x + y) - f(x) = 3f(y) - 3f(-y). \tag{8.90}$$

The general solution of the functional equation (8.90) is established in the following theorem.

Theorem 8.7 ([333]). *A function $f : X \to Y$ satisfies the functional equation (8.90) if and only if there exist a quadratic function $A^2 : X \to Y$, a cubic function $A^3 : X \to Y$ and a constant A^0 such that*

$$f(x) = A^0 + A^2(x) + A^3(x) \tag{8.91}$$

for all $x \in X$.

Proof. Assume that a function $f : X \to Y$ satisfies equation (8.90). Replacing x by $x + y$ in (8.90), and taking the difference of the previous result and (8.90), we then obtain

$$f(x + 4y) - 4f(x + 3y) + 6f(x + 2y) - 4f(x + y) + f(x) = 0. \tag{8.92}$$

Hence, by Theorem 8.6, we see that, f is a polynomial function of order 3, and takes the form following

$$f(x) = A^0 + A^1(x) + A^2(x) + A^3(x) \tag{8.93}$$

for all $x \in X$. Substituting (8.93) into (8.90), one obtains that

$$6A^3(y) = 6A^1(y) + 6A^3(y).$$

Thus, it yields $A^1(y) = 0$ for all $y \in X$, which completes the proof of the theorem.
□

8.6. Mixed type quadratic-quartic functional equations

This section deals with the investigation of general solution of the functional equation

$$f(2x + y) + f(2x - y) = 4[f(x + y) + f(x - y)] + 2f[f(2x) - 4f(x)] - 6f(y). \tag{8.94}$$

It is easy to see that the function $f(x) = ax^2 + bx^4$ is a solution of (8.94). We need the following lemma for obtaining the solution of the functional equation (8.94).

8.6.1. *Mixed type functional equation* (8.94)

Lemma 8.6 ([99]). *If a mapping $f : X \to Y$ satisfies (8.94) for all $x, y \in X$, then f is quadratic-quartic.*

Proof. We show that the mappings $g : X \to Y$ defined by $g(x) := f(2x) - 16f(x)$ and $h : X \to Y$ defined by $h(x) := f(2x) - 4f(x)$ are quadratic and quartic, respectively.

Letting $x = y = 0$ in (8.94), we have $f(0) = 0$. Putting $x = 0$ in (8.94), we get $f(-y) = f(y)$. Thus the mapping f is even. Replacing y by $2y$ in (8.94), we obtain

$$f(2x + 2y) + f(2x - 2y) = 4[f(x + 2y) + f(x - 2y)] + 2[f(2x) - 4f(x)] - 6f(2y) \tag{8.95}$$

for all $x, y \in X$. Interchanging x with y in (8.94), we obtain

$$f(2y + x) + f(2y - x) = 4[f(y + x) + f(y - x)] + 2[f(2y) - 4f(y)] - 6f(x) \tag{8.96}$$

for all $x, y \in X$. Since f is even, by (8.96), one gets

$$f(x + 2y) + f(x - 2y) = 4[f(x + y) + f(x - y)] + 2[f(2y) - 4f(y)] - 6f(x) \tag{8.97}$$

for all $x, y \in X$. It follows from (8.95) and (8.97) that

$$[f(2(x+y)) - 16f(x+y)] + [f(2(x-y)) - 16f(x-y)]$$
$$= 2[f(2x) - 16f(x)] + 2[f(2y) - 16f(y)] \tag{8.98}$$

for all $x, y \in X$. It means that

$$g(x+y) + g(x-y) = 2g(x) + 2g(y) \tag{8.99}$$

for all $x, y \in X$. Therefore, the mapping $g : X \to Y$ is quadratic.

To prove that $h : X \to Y$ is quartic, we have to show that

$$h(x+2y) + h(x-2y) = 4[h(x+y) + h(x-y) + 6h(y)] - 6h(x) \tag{8.100}$$

for all $x, y \in X$. Since f is even, the mapping h is even. Now if we interchange x with y in the last equation, we obtain

$$h(2x+y) + h(2x-y) = [h(x+y) + h(x-y) + 6h(x)] - 6h(y) \tag{8.101}$$

for all $x, y \in X$. Thus, it is enough to prove that h satisfies (8.101). Replacing x and y by $2x$ and $2y$ in (8.94), respectively, we obtain

$$f(2(2x+y)) + f(2(2x-y)) = 4[f(2(x+y)) + f(2(x-y))]$$
$$+ 2[f(4x) - 4f(2x)] - 6f(2y) \tag{8.102}$$

for all $x, y \in X$. Since $g(2x) = 4g(x)$ for all $x \in X$,

$$f(4x) = 20f(2x) - 64f(x) \tag{8.103}$$

for all $x \in X$. By (8.102) and (8.103), we obtain

$$f(2(2x+y)) + f(2(2x-y)) = 4[f(2(x+y)) + f(2(x-y))]$$
$$+ 32[f(2x) - 4f(x)] - 6f(2y) \tag{8.104}$$

for all $x, y \in X$. By multiplying both sides of (8.94) by 4, we obtain

$$4[f(2x+y) + f(2x-y)] = 16[f(x+y) + f(x-y)] + 8[f(2x) - 4f(x)] - 24f(y) \tag{8.105}$$

for all $x, y \in X$. If we subtract the last equation from (8.104), and then simplifying, we obtain

$$h(2x+y) + h(2x-y) = 4[h(x+y) + h(x-y) + 6h(x)] - 6h(y) \tag{8.106}$$

for all $x, y \in X$. Therefore, the mapping $h : X \to Y$ is quartic. This completes the proof of the lemma. $\qquad\square$

Theorem 8.8. *A mapping $f : X \to Y$ satisfies (8.94) for all $x, y \in X$ if and only if there exist a unique symmetric multiadditive mapping $M : X^4 \to Y$ and a unique symmetric bi-additive mapping $B : X \times X \to Y$ such that*

$$f(x) = M(x, x, x, x) + B(x, x) \tag{8.107}$$

for all $x \in X$.

Proof. Let f satisfy (8.94), and assume that $g : x \to y$ and $h : X \to Y$ are mappings defined by

$$g(x) = f(2x) - 16f(x) \quad \text{and} \quad h(x) = f(2x) - 4f(x), \tag{8.108}$$

respectively, for all $x \in X$. By Lemma 8.6, we obtain that the mappings g and h are quadratic and quartic, respectively, and

$$f(x) = \frac{1}{12}h(x) - \frac{1}{12}g(x) \tag{8.109}$$

for all $x \in X$.

Therefore, there exist a unique symmetric multiadditive mapping $M : X^4 \to Y$ and a unique symmetric bi-additive mapping $B : X \times X \to Y$ such that

$$\frac{1}{12}h(x) = M(x, x, x, x) \quad \text{and} \quad -\frac{1}{12}g(x) = B(x, x)$$

for all $x \in X$ (see [2, 246]). So $f(x) = M(x, x, x, x) + B(x, x)$ for all $x \in X$. The converse part of the proof is obvious. This completes the proof of the theorem. $\square$

8.7. Mixed type cubic-quartic functional equations

The general solution of mixed type cubic-quartic functional equation

$$f(x + 2y) + f(x - 2y) = 4(f(x + y) + f(x - y)) - 24f(y) - 6f(x) + 3f(2y) \tag{8.110}$$

is obtained in the following theorem. It is easy to see that the function $f(x) = ax^3 + bx^4$ is a solution of the functional equation (8.110).

8.7.1. *Mixed type functional equation (8.110)*

Lemma 8.7 ([96]). *If an even function $f : X \to Y$ satisfies (8.110), then f is a quartic function.*

Proof. Putting $x = y = 0$ in (8.110), we have $f(0) = 0$. Setting $x = 0$ in (8.110), then by evenness of f, we have

$$f(2y) = 16f(y) \tag{8.111}$$

for all $y \in X$. Hence equation (8.110) can be written as

$$f(x + 2y) + f(x - 2y) = f(f(x + y) + f(x - y)) + 24f(y) - 6f(x). \tag{8.112}$$

It means that f is quartic function, which completes the proof of the lemma. $\square$

Lemma 8.8 ([96]). *If an odd function $f : X \to Y$ satisfies (8.110), then f is a cubic function.*

Proof. Setting $x = y = 0$ in (8.110), we have $f(0) = 0$. Putting $x = 0$ in (8.110), then by oddness of f, we have

$$f(2y) = 8f(y). \tag{8.113}$$

Hence equation (8.110) can be written as

$$f(x + 2y) + f(x - 2y) = 4f(x + y) + 4f(x - y) - 6f(x). \tag{8.114}$$

Replacing x by $x + y$ in (8.114), we obtain

$$f(x + 3y) + f(x - y) = 4f(x + 2y) - 6f(x + y) + 4f(x). \tag{8.115}$$

Substituting $-y$ for y in (8.115) gives

$$f(x - 3y) + f(x + y) = 4f(x - 2y) - 6f(x - y) + 4f(x). \tag{8.116}$$

If we subtract (8.115) from (8.116), we obtain

$$f(x + 3y) - f(x - 3y) = 4f(x + 2y) - 4f(x - 2y) - 5f(x + y) + 5f(x - y). \tag{8.117}$$

Interchanging x and y in (8.117), then we see that

$$f(3x + y) + f(3x - y) = 4f(2x + y) + 4f(2x - y) - 5f(x + y) - 5f(x - y). \tag{8.118}$$

With the substitution $y = x + y$ in (8.114), we have

$$f(3x + 2y) - f(x + 2y) = 4f(2x + y) - 4f(y) - 6f(x). \tag{8.119}$$

From the substitution $y = -y$ in (8.119) it follows that

$$f(3x - 2y) - f(x - 2y) = 4f(2x - y) + 4f(y) - 6f(x). \tag{8.120}$$

Adding (8.119) with (8.120), we obtain

$$f(3x + 2y) + f(3x - 2y)$$
$$= 4f(2x + y) + 4f(2x - y) + f(x + 2y) + f(x - 2y) - 12f(x). \qquad (8.121)$$

Replacing x by $2x$ in (8.117) and using (8.113), we obtain

$$f(2x + 3y) - f(2x - 3y) = 32f(x + y) - 32f(x - y) - 5f(2x + y) + 5f(2x - y). \qquad (8.122)$$

Interchanging x with y in (8.122) gives the equation

$$f(3x + 2y) + f(3x - 2y) = 32f(x + y) + 32f(x - y) - 5f(x + 2y) - 5f(x - 2y). \qquad (8.123)$$

If we compare (8.121) and (8.123) and employ (8.114), we conclude that

$$f(2x + y) + f(2x - y) = 2f(x + y) + 2f(x - y) + 12f(x). \qquad (8.124)$$

This means that f is cubic function. This completes the proof of lemma. $\qquad \square$

Theorem 8.9 ([96]). *A function $f : X \to Y$ satisfies (8.90) for all $x, y \in X$ if and only if there exists a unique function $C : X \times X \times X \to Y$ and a unique symmetric multiadditive function $Q : X \times X \times X \times X \to Y$ such that $f(x) = C(x, x, x) + Q(x, x, x, x)$ for all $x \in X$, and that C is symmetric for each fixed one variable and is additive for fixed two variables.*

Proof. Let f satisfy equation (8.110). We decompose f into the even part and odd part by setting

$$f_e(x) = \frac{f(x) + f(-x)}{2} \quad \text{and} \quad f_o(x) = \frac{f(x) - f(-x)}{2}, \qquad (8.125)$$

respectively, for all $x \in X$. By (8.110), we have

$$f_e(x + 2y) + f_e(x - 2y)$$

$$= \frac{1}{2}[f(x + 2y) + f(-x - 2y) + f(x - 2y) + f(-x + 2y)]$$

$$= 4(f_e(x + y) + f_e(x - y)) - 24f_e(y) - 6f_e(y) + 3f_e(2y) \qquad (8.126)$$

for all $x, y \in X$. It means that f_e satisfies (8.110). Similarly, we can show that f_o satisfies (8.110). By Lemmas 8.7 and 8.8, f_e and f_o are quartic and cubic, respectively. Thus there exist a unique function $C : X \times X \times X \to Y$ and a unique symmetric multiadditive function $Q : X \times X \times X \times X \to Y$ such that $f_e(x) = Q(x, x, x, x)$ and that $f_o(x) = C(x, x, x)$ for all $x \in X$, and C is symmetric for each fixed one variable and is additive for fixed two variables. Thus $f(x) = C(x, x, x) + Q(x, x, x, x)$ for all $x \in X$. The proof of the converse is trivial. The proof of the theorem is completed. $\qquad \square$

Chapter 9

Mixed Type Functional Equations
(*continued*)

9.1. Introduction

This chapter is a continuation of Chapter 8, and in this chapter, we mainly concentrate and investigate the general solution of other mixed type functional equations like additive-quadratic-cubic, additive-cubic-quartic, additive-quadratic-quartic, quadratic-cubic-quartic, additive-quadratic-cubic-quartic functional equations. Throughout this chapter, let X and Y be vector spaces.

9.2. Mixed type additive-quadratic-cubic functional equation

9.2.1. *Mixed type functional equation* (9.1)

This section deals with the investigation of the general solution of the mixed type additive-quadratic-cubic functional equation

$$f(x + 2y) + f(x - 2y) + 6f(x) = 4f(x + y) + 4f(x - y). \qquad (9.1)$$

Theorem 9.1 ([138]). *A function* $f : X \to Y$ *satisfies the functional equation* (9.1) *if and only if there exist functions* $B : X \times X \times X \to Y$, $Q : X \times X \to Y$, $A : X \to Y$ *and a constant* c *in* Y *such that* $f(x) = B(x, x, x) + Q(x, x) + A(x) + c$ *for all* $x \in X$, *where* B *is symmetric for each fixed one variable and is additive for fixed two variables,* Q *is symmetric bi-additive and* A *is additive.*

Proof. First we assume that f is a solution of the functional equation (9.1). If we put $g(x) = f(x) - f(0)$, we get g is also a solution of (9.1), and $g(0) = 0$. So we may assume without loss of generality that f is a solution of (9.1), and $f(0) = 0$. Let $f_e(x) = \frac{f(x) + f(-x)}{2}$ and $f_o(x) = \frac{f(x) - f(-x)}{2}$ for all $x \in X$. Then $f_e(0) = 0 = f_o(0)$, f_e is even and f_o is odd. Since f is a solution of (9.1), f_e and f_o also satisfy equation (9.1). Replacing f by f_o, y by $y + z$ in (9.1), and then y by $y - z$ in (9.1), separately,

109

and then adding the resulting two relations, we have

$$f_o(x + 2y + 2z) + f_o(x + 2y - 2z) + f_o(x - 2y + 2z) + f_o(x - 2y - 2z) + 12f_o(x)$$
$$= 4f_o(x + y + z) + 4f_o(x + y - z) + 4f_o(x - y + z) + 4f_o(x - y - z). \quad (9.2)$$

Also, by virtue of equation (9.1), expanding the left-hand side of (9.2), we can rewrite (9.2) in the form

$$f_o(x + y + z) + f_o(x + y + -z) + f_o(x - y + z) + f_o(x - y - z) + 4f_o(x)$$
$$= 2f_o(x + y) + 2f_o(x - y) + 2f_o(x + z) + 2f_o(x - z). \quad (9.3)$$

Exchanging x with y in (9.3), and then adding the resulting relation to (9.3), we have

$$f_o(x + y + z) + f_o(x + y - z) + 2f_o(x) + 2f_o(y)$$
$$= 2f_o(x + y) + f_o(x + z) + f_o(x - z) + f_o(y + z) + f_o(y - z) \quad (9.4)$$

for all $x, y \in X$. Hence by using (8.65) of Chapter 8, $f_o(x) = B(x, x, x) + A(x) + c$ for all $x \in X$, where $c = f(0)$, B is symmetric for each fixed one variable and is additive for fixed two variables, and A is additive.

In turn, since f_e satisfies equation (9.1), we obtain

$$f_e(2x + y) + f_e(2x - y) + 6f_e(y) = 4f_e(x + y) + 4f_e(x - y). \quad (9.5)$$

Replacing x and y by $x + y$ and $x - y$ in (9.5), respectively, we have

$$f_e(3x + y) + f_e(x + 3y) + 6f_e(x - y) = 16f_e(x) + 16f_e(y). \quad (9.6)$$

Replacing $x + y$ instead of y in (9.5), one obtains

$$f_e(3x + y) + f_e(x - y) + 6f_e(x + y) = 4f_e(2x + y) + 4f_e(y). \quad (9.7)$$

Interchange x and y in (9.7) to get the relation

$$f_e(x + 3y) + f_e(x - y) + 6f_e(x + y) = 4f_e(2y + x) + 4f_e(x). \quad (9.8)$$

Adding (9.8) to (9.7) and using (9.6), we have

$$12f_e(x) + 12f_e(y) - 4f_e(x - y) + 12f_e(x + y) = 4f_e(2x + y) + 4f_e(2y + x). \quad (9.9)$$

Replacing y by $-y$ in (9.9) and then adding the resulting relation to (9.9) together with (9.5), we have

$$f_e(x + y) + f_e(x - y) = 2f_e(x) + 2f_e(y) \quad (9.10)$$

for all $x, y \in X$. Therefore, $f_e(x) = Q(x, x)$, where Q is a symmetric bi-additive function. As a result, $f(x) = f_e(x) + f_o(x) = B(x, x, x) + Q(x, x) + A(x) + c$ for all $x \in X$.

Conversely, if there exist functions $B : X \times X \times X \to Y$, $Q : X \times X \to Y$, $A : X \to Y$ and a constant c such that $f(x) = B(x, x, x) + Q(x, x) + A(x) + c$ for all $x \in X$, where A is additive, Q is symmetric bi-additive, and B is symmetric for fixed one variable and is additive for fixed two variables, then it is obvious that f satisfies equation (9.1). This completes the proof. $\qquad\square$

9.2.2. *Mixed type functional equation* (9.11)

In the following theorem, we obtain the general solution of the mixed type additive, quadratic and cubic functional equation

$$f(x + 2y) - f(x - 2y) = 2(f(x + y) - f(x - y)) + 2f(3y) - 6f(2y) + 6f(y). \quad (9.11)$$

It is easy to see that the function $f(x) = ax + bx^2 + cx^3$ is a solution of the functional equation (9.11).

Theorem 9.2 ([95]). *Let $f : X \to Y$ be a function. Then f satisfies (9.11) if and only if there exists a unique additive function $A : X \to Y$, a unique symmetric and bi-additive function $Q : X \times X \to Y$, and a unique symmetric and 3-additive function $C : X \times X \times X \to Y$ such that $f(x) = A(x) + Q(x, x) + C(x, x, x)$ for all $x \in X$.*

Proof. Suppose that $f(x) = A(x) + Q(x, x) + C(x, x, x)$ for all $x \in X$, where $A : X \to Y$ is additive, $Q : X \times X \to Y$ is symmetric and bi-additive, and $C : X \times X \times X \to Y$ is symmetric and 3-additive. Then it is easy to see that f satisfies equation (9.11).

For the converse let f satisfy (9.11). We decompose f into the even part and odd part by setting

$$f_e(x) = \frac{1}{2}(f(x) + f(-x)) \quad \text{and} \quad f_o(x) = \frac{1}{2}(f(x) - f(-x)), \quad (9.12)$$

respectively, for all $x \in X$. By (9.11), we have

$$f_e(x + 2y) - f_e(x - 2y)$$

$$= \frac{1}{2}[f(x + 2y) + f(-x - 2y) - f(x - 2y) - f(-x + 2y)]$$

$$= 2(f_e(x + y) - f_e(x - y) + 2f_e(3y) - 6f_e(2y) + 6f_e(y)), \quad (9.13)$$

for all $x, y \in X$. This means that f_e satisfies (9.11), that is

$$f_e(x+2y) - f_e(x-2y) = 2(f_e(x+y) - f_e(x-y) + 2f_e(3y) - 6f_e(2y) + 6f_e(y)). \quad (9.14)$$

Now, putting $x = y = 0$ in (9.14), we get $f_e(0) = 0$. Setting $x = 0$ in (9.14), by evenness of f_e we obtain

$$3f_e(2y) = f_e(3y) + 3f_e(y). \quad (9.15)$$

Replacing x by y in (9.14), we obtain

$$4f_e(2y) = f_e(3y) + 7f_e(y). \tag{9.16}$$

Comparing (9.15) with (9.16), we get

$$f_e(3y) = 9f_e(y). \tag{9.17}$$

By utilizing (9.16) with (9.17), we obtain

$$f_e(2y) = 4f_e(y). \tag{9.18}$$

Hence, according to (9.17) and (9.18), the relation (9.14) can be written as

$$f_e(x + 2y) - f_e(x - 2y) = 2f_e(x + y) - 2f_e(x - y). \tag{9.19}$$

With the substitution $x = x + y$, $y = x - y$ in (9.19), we have

$$f_e(3x - y) - f_e(x - 3y) = 8f_e(x) - 8f_e(y). \tag{9.20}$$

Replacing y by $-y$ in the above relation, we obtain

$$f_e(3x + y) - f_e(x + 3y) = 8f_e(x) - 8f_e(y). \tag{9.21}$$

Setting $x + y$ instead of x in (9.19), we obtain

$$f_e(x + 3y) - f_e(x - y) = 2f_e(x + 2y) - 2f_e(x). \tag{9.22}$$

Interchanging x and y in (9.20), we obtain

$$f_e(3x + y) - f_e(x - y) = 2f_e(2x + y) - 2f_e(y). \tag{9.23}$$

Subtracting (9.23) from (9.22), and then using (9.21), we obtain

$$f_e(x + 2y) - f_e(2x + y) = 3f_e(y) - 3f_e(x), \tag{9.24}$$

which, by putting $y = 2y$, and using (9.18), leads to

$$f_e(x + 4y) - 4f_e(x + y) = 12f_e(y) - 3f_e(x). \tag{9.25}$$

Interchanging x with y in (9.25), then we see that

$$f_e(4x + y) - 4f_e(x + y) = 12f_e(x) - 3f_e(y), \tag{9.26}$$

and by adding (9.25) and (9.26), we arrive at

$$f_e(x + 4y) + f_e(4x + y) = 8f_e(x + y) + 9f_e(x) + 9f_e(y). \tag{9.27}$$

Replacing y by $x + y$ in (9.19), we obtain

$$f_e(3x + 2y) - f_e(x + 2y) = 2f_e(2x + y) - 2f_e(y). \tag{9.28}$$

Interchanging x with y in (9.28), then we see that

$$f_e(2x + 3y) - f_e(2x + y) = 2f_e(x + 2y) - 2f_e(x). \tag{9.29}$$

Thus by adding (9.28) and (9.29), we have

$$f_e(2x + 3y) + f_e(3x + 2y) = 3f_e(x + 2y) + 3f_e(2x + y) - 2f_e(x) - 2f_e(y). \quad (9.30)$$

Replacing x by $2x$ in (9.22) and using (9.18), we have

$$f_e(2x + 3y) - f_e(2x - y) = 8f_e(x + y) - 8f_e(x), \quad (9.31)$$

and interchanging x with y in (9.31) yields

$$f_e(3x + 2y) - f_e(x - 2y) = 8f_e(x + y) - 8f_e(y). \quad (9.32)$$

Adding (9.31) and (9.32), we have

$$f_e(2x + 3y) + f_e(3x + 2y)$$
$$= f_e(2x - y) + f_e(x - 2y) + 16f_e(x + y) - 8f_e(x) - 8f_e(y). \quad (9.33)$$

Interchanging x with y in (9.19), we obtain

$$f_e(2x + y) - f_e(2x - y) = 2f_e(x + y) - 2f_e(x - y), \quad (9.34)$$

and by adding the last equation, and (9.19) with (9.30), we obtain

$$f_e(2x + 3y) + f_e(3x + 2y) - f_e(2x - y) - f_e(x - 2y)$$
$$= 2f_e(x + 2y) + 2f_e(2x + y) + 4f_e(x + y)$$
$$- 4f_e(x - y) - 2f_e(x) - 2f_e(y). \quad (9.35)$$

Now, according to (9.33) and (9.34), it follows that

$$f_e(x + 2y) + f_e(2x + y) = 6f_e(x + y) + 2f_e(x - y) - 3f_e(x) - 3f_e(y). \quad (9.36)$$

From the substitution $y = -y$ in (9.36), it follows that

$$f_e(x - 2y) + f_e(2x - y) = 6f_e(x - y) + 2f_e(x + y) - 3f_e(x) - 3f_e(y). \quad (9.37)$$

Replacing y by $2y$ in (9.36), we have

$$f_e(x + 4y) + 4f_e(x + y) = 6f_e(x + 2y) + 2f_e(x - 2y) - 3f_e(x) - 12f_e(y), \quad (9.38)$$

and interchanging x with y yields

$$f_e(4x + y) + 4f_e(x + y) = 6f_e(2x + y) + 2f_e(2x - y) - 12f_e(x) - 3f_e(y). \quad (9.39)$$

By adding (9.38) and (9.39), and then using (9.36), and (9.37), we have

$$f_e(x + 4y) + f_e(4x + y) = 32f_e(x + y) + 24f_e(x - y) - 39f_e(x) - 39f_e(y). \quad (9.40)$$

If we compare (9.27) and (9.40), then we conclude that

$$f_e(x + y) + f_e(x - y) = 2f_e(x) + 2f_e(y). \quad (9.41)$$

This means that f_e is quadratic. Thus there exists a unique quadratic function $Q : X \times X \to Y$ such that $f_e(x) = Q(x, x)$, for all $x \in X$. On the other hand, we can show that f_0 satisfies (9.11), that is,

$$f_0(x + 2y) - f_0(x - 2y)$$
$$= 2\left(f_0(x - y) + f_0(x - y)\right) + 2f_0(3y) - 6f_0(2y) + 6f_0(y). \qquad (9.42)$$

Now, we show that the mapping $g : X \to Y$ defined by $g(x) := f_o(2x) - 8f_o(x)$ is additive and the mapping $h : X \to Y$ defined by $h(x) := f_o(2x) - 2f_o(x)$ is cubic. Putting $x = 0$ in (9.42), then by oddness of f_o, we have

$$4f_o(2y) = 5f_o(y) + f_o(3y). \qquad (9.43)$$

Hence (9.42) can be written as

$$f_o(x + 2y) - f_o(x - 2y) = 2f_o(x + y) - 2f_o(x - y) + 2f_o(2y) - 4f_o(y). \qquad (9.44)$$

From the substitution $y = -y$ in (9.44) it follows that

$$f_o(x - 2y) - f_o(x + 2y) = 2f_o(x - y) - 2f_o(x + y) - 2f_o(2y) + 4f_o(y). \qquad (9.45)$$

Interchanging x with y in (9.45), we have that

$$f_o(2x + y) + f_o(2x - y) = 2f_o(x + y) + 2f_o(x - y) + 2f_o(2x) - 4f_o(x). \qquad (9.46)$$

With the substitutions $x = x - y$, and $y = x + y$ in (9.46), we have

$$f_o(3x - y) + f_o(x - 3y) = 2f_o(2x - 2y) - 4f_o(x - y) + 2f_o(2x) - 2f_o(2y). \quad (9.47)$$

Replacing x by $x - y$ in (9.45), we have

$$f_o(x - 3y) - f_o(x + y) = 2f_o(x - 2y) - 2f_o(x) - 2f_o(2y) + 4f_o(y). \qquad (9.48)$$

Replacing y by $-y$ in (9.48) gives

$$f_o(x + 3y) - f_o(x - y) = 2f_o(x + 2y) - 2f_o(x) + 2f_o(2y) - 4f_o(y). \qquad (9.49)$$

Interchanging x with y in (9.49), we obtain

$$f_o(3x + y) + f_o(x - y) = 2f_o(2x + y) - 2f_o(y) + 2f_o(2x) - 4f_o(x). \qquad (9.50)$$

Adding (9.49) and (9.50), we have

$$f_o(x + 3y) + f_o(3x + y)$$
$$= 2f_o(x + 2y) + 2f_o(2x + y) + 2f_o(2x) + 2f_o(2y) - 6f_o(x) - 6f_o(y). \quad (9.51)$$

Replacing y by $-y$ in (9.47) gives

$$f_0(x + 3y) + f_o(3x + y) = 2f_o(2x + 2y) - 4f_o(x + y) + 2f_o(2x) + 2f_o(2y). \quad (9.52)$$

By comparing (9.51) with (9.52), we arrive at

$$f_o(x + 2y) + f_o(2x + y) = f_o(2x + 2y) - 2f_o(x + y) + 3f_o(x) + 3f_o(y). \quad (9.53)$$

Replacing y by $-y$ in (9.53) gives

$$f_o(x - 2y) + f_o(2x - y) = f_o(2x - 2y) - 2f_o(x - y) + 3f_o(x) - 3f_o(y). \quad (9.54)$$

With the substitution $y = x + y$ in (9.54), we have

$$f_o(x + y) - f_o(x + 2y) = -f_o(2y) - 3f_o(x + y) + 3f_o(x) + 2f_o(y), \quad (9.55)$$

and replacing $-y$ by y gives

$$f_o(x + y) - f_o(x - 2y) = f_o(2y) - 3f_o(x - y) + 3f_o(x) - 2f_o(y). \quad (9.56)$$

Interchanging x with y in (9.56), we see that

$$f_o(x + y) + f_o(2x - y) = f_o(2x) + 3f_o(x - y) - 2f_o(x) + 3f_o(y). \quad (9.57)$$

Adding (9.56) and (9.57), we have

$$f_o(2x - y) - f_o(x - 2y) = f_o(2x) - 2f_o(x + y) + f_o(x) + f_o(2y) + f_o(y). \quad (9.58)$$

Adding (9.53) to (9.58), and using (9.44) and (9.46), we obtain

$$f_o(2(x + y)) - 8f_o(x + y) = [f_o(2x) - 8f_o(x)] + [f_o(2y) - 8f_o(y)], \quad (9.59)$$

for all $x, y \in X$. The last inequality implies that

$$g(x + y) = g(x) + g(y), \quad (9.60)$$

for all $x, y \in X$. Therefore, the mapping $g : X \to Y$ is additive. With the substitutions $x := 2x$ and $y := 2y$ in (9.46), we have

$$f_o(4x + 2y) + f_o(4x - 2y) = 2f_o(2x + 2y) + 2f_o(2x - 2y) + 2f_o(4x) - 4f_o(2x). \quad (9.61)$$

Let $g : X \to Y$ be the additive mapping defined above. It is easy to show that f_o is cubic-additive function. Then there exist a unique function $C : X \times X \times X \to Y$ and a unique additive function $A : X \to Y$ such that $f_o(x) = C(x, x, x) + A(x)$, for all $x \in X$, and C is symmetric and 3-additive. Thus for all $x \in X$, we have

$$f(x) = f_e(x) + f_o(x) = Q(x, x) + C(x, x, x) + A(x). \quad (9.62)$$

This completes the proof of the theorem. $\qquad\qquad\qquad\qquad\qquad\square$

9.3. Mixed type additive-quadratic-quartic functional equation

9.3.1. *Mixed type functional equation* (9.63)

This section contains the general solution of mixed type additive-quadratic-quartic functional equation

$$
\begin{aligned}
f(x+2y) + f(x-2y) = {} & 2f(x+y) + 2f(-x-y) + 2f(x-y) \\
& + 2f(y-x) - 4f(-x) - 2f(x) + f(2y) \\
& + f(-2y) - 4f(y) - 4f(-y).
\end{aligned} \tag{9.63}
$$

Theorem 9.3 ([240]). *If a mapping $f : X \to Y$ satisfies* (9.63), *then f is additive-quadratic-quartic.*

Proof. It is easy to show that an odd mapping $f : X \to Y$ satisfies (9.63) if and only if the odd mapping $f : X \to Y$ is an additive mapping, that is

$$
f(x+2y) + f(x-2y) = 2f(x).
$$

It is also easy to show that an even mapping $f : X \to Y$ satisfies (9.63) if and only if the even mapping $f : X \to Y$ is a quadratic-quartic mapping, that is

$$
\begin{aligned}
f(x+2y) + f(x-2y) = {} & 4f(x+y) + 4f(x-y) - 6f(x) \\
& + 2f(2y) - 8f(y).
\end{aligned}
$$

It was shown in [240] that $g(x) := f(2x) - 4f(x)$ and $h(x) := f(2x) - 16f(x)$ are quadratic and quartic, respectively, and that $f(x) = \frac{1}{12}g(x) - \frac{1}{12}h(x)$. This completes the proof of the theorem. $\square$

9.4. Mixed type additive-cubic-quartic functional equation

9.4.1. *Mixed type functional equation* (9.65)

This section is devoted to the general solution of mixed type additive-cubic-quartic functional equation

$$
\begin{aligned}
11[f(x+2y) + f(x-2y)] = {} & 44[f(x+y) + f(x-y)] + 12f(3y) \\
& - 48f(2y) + 60f(y) - 66f(x).
\end{aligned} \tag{9.64}
$$

It is easy to show that the function $f(x) = ax + bx^3 + cx^4$ satisfies the functional equation (9.64).

Lemma 9.1 ([97]). *If an even mapping $f : X \to Y$ satisfies* (9.64), *then f is quartic.*

Proof. Putting $x = y = 0$ in equation (9.64), we get $f(0) = 0$. Setting $x = 0$ in (9.64), by the evenness of f, we obtain

$$6f(3y) = 35f(2y) - 74f(y) \qquad (9.65)$$

for all $y \in X$. Hence (9.64) can be written as

$$f(x + 2y) + f(x - 2y) = 4[f(x + y) + f(x - y)] + 2f(2y) - 8f(y) - 6f(x) \quad (9.66)$$

for all $x, y \in X$. Replacing x by y in (9.64), we obtain

$$f(3y) = 4f(2y) + 17f(y) \qquad (9.67)$$

for all $y \in X$. By (9.65) and (9.67), we obtain

$$f(2y) = 16f(y) \qquad (9.68)$$

for all $y \in X$. According to (9.68), equation (9.66) can be written as

$$f(x + 2y) + f(x - 2y) = 4f(x + y) + 4f(x - y) + 24f(y) - 6f(x)$$

for all $x, y \in X$. This shows that f is quartic, which completes the proof of lemma. $\square$

Lemma 9.2 ([97]). *If an odd mapping $f : X \to Y$ satisfies (9.64), then f is cubic-additive.*

Proof. We show that the mappings $g : X \to Y$ and $h : X \to Y$, respectively, defined by $g(x) = f(2x) - 8f(x)$ and $h(x) = f(2x) - 2f(x)$, are additive and cubic, respectively.

Since f is odd, $f(0) = 0$. Letting $x = 0$ in (9.64), we obtain

$$f(3y) = 4f(2y) - 5f(y) \qquad (9.69)$$

for all $y \in X$. Hence (9.64) can be written as

$$f(x + 2y) + f(x - 2y) = 4[f(x + y) + f(x - y)] - 6f(x) \qquad (9.70)$$

for all $x, y \in X$. Replacing x by $x + y$ and y by $x - y$ in (9.70), respectively, we get

$$f(3x - y) - f(x - 3y) = -6f(x + y) + 4f(2x) + 4f(2y) \qquad (9.71)$$

for all $x, y \in X$. Replacing x by $x + y$ in (9.70), we obtain

$$f(x + 3y) + f(x - y) = 4f(x + 2y) - 6f(x + y) + 4f(x) \qquad (9.72)$$

for all $x, y \in X$. Replacing y by $-y$ in (9.72), we obtain

$$f(x - 3y) + f(x + y) = 4f(x - 2y) - 6f(x - y) + 4f(x) \qquad (9.73)$$

for all $x, y \in X$. Replacing x by y in (9.72), we obtain

$$f(3x + y) - f(x - y) = 4f(2x + y) - 6f(x + y) + 4f(y) \qquad (9.74)$$

for all $x, y \in X$. Replacing $-y$ by y in (9.74), we obtain

$$f(3x - y) - f(x + y) = 4f(2x - y) - 6f(x - y) - 4f(y) \qquad (9.75)$$

for all $x, y \in X$.

Subtracting (9.75) from (9.73), we obtain

$$f(3x - y) - f(x - 3y) = 4f(2x - y) - 4f(x - 2y)$$
$$+ 2f(x + y) - 4f(x) - 4f(y) \qquad (9.76)$$

for all $x, y \in X$. By (9.71) and (9.76), we obtain

$$f(x - 2y) = f(2x - y) + 2f(x + y) - f(2x) - f(2y) - f(x) - f(y) \qquad (9.77)$$

for all $x, y \in X$.

Replacing y by $-y$ in (9.77), we obtain

$$f(x + 2y) = f(2x + y) + 2f(x - y) - f(2x) + f(2y) - f(x) + f(y) \qquad (9.78)$$

for all $x, y \in X$.

By (9.77) and (9.78), we obtain

$$f(x + 2y) + f(x - 2y) = f(2x + y) + f(2x - y) + 2f(x + y)$$
$$+ 2f(x - y) - 2f(2x) - 2f(x) \qquad (9.79)$$

for all $x, y \in X$.

By (9.70) and (9.79), we have

$$f(2x + y) + f(2x - y) = 2f(x + y) + 2f(x - y) + 2f(2x) - 4f(x) \qquad (9.80)$$

for all $x, y \in X$. Replacing y by $x + y$ in (9.80), we obtain

$$f(3x + y) + f(x - y) = 2f(2x + y) - 2f(y) + 2f(2x) - 4f(x) \qquad (9.81)$$

for all $x, y \in X$. Interchanging x with y in (9.81), we obtain

$$f(x + 3y) - f(x - y) = 2f(x + 2y) - 2f(x) + 2f(2y) - 4f(y) \qquad (9.82)$$

for all $x, y \in X$.

By (9.81) and (9.82), we obtain

$$f(3x + y) + f(x + 3y) = 2f(2x + y) + 2f(x + 2y) + 2f(2x)$$
$$+ 2f(2y) - 6f(x) - 6f(y) \qquad (9.83)$$

for all $x, y \in X$. Replacing x by $x + y$ and y by $x - y$ in (9.80), respectively, we obtain

$$f(3x + y) + f(x + 3y) = 2f(2x + 2y) - 4f(x + y) + 2f(2x) + 2f(2y) \qquad (9.84)$$

for all $x, y \in X$. Thus, it follows from (9.81) and (9.84) that

$$f(2x + y) + f(x + 2y) = f(2x + 2y) - 2f(x + y) + 3f(x) + 3f(y) \qquad (9.85)$$

for all $x, y \in X$. Replacing x by $x - y$ in (9.85), we obtain

$$f(2x - y) + f(x + y) = 3f(x - y) + f(2x) - 2f(x) + 3f(y) \qquad (9.86)$$

for all $x, y \in X$. Interchanging x with y in (9.86), we obtain

$$f(2y - x) + f(x + y) = 3f(y - x) + f(2y) - 2f(y) + 3f(x) \qquad (9.87)$$

for all $x, y \in X$. By (9.86) and (9.87), we obtain

$$f(2x - y) + f(2y - x) = -2f(x + y) + f(x) + f(y) + f(2x) + f(2y) \qquad (9.88)$$

for all $x, y \in X$. Adding (9.85) to (9.88), and using (9.80), we obtain

$$f(2x + 2y) - 8f(x + y) = [f(2x) - 8f(x)] + [f(2y) - 8f(y)] \qquad (9.89)$$

for all $x, y \in X$. The last equality means that

$$g(x + y) = g(x) + g(y)$$

for all $x, y \in X$. Thus the mapping $g : X \to Y$ is additive.

Replacing x by $2x$ and y by $2y$ in (9.80), respectively, we get

$$f(4x + 2y) + f(4x - 2y) = 2f(2x + 2y) + 2f(2x - 2y) + 2f(4x) - 4f(2x) \quad (9.90)$$

for all $x, y \in X$. Since $g(2x) = 2g(x)$ for all $x \in X$,

$$f(4x) = 10f(2x) - 16f(x)$$

for all $x, y \in X$. Hence it follows from (9.80), (9.89), and (9.90) that

$$
\begin{aligned}
h(2x + y) + h(2x - y) &= [f(2(2x + y)) - 2f(2x + y)] + [f(2(2x - y)) - 2f(2x - y)] \\
&= 2[f(2(x + y)) - 2f(x + y)] \\
&\quad + 2[f(2(x - y)) - 2f(x - y)] + 12[f(2x) - 2f(x)] \\
&= 2h(x + y) + 2h(x - y) + 12h(x)
\end{aligned}
$$

for all $x, y \in X$. Thus the mapping $h : X \to Y$ is cubic.

On the other hand, we have $f(x) = \frac{1}{6}h(x) - \frac{1}{6}g(x)$ for all $x \in X$. This means that f is cubic-additive. This completes the proof of the lemma. $\qquad \square$

Theorem 9.4 ([97]). *A mapping $f : X \to Y$ satisfies (9.64) for all $x, y \in X$ if and only if there exist a unique additive mapping $A : X \to Y$, a unique mapping $C : X \times X \times X \to Y$ and a unique symmetric multiadditive mapping $Q : X \times X \times X \times X \to Y$ such that $f(x) = A(x) + C(x, x, x) + Q(x, x, x, x)$ for all $x \in X$, and that C is symmetric for each fixed one variable and is additive for fixed two variables.*

Proof. Let f satisfy equation (9.64). We decompose f into the even part, and the odd part by setting

$$f_e(x) = \frac{f(x) + f(-x)}{2} \quad \text{and} \quad f_o(x) = \frac{f(x) - f(-x)}{2}$$

for all $x \in X$. By (9.64), we have

$$11 \left[f_e(x + 2y) + f_e(x - 2y) \right]$$
$$= \frac{1}{2} [11f(x + 2y) + 11f(-x - 2y) + 11f(x - 2y) + 11f(-x + 2y)]$$
$$= 44 \left(f_e(x + y) + f_e(x - y) \right) + 12f_e(3y) - 48f_e(2y) + 60f_e(y) - 66f_e(x)$$

for all $x, y \in X$. This means that f_e satisfies (9.64). Similarly we can show that f_o satisfies (9.64). By Lemmas 9.1 and 9.2, f_e and f_o are quartic and cubic-additive, respectively. Thus there exist a unique additive mapping $A : X \to Y$, a unique mapping $C : X \times X \times X \times X \to Y$ such that $f_e(x) = Q(x, x, x, x)$ and that $f_o(x) = A(x) + C(x, x, x)$ for all $x \in X$, and C is symmetric for each fixed one variable and is additive for fixed two variables. Thus $f(x) = A(x) + C(x, x, x) + Q(x, x, x, x)$ for all $x \in X$. The proof of the converse is obvious. This completes the proof of the theorem. $\square$

9.4.2. *Mixed type functional equation* (9.91)

The following theorem provides the general solution of the functional equation derived from additive, cubic and quartic functions

$$f(x + ky) + f(x - ky) = k^2 f(x + y) k^2 f(x - y)$$
$$+ (k^2 - 1) \left(k^2 f(y) + k^2 f(-y) - 2f(x) \right), \quad (9.91)$$

for fixed integers k with $k \neq 0, \pm 1$. It is easy to see that the function $f(x) = ax + bx^3 + cx^4$ is a solution of the functional equation (9.91).

Theorem 9.5 ([112]). *A function $f : X \to Y$ with $f(0) = 0$ satisfies the equation* (9.91) *for all $x, y \in X$ if and only if there exist a unique symmetric bi-quadratic function $B : X \times X \to Y$, a unique function $C : X \times X \times X \to Y$ and a unique additive function $A : X \to Y$ such that*

$$f(x) = B(x, x) + C(x, x, x) + A(x),$$

for all $x \in X$ and C is symmetric for each fixed one variable and is additive for fixed two variables.

Proof. Let f satisfy equation (9.91). We decompose f into the even part, and odd part by putting

$$f_e(x) = \frac{1}{2}(f(x) + f(-x)), \quad \text{and} \quad f_o(x) = \frac{1}{2}(f(x) - f(-x)), \ \forall x \in X.$$

It is clear that $f(x) = f_e(x) + f_o(x)$ for all $x \in X$. It is easy to show that the functions f_e, and f_o satisfy equation (9.91).

Now, we show that the function $f_e : X \to Y$ is quartic. In fact, it follows from equation (9.91) that

$$f_e(x + ky) + f_e(x - ky) = k^2 f_e(x + y) + k^2 f_e(x - y)$$
$$+ 2(1 - k^2)f_e(x) + 2k^2(k^2 - 1)f_e(y), \qquad (9.92)$$

for all $x, y \in X$. Letting $x = y = 0$ in (9.92), we have $f(0) = 0$. Putting $x = 0$ in (9.92), we obtain

$$f_e(ky) = k^4 f_e(y) \qquad (9.93)$$

for all $y \in X$. Replacing x by $2x$ in (9.92), we obtain

$$f_e(2x + ky) + f_e(2x - ky) = k^2 f_e(2x + y) + k^2 f_e(2x - y)$$
$$+ 2(1 - k^2)f_e(2x) + 2k^2(k^2 - 1)f_e(y), \qquad (9.94)$$

for all $x, y \in X$. If we put $y = x + y$ in equation (9.92) and then $y = x - y$ in equation (9.92) again, then it follows from the evenness of f_e that

$$f_e(k(x + y) + x) + f_e(k(x + y) - x)) = k^2 f_e(2x + y) + k^2 f_e(y) + 2(1 - k^2)f_e(x)$$
$$+ 2k^2(k^2 - 1)f_e(x + y), \qquad (9.95)$$

and

$$f_e(k(x - y) + x) + f_e(k(x - y) - x) = k^2 f_e(2x - y) + k^2 f_e(y) + 2(1 - k^2)f_e(x)$$
$$+ 2k^2(k^2 - 1)f_e(x - y), \qquad (9.96)$$

for all $x, y \in X$. Adding equations (9.95) and (9.96), we have

$$f_e(k(x + y) + x) + f_e(k(x + y) - x) + f_e(k(x - y) + x) + f_e(k(x - y) - x)$$
$$= k^2 f_e(2x + y) + k^2 f_e(2x - y) + 2k^2 f_e(y) + 4(1 - k^2)f_e(x)$$
$$+ 2k^2(k^2 - 1)(f_e(x + y) + f_e(x - y)), \qquad (9.97)$$

for all $x, y \in X$. Interchanging x with y in (9.92) and using the evenness of f_e, we obtain

$$f_e(kx + y) + f_e(kx - y) = k^2 f_e(x + y) + k^2 f_e(x - y) + 2k^2(k^2 - 1)f_e(x)$$
$$+ 2(1 - k^2)f_e(y), \qquad (9.98)$$

for all $x, y \in X$. We substitute $y = x + ky$ in (9.98), and then $y = x - ky$ in (9.98) and use (9.94), we obtain

$$f_e(k(x + y) + x) + f_e(k(x - y) - x)$$
$$= k^2 f_e(2x + ky) + k^6 f_e(y) + 2k^2(k^2 - 1)f_e(x) + 2(1 - k^2)f_e(x + ky), \quad (9.99)$$

and

$$f_e(k(x - y) + x) + f_e(k(x + y) - x)$$
$$= k^2 f_e(2x - ky) + k^6 f_e(y) + 2k^2(k^2 - 1)f_e(x) + 2(1 - k^2)f_e(x - ky)$$
$$(9.100)$$

for all $x, y \in X$. Adding equations (9.99) and (9.100), we have

$$f_e(k(x + y) + x) + f_e(k(x + y) - x)$$
$$+ f_e(k(x - y) + x) + f_e(k(x - y) - x)$$
$$= k^2 f_e(2x + ky) + k^2 f_e(2x - ky) + 2k^6 f_e(y) + 4k^2(k^2 - 1)f_e(x)$$
$$+ 2(1 - k^2)(f_e(x + ky) + f_e(x - ky))$$
$$(9.101)$$

for all $x, y \in X$. It follows from equations (9.93), (9.95), (9.98), and (9.99) that

$$f_e(2x + y) + f_e(2x - y)$$
$$= 4(f_e(x + y) + f_e(x - y)) + 2f_e(2x) - 8f_e(x) - 6f_e(y), \quad (9.102)$$

for all $x, y \in X$. Letting $y = x$ in (9.102), we have $f_e(3x) = 6f_e(2x) - 15f_e(x)$ and letting $y = 2x$ in (9.102), we have $f_e(4x) = 20f_e(2x) - 64f_e(x)$. Thus, by induction, we get

$$f_e(mx) = \frac{m(m^2 - 1)}{12} f_e(2x) + \frac{m^2(4 - m^2)}{3} f_e(x), \quad (9.103)$$

for each fixed integer $m \neq 0, \pm 1, \pm 2$ and $x \in X$. But, $k \neq 0, \pm 1$, and also if $k \neq 2$, then it follows from equation (9.99) that f_e is quartic. Otherwise, if we use equation (9.103) for $m = k$ and equation (9.94), then we obtain $f_e(2x) = 16f_e(x)$ and so it follows from equation (9.103) that

$$f_e(2x + y) + f_e(2x - y) = 4f_e(x + y) + 4f_e(x - y) + 24f_e(x) - 6f_e(y),$$

for all $x, y \in X$. This shows that f_e is quartic, and so there exists a unique symmetric bi-quadratic function $B : X \times X \to Y$ such that

$$f_e(x) = B(x, x), \quad \text{for all } x \in X. \quad (9.104)$$

On the other hand, we show that the function $f_0 : X \to Y$ is cubic-additive. In fact, it follows from equation (9.92) that

$$f_o(x + ky) + f_o(x - ky) = k^2 f_o(x + y) + k^2 f_o(x - y) + 2(1 - k^2)f_o(x)$$

for all $x, y \in X$. By the same method as in [102, Lemma 2.2], we can show that f_o is cubic-additive. Therefore, it follows that

$$f_o(x) = C(x, x, x) + A(x), \tag{9.105}$$

for all $x, y \in X$, C is symmetric for each fixed one variable and is additive for fixed two variables, and A is additive. Hence, from equations (9.104) and (9.105), it follows that

$$f(x) = f_e(x) + f_0(x) = B(x, x) + C(x, x, x) + A(x), \quad \text{for all } x \in X.$$

Conversely, let $f(x) = B(x, x) + C(x, x, x) + A(x)$ for all $x \in X$, where the function B is symmetric bi-quadratic, C is symmetric for each fixed one variable and is additive for fixed two variables, and A is additive. By a simple computation, we can show that the functions $x \mapsto B(x, x)$, $x \mapsto C(x, x, x)$, and $x \mapsto A(x)$ satisfy the functional equation (9.92). Therefore, the function f satisfies equation (9.92). This completes the proof. $\qquad\square$

9.5. Mixed type quadratic-cubic-quartic functional equations

9.5.1. *Mixed type functional equation* (9.106)

This section comprises the general solution of the mixed type quadratic-cubic-quartic functional equation for fixed integers $k \neq \pm 1$,

$$f(x + ky) + f(x - ky) = k^2 f(x + y) + k^2 f(x - y) + 2(1 - k^2) f(x)$$
$$+ \frac{k^2(k^2 - 1)}{6}(f(2y) + 2f(-y) - 6f(y)). \tag{9.106}$$

It is easy to see that the function $f(x) = ax^2 + bx^3 + cx^4$ is a solution of the functional equation (9.106).

Theorem 9.6 ([104]). *A function $f : X \to Y$ satisfies (9.106) for all $x, y \in X$ if and only if there exist a unique symmetric bi-quadratic function $B_2 : X \times X \to Y$, a unique function $C : X \times X \times X \to Y$ and a unique symmetric bi-additive function $B_1 : X \times X \to Y$ such that*

$$f(x) = B_2(x, x) + C(x, x, x) + B_1(x, x)$$

for all $x \in X$, and that C is symmetric for each fixed one variable and is additive for fixed two variables.

Proof. Let f satisfy equation (9.106). We decompose f into the even part and odd part by setting

$$f_e(x) = \frac{1}{2}(f(x) + f(-x)) \quad \text{and} \quad f_o(x) = \frac{1}{2}(f(x) - f(-x)),$$

respectively, for all $x \in X$. It is clear that $f(x) = f_e(x) + f_o(x)$ for all $x \in X$. It is easy to show that the functions f_e and f_o satisfy (9.106). Now, let $g, h : X \to Y$ be functions defined by

$$g(x) = f_e(2x) - 16f_e(x), \quad \text{and} \quad h(x) = f_e(2x) - 4f_e(x)$$

for all $x \in X$. We show that the functions g, and h are quadratic and quartic, respectively.

Interchanging x with y in (9.106) and then using $f_e(-x) = f_e(x)$, we have

$$f_e(kx + y) + f_e(kx - y) = k^2 f_e(x + y) + k^2 f_e(x - y)$$
$$+ \frac{k^2(k^2 - 1)}{6} (f_e(2x) - 4f_e(x)) + (1 - k^2)f_e(y)$$
$$\tag{9.107}$$

for all $x, y \in X$. Letting $x = y = 0$ in (9.107), we have $f_e(0) = 0$. Putting $y = x + y$ in (9.107), gives

$$f_e((k + 1)x + y) + f_e((k - 1)x - y)$$
$$= k^2 f_e(2x + y) + k^2 f_e(-y)$$
$$+ \frac{k^2(k^2 - 1)}{6} (f_e(2x) - 4f_e(x)) + 2(1 - k^2)f_e(x + y) \tag{9.108}$$

for all $x, y \in X$. Replacing y by $-y$ in (9.108), we obtain

$$f_e((k + 1)x - y) + f_e((k - 1)x + y)$$
$$= k^2 f_e(2x - y) + k^2 f_e(y)$$
$$+ \frac{k^2(k^2 - 1)}{6} (f_e(2x) - 4f_e(x)) + 2(1 - k^2)f_e(x - y) \tag{9.109}$$

for all $x, y \in X$. Adding (9.108) and (9.109), we obtain

$$f_e((k + 1)x + y) + f_e((k + 1)x - y) + f_e((k - 1)x + y) + f_e((k - 1)x - y)$$
$$= k^2 (f_e(2x + y) + f_e(2x - y)) + \frac{2k^2(k^2 - 1)}{6} (f_e(2x) - 4f_e(x))$$
$$+ 2(1 - k^2) (f_e(x + y) + f_e(x - y)) + 2k^2 f_e(y) \tag{9.110}$$

for all $x, y \in X$. From the substitution $y = kx + y$ in (9.107), we have

$$f_e(2kx + y) + f_e(y) = k^2 f_e((k + 1)x + y) + k^2 f_e((k - 1)x + y)$$
$$+ \frac{k^2(k^2 - 1)}{6} (f_e(2x) - 4f_e(x)) + 2(1 - k^2)f_e(kx + y) \tag{9.111}$$

for all $x, y \in X$. Replacing y by $-y$ in (9.111), we obtain

$$f_e(2kx - y) + f_e(-y) = k^2 f_e((k+1)x - y) + k^2 f_e((k-1)x - y)$$
$$+ \frac{k^2(k^2 - 1)}{6}(f_e(2x) - 4f_e(x)) + 2(1 - k^2)f_e(kx - y)$$
$$\tag{9.112}$$

for all $x, y \in X$. Adding (9.111) and (9.112), we obtain

$$f_e(2kx + y) + f_e(2kx - y)$$
$$= k^2(f_e((k+1)x + y) + f_e((k+1)x - y)$$
$$+ f_e((k-1)x + y) + f_e((k+1)x - y))$$
$$+ 2(1 - k^2)(f_e(kx + y) + f_e(kx - y))$$
$$+ \frac{2k^2(k^2 - 1)}{6}(f_e(2x) - 4f_e(x)) - 2f_e(y) \tag{9.113}$$

for all $x, y \in X$. It follows from (9.113), by using (9.107), and (9.110) that

$$f_e(2kx + y) + f_e(2kx - y)$$
$$= k^2 \left[k^2(f_e(2x + y) + f_e(2x - y)) + \frac{2k^2(k^2 - 1)}{6}(f_e(2x) - 4f_e(x)) \right.$$
$$\left. + 2(1 - k^2)(f_e(x + y) + f_e(x - y)) + 2k^2 f_e(y) \right]$$
$$+ 2(1 - k^2) \left[k^2 f_e(x + y) + k^2 f_e(x - y) \right.$$
$$+ \frac{k^2(k^2 - 1)}{6}(f_e(2x) - 4f_e(x)) + 2(1 - k^2)f_e(y) \right]$$
$$+ \frac{2k^2(k^2 - 1)}{6}(f_e(2x) - 4f_e(x)) - 2f_e(y) \tag{9.114}$$

for all $x, y \in X$. If we replace x by $2x$ in (9.107), we obtain

$$f_e(2kx + y) + f_e(2kx - y) = k^2 f_e(2x + y) + k^2 f_e(2x - y)$$
$$+ \frac{k^2(k^2 - 1)}{6}(f_e(4x) - 4f_e(2x)) + 2(1 - k^2)f_e(y)$$
$$\tag{9.115}$$

for all $x, y \in X$. It follows from (9.114) and (9.115) that

$$k^2 \left[k^2 \left(f_e(2x + y) + f_e(2x - y) \right) + \frac{2k^2(k^2 - 1)}{6} \left(f_e(2x) - 4f_e(x) \right) \right.$$

$$\left. + 2(1 - k^2) \left(f_e(x + y) + f_e(x - y) \right) + 2k^2 f_e(y) \right]$$

$$+ 2(1 - k^2) \left[k^2 f_e(x + y) + k^2 f_e(x - y) \right.$$

$$+ \frac{k^2(k^2 - 1)}{6} \left(f_e(2x) - 4f_e(x) \right) + 2(1 - k^2) f_e(y) \right]$$

$$+ \frac{2k^2(k^2 - 1)}{6} \left(f_e(2x) - 4f_e(x) \right) + 2(1 - k^2) f_e(y)$$

$$+ \frac{2k^2(k^2 - 1)}{6} \left(f_e(2x) - 4f_e(x) \right) - 2f_e(y)$$

$$= k^2 f_e(2x + y) + k^2 f_e(2x - y) + \frac{k^2(k^2 - 1)}{6} \left(f_e(4x) - 4f_e(2x) \right)$$

$$+ 2(1 - k^2) f_e(y) \tag{9.116}$$

for all $x, y \in X$. Also, putting $y = 0$ in (9.107), we obtain

$$f_e(kx) = k^2 f_e(x) + \frac{k^2(k^2 - 1)}{12} \left(f_e(2x) - 4f_e(x) \right) \tag{9.117}$$

for all $x \in X$. Setting $y = x$ in (9.107), we obtain

$$f_e((k + 1)x) + f_e((k - 1)x) = k^2 f_e(2x) + \frac{k^2(k^2 - 1)}{6} \left(f_e(2x) - 4f_e(x) \right)$$

$$+ 2(1 - k^2) f_e(x) \tag{9.118}$$

for all $x, y \in X$. Putting $y = kx$ in (9.107), we obtain

$$f_e(2kx) = k^2 f_e((k + 1)x) + f_e((k - 1)x) + \frac{k^2(k^2 - 1)}{6} \left(f_e(2x) - 4f_e(x) \right)$$

$$+ 2(1 - k^2) f_e(kx) \tag{9.119}$$

for all $x \in X$. Letting $y = 0$ in (9.115), we have

$$f_e(2kx) = k^2 f_e(2x) + \frac{k^2(k^2 - 1)}{12} \left(f_e(4x) - 4f_e(2x) \right) \tag{9.120}$$

for all $x \in X$. It follows from (9.119) and (9.120) that

$$\frac{k^2(k^2-1)}{12}\left(f_e(4x)-4f_e(2x)\right)$$

$$= k^2 f_e((k+1)x) + f_e((k-1)x)$$

$$+ \frac{k^2(k^2-1)}{6}\left(f_e(2x)-4f_e(x)\right) + 2(1-k^2)f_e(kx) - k^2 f_e(2x) \quad (9.121)$$

for all $x \in X$. Now, using (9.117), (9.118), and (9.121), we have

$$\frac{k^2(k^2-1)}{12}\left(f_e(4x)-4f_e(2x)\right)$$

$$= k^2\left[k^2 f_e(2x) + \frac{k^2(k^2-1)}{6}\left(f_e(2x)-4f_e(x)\right) + 2(1-k^2)f_e(x)\right]$$

$$+ 2(1-k^2)\left[k^2 f_e(x) + \frac{k^2(k^2-1)}{12}\left(f_e(2x)-4f_e(x)\right)\right]$$

$$+ \frac{k^2(k^2-1)}{6}\left(f_e(2x)-4f_e(x)\right) - k^2 f_e(2x) \quad (9.122)$$

for all $x \in X$. Finally, if we compare (9.116) with (9.122), then we conclude that

$$f_e(2x+y) + f_e(2x-y) = 4f_e(x+y) + 4f_e(x-y) + 2\left(f_e(2x)-4f_e(x)\right) - 6f_e(y) \quad (9.123)$$

for all $x, y \in X$. Replacing y by $2y$ in (9.123), we obtain

$$f_e(2x+2y) + f_e(2x-2y) = 4f_e(x+2y) + 4f_e(x-2y)$$

$$+ 2\left(f_e(2x)-4f_e(x)\right) - 6f_e(2y) \quad (9.124)$$

for all $x, y \in X$. Interchanging x with y in (9.123), we obtain

$$f_e(2y+x) + f_e(2y-x) = 4f_e(y+x) + 4f_e(y-x) + 2\left(f_e(2y)-4f_e(y)\right) - 6f_e(x) \quad (9.125)$$

for all $x, y \in X$, which implies that

$$f_e(x+2y) + f_e(x-2y) = 4f_e(x+y) + 4f_e(x-y) + 2\left(f_e(2y)-4f_e(y)\right) - 6f_e(x) \quad (9.126)$$

for all $x, y \in X$. It follows from (9.124) and (9.126) that

$$f_e(2(x+y)) - 16f_e(x+y) + f_e(2(x-y)) - 16f_e(x-y)$$

$$= 2\left(f_e(2x)-16f_e(x)\right) + 2\left(f_e(2y)-16f_e(y)\right)$$

for all $x, y \in X$. This means that

$$g(x + y) + g(x - y) = 2g(x) + 2g(y)$$

for all $x, y \in X$. So the function $g : X \to Y$ defined by $g(x) = f_e(2x) - 16f_e(x)$ is quadratic.

To prove that $h : X \to Y$ defined by $h(x) = f_e(2x) - 4f_e(x)$ is quartic, we have to show that

$$h(2x + y) + h(2x - y) = 4h(x + y) + 4h(x - y) + 24h(x) - 6h(y) \tag{9.127}$$

for all $x, y \in X$. Replacing x by $2x$ and y by $2y$ in (9.123), respectively, we obtain

$$f_e(2(2x + y)) + f_e(2(2x - y)) = 4f_e(2(x + y)) + 4f_e(2(x - y))$$
$$+ 2\left(f_e(4x) - 4f_e(2x) \right) - 6f_e(2y) \tag{9.128}$$

for all $x, y \in X$. But, since $g(2x) = 4g(x)$ for all $x \in X$, with $g : X \to Y$ is the quadratic function defined above, thus we see that

$$f_e(4x) = 20f_e(2x) - 64f_e(x) \tag{9.129}$$

for all $x \in X$. Hence, according to (9.128) and (9.129), we obtain

$$f_e(2(2x + y)) + f_e(2(2x - y)) = 4f_e(2(x + y)) + 4f_e(2(x - y))$$
$$+ 32\left(f_e(2x) - 4f_e(x) \right) - 6f_e(2y) \tag{9.130}$$

for all $x, y \in X$. By multiplying both sides of (9.123) by 4, we obtain

$$4f_e(2x + y) + 4f_e(2x - y) = 16f_e(x + y) + 16f_e(x - y)$$
$$+ 8\left(f_e(2x) - 4f_e(x) \right) - 24f_e(y) \tag{9.131}$$

for all $x, y \in X$. If we subtract the last equation from (9.130), we arrive at

$$f_e(2(2x + y)) - 4f_e(2x + y) + f_e(2(2x - y)) - 4f_e(2x - y)$$
$$= 4\left(f_e(2(x + y)) - 4f_e(x + y) \right) + 4\left(f_e(2(x - y)) - 4f_e(x - y) \right)$$
$$+ 24\left(f_e(2x) - 4f_e(x) \right) - 6\left(f_e(2y) - 4f_e(y) \right)$$

for all $x, y \in X$. This means that h satisfies equation (9.127), and hence the function $h : X \to Y$ is quartic. But, since $f_e(x) = \frac{1}{12}h(x) - \frac{1}{2}g(x)$ for all $x \in X$, there exist a unique symmetric bi-quadratic function $B_2 : X \times X \to Y$, and a unique symmetric bi-additive function $B_1 : X \times X \to Y$ such that $h(x) = 12B_2(x, x)$, and $g(x) = -12B_1(x, x)$ for all $x \in X$ (see [2, 178]). So

$$f_e(x) = B_2(x, x) + B_1(x, x) \tag{9.132}$$

for all $x \in X$. On the other hand, we show that the function $f_o : X \to Y$ is cubic.

It follows from (9.106), and $f_o(-x) = -f_o(x)$ that

$$f_o(x + ky) + f_o(x - ky) = k^2 f_o(x + y) + k^2 f_o(x - y) + 2(1 - k^2) f_o(x)$$
$$+ \frac{k^2(k^2 - 1)}{6} (f_o(2y) - 8f_o(y)) \qquad (9.133)$$

for all $x, y \in X$. Putting $x = 0$ in (9.133), to get $f_o(2y) = 8f_o(y)$ for all $y \in X$, so we get from (9.133) that

$$f_o(x + ky) + f_o(x - ky) = k^2 f_o(x + y) + k^2 f_o(x - y) + 2(1 - k^2) f_o(x) \qquad (9.134)$$

for all $x, y \in X$. Setting $x = x - y$ in (9.134), we have

$$f_o(x + (k-1)y) + f_o(x - (k+1)y) = k^2 f_o(x) + k^2 f_o(x - 2y)$$
$$+ 2(1 - k^2) f_0(x - y) \qquad (9.135)$$

for all $x, y \in X$. Replacing y by $-y$ in (9.135), gives

$$f_o(x - (k-1)y) + f_o(x + (k+1)y) = k^2 f_o(x) + k^2 f_o(x + 2y)$$
$$+ 2(1 - k^2) f_o(x + y) \qquad (9.136)$$

for all $x, y \in X$. Adding (9.135) and (9.136), we obtain

$$f_o(x + (k-1)y) + f_o(x - (k-1)y) + f_o(x + (k+1)y) + f_o(x - (k+1)y)$$
$$= k^2 (f_o(x + 2y) + f_o(x - 2y)) + 2k^2 f_o(x)$$
$$+ 2(1 - k^2) (f_o(x + y) + f_o(x - y)) \qquad (9.137)$$

for all $x, y \in X$. Setting $x = x + ky$ in (9.134), we have

$$f_o(x + 2ky) + f_o(x) = k^2 (f_o(x + (k-1)y) + f_o(x + (k+1)y))$$
$$+ 2(1 - k^2) f_o(x + ky) \qquad (9.138)$$

for all $x, y \in X$. Replacing y by $-y$ in (9.138), we have

$$f_o(x - 2ky) + f_o(x) = k^2 (f_o(x - (k-1)y) + f_o(x - (k+1)y))$$
$$+ 2(1 - k^2) f_o(x - ky) \qquad (9.139)$$

for all $x, y \in X$. Adding (9.138) and (9.139), one obtains

$$f_o(x + 2ky) + f_o(x - 2ky)$$
$$= k^2 \Big[f_o(x + (k-1)y) + f_o(x + (k+1)y) + f_o(x - (k-1)y)$$
$$+ f_o(x - (k+1)y) \Big] + 2(1 - k^2) (f_o(x + ky) + f_o(x - ky)) - 2f_o(x)$$
$$(9.140)$$

for all $x, y \in X$. Using (9.134), (9.137) and (9.138), we have

$$f_o(x + 2ky) + f_o(x - 2ky)$$
$$= 4k^2(1 - k^2)\left(f_o(x + y) + f_o(x - y)\right) + (6k^4 - 8k^2 + 2)f_o(x)$$
$$+ k^4\left(f_o(x + 2y) + f_o(x - 2y)\right) \tag{9.141}$$

for all $x, y \in X$. Replacing y by $2y$ in (9.134), we obtain

$$f_o(x + 2ky) + f_o(x - 2ky) = k^2\left(f_o(x + 2y) + f_o(x - 2y)\right) + 2(1 - k^2)f_o(x) \tag{9.142}$$

for all $x, y \in X$. If we compare (9.141) with (9.142), then we conclude that

$$f_o(x + 2y) + f_o(x - 2y) = 4\left(f_o(x + y) + f_o(x - y)\right) - 6f_o(x) \tag{9.143}$$

for all $x, y \in X$. Replacing x by $2x$ in (9.143) gives

$$f_o(2(x + y)) + f_o(2(x - y)) = 4\left(f_o(2x + y) + f_o(2x - y)\right) - 6f_o(2x) \tag{9.144}$$

for all $x, y \in X$, which by considering $f_o(2x) = 8f_o(x)$ and (9.144) gives

$$f_o(2x + y) + f_o(2x - y) = 2f_o(x + y) + 2f_o(x - y) - 12f_o(x)$$

for all $x, y \in X$. This means that f_0 is cubic. So

$$f_0(x) = C(x, x, x) \tag{9.145}$$

for all $x \in X$, where C is symmetric for each fixed one variable, and is additive for fixed two variables (see [137]).

Hence, according to (9.132) and (9.145), we obtain that

$$f(x) = f_e(x) + f_o(x) = B_2(x, x) + C(x, x, x) + B_1(x, x)$$

for all $x \in X$. The proof of the converse is trivial. This completes the proof of the theorem. $\qquad\square$

9.6. Mixed type additive-quadratic-cubic-quartic functional equation

9.6.1. *Mixed type functional equation* (9.146)

The general solution of a mixed type functional equation of the form

$$f(x + ay) + f(x - ay) = a^2[f(x + y) + f(x - y)] + 2\left(1 - a^2\right)f(x)$$
$$+ \frac{\left(a^4 - a^2\right)}{12}[f(2y) + f(-2y) - 4f(y) - 4f(-y)], \tag{9.146}$$

which is additive, quadratic, cubic and quartic, where a is an integer with $a \neq 0, \pm 1$, is obtained in the following theorem.

Theorem 9.7 ([287]). *Let* $f : X \rightarrow Y$ *be a function satisfying* (9.146) *for all* $x, y \in X$. *If* f *is even, then* f *is quadratic-quartic.*

Proof. Let f be an even function. Then equation (9.146) becomes

$$f(x + ay) + f(x - ay) = a^2[f(x + y) + f(x - y)] + 2\left(1 - a^2\right) f(x)$$

$$+ \frac{\left(a^4 - a^2\right)}{6}[f(2y) - 4f(y)] \tag{9.147}$$

for all $x, y \in X$. Interchanging x with y in (9.147), and using the evenness of f, we obtain

$$f(ax + y) + f(ax - y) = a^2[f(x + y) + f(x - y)] + 2\left(1 - a^2\right) f(y)$$

$$+ \frac{\left(a^4 - a^2\right)}{6}[f(2x) - 4f(x)] \tag{9.148}$$

for all $x, y \in X$. Setting (x, y) as $(0, 0)$ in (9.148), we obtain $f(0) = 0$. Replacing y by $x + y$ in (9.143), and using the evenness of f, we have

$$f((a + 1)x + y) + f((a - 1)x - y)$$

$$= a^2[f(2x + y) + f(y)] + 2\left(1 - a^2\right) f(x + y)$$

$$+ \frac{\left(a^4 - a^2\right)}{6}[f(2x) - 4f(x)] \tag{9.149}$$

for all $x, y \in X$. Replacing y by $x - y$ in (9.147), we obtain

$$f((a + 1)x - y) + f((a - 1)x + y)$$

$$= a^2[f(2x - y) + f(y)] + 2\left(1 - a^2\right) f(x - y)$$

$$+ \frac{\left(a^4 - a^2\right)}{6}[f(2x) - 4f(x)] \tag{9.150}$$

for all $x, y \in X$. Adding (9.149) and (9.150), we obtain

$$f((a + 1)x + y) + f((a - 1)x - y) + f((a + 1)x - y) + f((a - 1)x + y)$$

$$= a^2[f(2x + y) + f(2x - y) + 2f(y)] + 2\left(1 - a^2\right)[f(x + y) + f(x - y)]$$

$$+ \frac{\left(a^4 - a^2\right)}{6}[2f(2x) - 8f(x)] \tag{9.151}$$

for all $x, y \in X$. Replacing y by $ax + y$ in (9.148), we obtain

$$f(2ax + y) + f(y) = a^2[f((a + 1)x + y) + f((1 - a)x - y)]$$

$$+ 2\left(1 - a^2\right) f(ax + y) + \frac{\left(a^4 - a^2\right)}{6}[f(2x) - 4f(x)]$$

$$\tag{9.152}$$

for all $x, y \in X$. Replacing y by $ax - y$ in (9.149), we obtain

$$f(2ax - y) + f(y) = a^2[f((a+1)x - y) + f((1-a)x + y)]$$
$$+ 2\left(1 - a^2\right) f(ax - y) + \frac{\left(a^4 - a^2\right)}{6}[f(2x) - 4f(x)]$$

(9.153)

for all $x, y \in X$. Adding (9.152) and (9.153), we obtain

$$f(2ax + y) + f(2ax - y) + 2f(y)$$
$$= a^2[f((a+1)x + y) + f((a+1)x - y) + f((a-1)x + y) + f((a-1)x - y)]$$
$$+ 2\left(1 - a^2\right)[f(ax + y) + f(ax - y)] + \frac{\left(a^4 - a^2\right)}{3}[f(2x) - 4f(x)]$$

(9.154)

for all $x, y \in X$. Using (9.151) in (9.154), we arrive at

$$f(2ax + y) + f(2ax - y) + 2f(y)$$
$$= a^4[f(2x + y) + f(2x - y)] + 2a^4 f(y) + 2a^2\left(1 - a^2\right)[f(x + y) + f(x - y)]$$
$$+ \frac{a^2\left(a^4 - a^2\right)}{3}[f(2x) - 4f(x)] + 2\left(1 - a^2\right)[f(ax + y) + f(ax - y)]$$
$$+ \frac{\left(a^4 - a^2\right)}{3}[f(2x) - f(x)]$$

(9.155)

for all $x, y \in X$. Replacing x by $2x$ in (9.148), we obtain

$$f(2ax + y) + f(2ax - y)$$
$$= a^2[f(2x + y) + f(2x - y)] + 2\left(1 - a^2\right) f(y)$$
$$+ \frac{\left(a^4 - a^2\right)}{6}[f(4x) - 4f(2x)]$$

(9.156)

for all $x, y \in X$. Using (9.156) in (9.155), we obtain

$$a^2[f(2x + y) + f(2x - y)] + 2\left(1 - a^2\right) f(y) + \frac{\left(a^4 - a^2\right)}{6}[f(4x) - 4f(2x)] + 2f(y)$$
$$= a^4[f(2x + y) + f(2x - y)] + 2a^2\left(1 - a^2\right)[f(x + y) + f(x - y)]$$
$$+ \frac{a^2\left(a^4 - a^2\right)}{3}[f(2x) - 4f(x)] + 2\left(1 - a^2\right)[f(ax + y) + f(ax - y)]$$
$$+ \frac{\left(a^4 - a^2\right)}{3}[f(2x) - 4f(x)] + 2a^4 f(y)$$

(9.157)

for all $x, y \in X$. Using (9.148) in (9.157), we obtain

$$a^2[f(2x+y) + f(2x-y)] + 2\left(1 - a^2\right) f(y) + \frac{(a4 - a^2)}{6}[f(4x) - 4f(2x)] + 2f(y)$$

$$= a^4[f(2x+y) + f(2x-y)] + 2a^2\left(1 - a^2\right)[f(x+y) + f(x-y)]$$

$$+ \frac{a^2\left(a^4 - a^2\right)}{3}[f(2x) - 4f(x)] + \frac{(a^4 - a^2)}{3}[f(2x) - 4f(x)] + 2a^4 f(y)$$

$$+ 2\left(1 - a^2\right)\left[a^2(f(x+y) + f(x-y)) + 2\left(1 - a^2\right) f(y) \right.$$

$$\left. + \frac{(a^4 - a^2)}{6}[f(2x) - 4f(x)] \right] \tag{9.158}$$

for all $x, y \in X$. Letting $y = 0$ in (9.148), we obtain

$$2f(ax) = 2a^2 f(x) + \frac{(a^4 - a^2)}{6}[f(2x) - 4f(x)] \tag{9.159}$$

for all $x, y \in X$. Replacing y by x in (9.148), we obtain

$$f((a+1)x) + f((a-1)x)$$

$$= a^2 f(2x) + 2\left(1 - a^2\right) f(x) + \frac{(a^4 - a^2)}{6}[f(2x) - 4f(x)] \tag{9.160}$$

for all $x \in X$. Replacing y by ax in (9.148), we obtain

$$f(2ax) = a^2[f((1+a)x) + f((1-a)x)]$$

$$+ 2\left(1 - a^2\right) f(ax) + \frac{(a^4 - a^2)}{6}[f(2x) - 4f(x)] \tag{9.161}$$

for all $x \in X$. Letting $y = 0$ in (9.156), we obtain

$$f(2ax) = a^2 f(2x) + \frac{(a^4 - a^2)}{12}[f(4x) - 4f(2x)] \tag{9.162}$$

for all $x \in X$. From (9.161) and (9.162), we arrive at

$$a^2 f(2x) + \frac{(a^4 - a^2)}{12}[f(4x) - 4f(2x)]$$

$$= a^2[f((1+a)x) + f((1-a)x)] + 2\left(1 - a^2\right) f(ax)$$

$$+ \frac{(a^4 - a^2)}{6}[f(2x) - 4f(x)] \tag{9.163}$$

for all $x \in X$. Using (9.159) and (9.160) in (9.163), we obtain

$$a^2 f(2x) + \frac{(a^4 - a^2)}{12}[f(4x) - 4f(2x)]$$

$$= a^2 \left[a^2 f(2x) + 2(1 - a^2) f(x) + \frac{(a^4 - a^2)}{6}[f(2x) - 4f(x)] \right]$$

$$+ (1 - a^2) \left[2a^2 f(x) + \frac{(a^4 - a^2)}{6}[f(2x) - 4f(x)] \right]$$

$$+ \frac{(a^4 - a^2)}{6}[f(2x) - 4f(x)] \tag{9.164}$$

for all $x \in X$. Comparing (9.158) with (9.164), we arrive at

$$f(2x + y) + f(2x - y) = 4[f(x + y) + f(x - y)] - 8f(x) + 2f(2x) - 6f(y) \tag{9.165}$$

for all $x, y \in X$. Replacing y by $2y$ in (9.165), we obtain

$$f(2x + 2y) + f(2x - 2y) = 4[f(x + 2y) + f(x - 2y)] - 8f(x) + 2f(2x) - 6f(2y) \tag{9.166}$$

for all $x, y \in X$. Interchanging x with y in (9.165), and using the evenness of f, we obtain

$$f(x + 2y) + f(x - 2y) = 4[f(x + 2y) + f(x - 2y)] - 8f(y) + 2f(2y) - 6f(x) \tag{9.167}$$

for all $x, y \in X$. Using (9.167) in (9.166), we obtain

$$f(2x + 2y) + f(2x - 2y)$$
$$= 16[f(x + y) + f(x - y)] + 2f(2y) - 32f(y) + 2f(2x) - 32f(x) \tag{9.168}$$

for all $x, y \in X$. Rearranging equation (9.168), we have

$$[f(2x + 2y) - 16f(x + y)] + [f(2x - 2y) - 16f(x - y)]$$
$$= 2[f(2x) - 16f(x)] + 2[f(2y) - 16f(y)] \tag{9.169}$$

for all $x, y \in X$. Let $\alpha : X \to Y$ be defined by

$$\alpha(x) = f(2x) - 16f(x), \qquad \forall x \in X. \tag{9.170}$$

Applying (9.171) in (9.169), we arrive at

$$\alpha(x + y) + \alpha(x - y) = 2\alpha(x) + 2\alpha(y), \qquad \forall x \in X. \tag{9.171}$$

Hence $\alpha : X \to Y$ is quadratic mapping.

Since α is quadratic, we have $\alpha(2x) = 4\alpha(x)$ for all $x \in X$. Then

$$f(4x) = 20f(2x) - 64f(x) \tag{9.172}$$

for all $x \in X$. Replacing x by $2x$ and y by $2y$ in (9.165), respectively, we get

$$f(2(2x + y)) + f(2(2x - y))$$
$$= 4[f(2(x + y)) + f(2(x - y))] - 8f(2x) + 2f(4x) - 6f(2y) \tag{9.173}$$

for all $x, y \in X$. Using (9.172) in (9.173), we obtain

$$f(2(2x + y)) + f(2(2x - y))$$
$$= 4[f(2(x + y)) + f(2(x - y))] + 32[f(2x) - 4f(x)] - 6f(2y) \tag{9.174}$$

for all $x, y \in X$. Multiplying equation (9.165) by 4, and then subtracting the resulting equation from (9.174), we have

$$[f(2(2x + y)) - 4f(2x + y)] + [f(2(2x - y)) - 4f(2x - y)]$$
$$= 4[f(2(x + y)) - 4f(x + y)] + 4[f(2(x - y)) - 4f(x - y)]$$
$$+ 24[f(2x) - 4f(x)] - 6[f(2y) - 4f(y)] \tag{9.175}$$

for all $x, y \in X$. Let $\beta : X \to Y$ be defined by

$$\beta(x) = f(2x) - 4f(x), \qquad \forall\, x \in X. \tag{9.176}$$

Applying (9.175) in (9.176), we arrive at

$$\beta(2x + y) + \beta(2x - y) = 4[\beta(x + y) + \beta(x - y)] + 24\beta(x) - 6\beta(y) \tag{9.177}$$

for all $x, y \in X$. Hence $\beta : X \to Y$ is quartic mapping.

On the other hand, we have

$$f(x) = \frac{\beta(x) - \alpha(x)}{12}, \qquad \forall\, x \in X. \tag{9.178}$$

This means that f is quadratic-quartic function. This completes the proof of the theorem. $\qquad\square$

Theorem 9.8 ([287]). *Let $f : X \to Y$ be a function satisfying (9.146) for all $x, y \in X$. If f is odd, then f is additive-cubic.*

Proof. Let f be an odd function. Then equation (9.146) becomes

$$f(x + ay) + f(x - ay) = a^2[f(x + y) + f(x - y)] + 2\left(1 - a^2\right)f(x) \tag{9.179}$$

for all $x, y \in X$. By [124, Lemma 2.2], f is additive-cubic. This completes the proof of the theorem. $\qquad\square$

Theorem 9.9 ([287]). *Let $f : X \to Y$ be a function satisfying (9.146) for all $x, y \in X$ if and only if there exist functions $A : X \to Y$, $B : X \times X \to Y$, $C : X \times X \times X \to Y$ and $D : X \times X \times X \times X \to Y$ such that*

$$f(x) = A(x) + B(x, x) + C(x, x, x) + D(x, x, x, x) \tag{9.180}$$

for all $x \in X$, where A is additive, B is symmetric bi-additive, C is symmetric for each fixed one variable and is additive for fixed two variables, and D is symmetric multiadditive.

Proof. Let $f : X \to Y$ be a function satisfying equation (9.146). We decompose f into even and odd parts by setting

$$f_e(x) = \frac{f(x) + f(-x)}{2} \quad \text{and} \quad f_o(x) = \frac{f(x) - f(-x)}{2},$$

respectively, for all $x \in X$. It is clear that $f(x) = f_e(x) + f_o(x)$ for all $x \in X$. It is easy to show that the functions f_e and f_o satisfy (9.146). Hence by Theorems 9.7 and 9.8, we see that the function f_e is quadratic-quartic, and f_o is additive-cubic, respectively. Thus there exist a symmetric bi-additive function $B : X \times X \to Y$, and a symmetric multiadditive function $D : X \times X \times X \times X \to Y$ such that $f_e(x) = B(x, x) + D(x, x, x, x)$ for all $x \in X$, and the function $A : X \to Y$ is additive, and $C : X \times X \times X \to Y$ such that $f_o(x) = A(x) + C(x, x, x)$, where C is symmetric for each fixed one variable and is additive for fixed two variables. Hence we get (9.180) for all $x \in X$.

Conversely, let $f(x) = A(x) + B(x, x) + C(x, x, x) + D(x, x, x, x)$ for all $x \in X$, where A is additive, B is symmetric bi-additive, C is symmetric for each fixed one variable and is additive for fixed two variables, and D is symmetric multiadditive. Then it is easy to show that f satisfies (9.146). This completes the proof of the theorem. $\square$

Chapter 10

Two-Variable and Three-Variable Functional Equations

10.1. Introduction

In the previous chapters, we have discussed about the general solution of one-variable functional equations of various types. This chapter concerns the study of the general solution of two-variable additive, two-variable Jensen's, two-variable quadratic, two-variable cubic and three-variable quadratic functional equations. Let X and Y be vector spaces throughout this chapter.

10.2. Two-variable additive functional equation

This section is devoted to the general solution of two-variable additive functional equation of the type

$$f(x + u, y + v) = f(x, y) + f(u, v). \tag{10.1}$$

It is easy to verify that the function $f(x, y) = ax + by$ is a solution of the functional equation (10.1).

Definition 10.1 ([289]). A mapping $f : X \times X \to Y$ is called a *two-variable additive mapping* if there exist $a, b \in \mathbb{R}$ such that $f(x, y) = ax + by$, for all $x, y \in X$.

10.2.1. *Relation between two-variable additive functional equation and Cauchy additive functional equation*

The two variable additive functional equation (10.1) induces the additive functional equation (1.1).

Theorem 10.1 ([289]). *Let $f : X \times X \to Y$ be a mapping satisfying (10.1) and let $A : X \to Y$ be a mapping given by*

$$A(x) = f(x, x) \tag{10.2}$$

for all $x \in X$. Then A satisfies equation (1.1).

Proof. From (10.1) and (10.2), we have

$$A(x + y) = f(x + y, x + y) = f(x, x) + f(y, y) = A(x) + A(y)$$

for all $x, y \in X$. This completes the proof. $\square$

Theorem 10.2 ([289]). *Let $a, b \in \mathbb{R}$, and $A : X \to Y$ be a mapping satisfying (1.1). If $f : X \times X \to Y$ is a mapping given by*

$$f(x, y) = aA(x) + bA(y) \qquad (10.3)$$

for all $x, y \in X$, then f satisfies equation (10.1).

Proof. From (1.1) and (10.3), we have

$$\begin{aligned}
f(x + u, y + v) &= aA(x + u) + bA(y + v) \\
&= a[A(x) + A(u)] + b[A(y) + A(v)] \\
&= f(x, y) + f(u, v)
\end{aligned}$$

for all $x, u, y, v \in X$. This completes the proof. $\square$

10.2.2. *General solution of two-variable additive functional equation*

Theorem 10.3 ([289]). *Let $f : X \times X \to Y$ be a mapping satisfying (10.1) if and only if there exist two additive mappings $s_1, s_2 : X \to Y$ such that*

$$f(x, y) = s_1(x) + s_2(y) \qquad (10.4)$$

for all $x, y \in X$.

Proof. Assume that f is a solution of equation (10.1). Define $f_1(x) = f(x, 0)$, and $f_2(y) = f(0, y)$, for all $x, y \in X$. It is easy to verify that f_1, f_2 are additive functions. Let $f_1(x) = s_1(x)$ and $f_2(x) = s_2(x)$, for all $x \in X$. Hence $s_1(x) + s_2(x) = ax + by = f(x, y)$, for all $x, y \in X$.

Conversely, assume that there exist two additive mappings $s_1, s_2 : X \to Y$ such that $f(x, y) = s_1(x) + s_2(y)$, for all $x, y \in X$.

$$\begin{aligned}
f(x + u, y + v) &= s_1(x + u) + s_2(y + v) \\
&= s_1(x) + s_1(u) + s_2(y) + s_2(v) \\
&= f(x, y) + f(u, v), \quad \text{for all } x, u, y, v \in X.
\end{aligned}$$

The proof is now completed. $\square$

10.3. Two-variable Jensen's functional equation

This section is devoted to the general solution of a two-variable Jensen's functional equation of the form

$$g\left(\frac{x+u}{2}, \frac{y+v}{2}\right) = \frac{1}{2}[g(x,y) + g(u,v)]. \tag{10.5}$$

The function $g(x,y) = ax + by + c$ becomes a solution of the functional equation (10.5).

Definition 10.2. A mapping $J : X \to Y$ is called Jensen's mapping if J satisfies the functional equation (with one variable)

$$J\left(\frac{x+y}{2}\right) = \frac{1}{2}[J(x) + J(y)]. \tag{10.6}$$

The function $J(x) = ax + c$ is a solution of the functional equation (10.6).

Definition 10.3 ([289]). A mapping $g : X \times X \to Y$ is called a two-variable Jensen's mapping if there exist $a, b, c \in \mathbb{R}$ such that $g(x,y) = ax + by + c$, for all $x, y \in X$.

10.3.1. *Relation between two-variable Jensen's functional equation and Jensen's functional equation of one variable*

The following theorem shows that the two-variable Jensen's functional equation (10.5) induces the Jensen's functional equation (10.6).

Theorem 10.4 ([289]). *Let $g : X \times X \to Y$ be a mapping satisfying (10.5), and let $J : X \to Y$ be a mapping given by*

$$J(x) = g(x,x) \tag{10.7}$$

for all $x \in X$. Then J satisfies equation (10.6).

Proof. From equations (10.5) and (10.7), we obtain

$$J\left(\frac{x+y}{2}\right) = g\left(\frac{x+y}{2}, \frac{x+y}{2}\right)$$

$$= \frac{1}{2}[g(x,x) + g(y,y)]$$

$$= \frac{1}{2}[J(x) + J(y)], \quad \text{for all } x, y \in X.$$

This completes the proof. $\qquad\square$

The following theorem establishes the relationship between the mappings satisfying the functional equations (10.1) and (10.5).

Theorem 10.5 ([289]). *Let $g : X \times X \to Y$ be a mapping satisfying the functional equation (10.5). Then there exists a two-variable additive function $f : X \times X \to Y$ such that $g(x, y) = f(x, y) + c$, for all $x, y \in X$ where $c = g(0, 0)$. Therefore, $g(x, y) = ax + by + c$ is a solution of the functional equation (10.5).*

Proof. Putting $u = v = 0$ in equation (10.5), we have

$$g\left(\frac{x}{2}, \frac{y}{2}\right) = \frac{1}{2}[g(x, y) + g(0, 0)] = \frac{1}{2}[g(x, y) + c].$$

From (10.5), we get

$$\frac{1}{2}[g(x, y) + g(u, v)] = g\left(\frac{x + u}{2}, \frac{y + v}{2}\right) = \frac{1}{2}[g(x + u, y + v) + c]$$

or

$$g(x + u, y + v) = g(x, y) + g(u, v) - c.$$

Taking $f(x, y) = g(x, y) - c$, we obtain

$$
\begin{aligned}
f(x + u, y + v) &= g(x + u, y + v) - c \\
&= [g(x, y) + g(u, v) - c] - c \\
&= f(x, y) + f(u, v).
\end{aligned}
$$

Therefore, $g(x, y) = f(x, y) + c$ is a solution of (10.5). Hence by Theorem 10.2, $g(x, y) = ax + by + c$ is a solution of (10.5). The proof is now completed. $\square$

10.3.2. *General solution of two-variable Jensen's functional equation*

The general solution of the two-variable Jensen's functional equation (10.5) is presented in the following theorem.

Theorem 10.6 ([289]). *Let $g : X \times X \to Y$ be a mapping satisfying (10.5) if and only if there exist two additive mappings $A_1, A_2 : X \to Y$ such that $g(x, y) = A_1(x) + A_2(y) + g(0, 0)$, for all $x, y \in X$.*

Proof. Assume that g is a solution of equation (10.5). Define $G_x, G'_y : X \to Y$ by $G_x(y) = G'_y(x) = g(x, y)$, for all $x, y \in X$. Then G_x, G'_y are Jensen mappings for all $x, y \in X$. By Theorem 10.5, there exist two additive mappings $F_x, F'_y : X \to Y$ such that $G_x(y) = F_x(y) + G_x(0)$ and $G'_y(x) = F'_y(x) + G'_y(0)$, for all $x, y \in X$. Define $A_1, A_2 : X \to Y$ by $A_1(x) = g(x, 0) - g(0, 0)$; $A_2(y) = g(0, y) - g(0, 0)$, for all $x, y \in X$. Then $A_1(x) = G'_0(x) - G'_0(0) = F'_0(x)$ and $A_2(y) = G_0(y) - G_0(0) = F_0(y)$, for all $x, y \in X$. Hence A_1, and A_2 are additive mappings.

Conversely, assume that there exist two additive mappings $A_1, A_2 : X \to Y$ such that $g(x, y) = A_1(x) + A_2(y) + g(0, 0)$, for all $x, y \in X$. Then

$$g\left(\frac{x+u}{2}, \frac{y+v}{2}\right) = A_1\left(\frac{x+u}{2}\right) + A_2\left(\frac{y+v}{2}\right) + g(0,0)$$

$$= \frac{1}{2}[A_1(x) + A_1(u)] + \frac{1}{2}[A_2(y) + A_2(v)] + g(0,0)$$

$$= \frac{1}{2}[g(x,y) + g(u,v)], \qquad \text{for all } x, u, y, v, \in X.$$

This completes the proof of Theorem 10.6. □

10.4. Two-variable quadratic functional equation

This section deals with the general solution of the two-variable quadratic functional equation

$$f(x + y, z + w) + f(x - y, z - w) = 2f(x, z) + 2f(y, w). \tag{10.8}$$

A mapping f is called a *quadratic form* if there exist $a, b, c \in \mathbb{R}$ such that $f : \mathbb{R} \times \mathbb{R} \to \mathbb{R}$ given by

$$f(x, y) = ax^2 + bxy + cy^2$$

for all $x, y \in X$, which is a solution of (10.8).

10.4.1. *Relation between two-variable quadratic functional equation and quadratic functional equation of one variable*

The following theorem shows that the two-variable quadratic functional equation (10.8) induces the quadratic functional equation (4.1) with one variable.

Theorem 10.7 ([18]). *Let $f : X \times X \to Y$ be a mapping satisfying equation (10.8), and let $g : X \to Y$ be a mapping given by*

$$g(x) = f(x, x) \tag{10.9}$$

for all $x \in X$. Then g satisfies equation (4.1).

Proof. By (10.8) and (10.9), we have

$$g(x + y) + g(x - y) = f(x + y, x + y) + f(x - y, x - y)$$
$$= 2f(x, x) + 2f(y, y)$$
$$= 2g(x) + 2g(y), \quad \text{for all } x, y \in X.$$

This completes the proof. □

Example 10.1 ([18]). Let $\mathcal{A}$ be a real algebra and $D : \mathcal{A} \to \mathcal{A}$ be a derivation on X. Define a mapping $f : \mathcal{A} \times \mathcal{A} \to \mathcal{A}$ by

$$f(x, y) = D(xy) = xD(y) + D(x)y$$

for all $x, y \in \mathcal{A}$. Then f satisfies (10.8). Define a mapping $g : \mathcal{A} \to \mathcal{A}$ by

$$g(x) = D(x^2) = xD(x) + D(x)x$$

for all $x \in \mathcal{A}$. Then g satisfies (10.9). By Theorem 10.7, g satisfies (4.1).

The following theorem proves that the quadratic functional equation (4.1) induces the two-variable quadratic functional equation (10.8) with an additional condition.

Theorem 10.8 ([18]). *Let $a, b, c \in \mathbb{R}$, and $g : X \to Y$ be a mapping satisfying equation (4.1). If $f : X \times X \to Y$ is the mapping given by*

$$f(x, y) = ag(x) + \frac{b}{4}[g(x + y) - g(x - y)] + cg(y) \tag{10.10}$$

for all $x, y \in X$, then f satisfies (10.8). Furthermore, (10.9) holds if $a + b + c = 1$.

Proof. By equations (4.1) and (10.10), we have

$$f(x + y, z + w) + f(x - y, z - w)$$

$$= ag(x + y) + \frac{b}{4}[g(x + y + z + w) - g(x + y - z - w)] + cg(z + w)$$

$$+ ag(x - y) + \frac{b}{4}[g(x - y + z - w) - g(x - y - z + w)] + cg(z - w)$$

$$= 2ag(x) + \frac{b}{2}[g(x + z) - g(x - z)] + 2cg(z) + 2ag(y)$$

$$+ \frac{b}{2}[g(y + w) - g(y - w)] + 2cg(w)$$

$$= 2f(x, z) + 2f(y, w), \quad \text{for all } x, y \in X.$$

Letting $x = y = 0$, and $y = x$ in equation (4.1), respectively, we get

$$g(0) = 0, \quad \text{and} \quad g(2x) = 4g(x)$$

for all $x \in X$. By (10.10), and the above two equalities,

$$f(x, x) = ag(x) + \frac{b}{4}[g(2x) - g(0)] + cg(x)$$

$$= (a + b + c)g(x)$$

$$= g(x), \quad \text{for all } x \in X.$$

This completes the proof. $\square$

Example 10.2 ([18]). Consider the function $g : \mathbb{R}^2 \to \mathbb{R}$ given by $g(x) := \mathbf{x}^T A \mathbf{x}$ for all $\mathbf{x} \in \mathbb{R}^2$, where A is a 2×2 real matrix. Then we can easily show that g satisfies (4.1). Let $a, b, c \in \mathbb{R}$ and define

$$f(\mathbf{x}, \mathbf{y}) := ag(\mathbf{x}) + \frac{b}{4}[g(\mathbf{x} + \mathbf{y}) - g(\mathbf{x} - \mathbf{y})] + cg(\mathbf{y})$$

for all $\mathbf{x}, \mathbf{y} \in \mathbb{R}^2$. Then one can obtain the function $f : \mathbb{R}^2 \times \mathbb{R}^2 \to \mathbb{R}$ given by

$$f(\mathbf{x}, \mathbf{y}) = \begin{pmatrix} \mathbf{x} \\ \mathbf{y} \end{pmatrix}^T \begin{pmatrix} a & \dfrac{b}{2} \\ \dfrac{b}{2} & c \end{pmatrix} \begin{pmatrix} A\mathbf{x} \\ A\mathbf{y} \end{pmatrix}$$

for all $\mathbf{x}, \mathbf{y} \in \mathbb{R}^2$. By Theorem 10.8, the function f satisfies equation (10.8).

Example 10.3 ([18]). Let M_n be the algebra of $n \times n$ real matrices. Consider the mapping $g : M_n \to M_n$ given by $g(A) := A^2$ for all $A \in M_n$. Then we can easily show that g satisfies (4.1). Let $a, b, c \in \mathbb{R}$ and define

$$f(A, B) := aA^2 + bA \circ B + cB^2,$$

where $A \circ B$ the Jordan product $\frac{1}{2}(AB + BA)$ of A and B for all $A, B \in M_n$. Then the mapping $f : M_n \times M_n \to M_n$ satisfies equation (10.10). By Theorem 10.8, the mapping f satisfies equation (10.8).

10.4.2. *General solution of two-variable quadratic functional equation*

The general solution of the two-variable quadratic functional equation (10.8) is presented in the following theorem.

Theorem 10.9 ([18]). *A mapping $f : X \times X \to Y$ satisfies (10.8) if and only if there exist two symmetric bi-additive mappings $S_1, S_2 : X \times X \to Y$, and a bi-additive mapping $B : X \times X \to Y$ such that*

$$f(x, y) = S_1(x, x) + B(x, y) + S_2(y, y)$$

for all $x, y \in X$.

Proof. We first assume that f is a solution of equation (10.8). Define $f_1, f_2 : X \to Y$ by $f_1(x) = f(x, 0)$, and $f_2(x) = f(0, x)$ for all $x \in X$. One can easily verify that f_1, f_2 are quadratic. By [2], there exist two symmetric bi-additive mappings $S_1, S_2 : X \times X \to Y$ such that $f_1(x) = S_1(x, x)$, and $f_2(x) = S_2(x, x)$ for all $x \in X$.

Define $B : X \times X \to Y$ by

$$B(x, y) := f(x, y) - [f(x, 0) + f(0, y)] \tag{10.11}$$

for all $x, y \in X$. Then B is bi-additive. Indeed, by equations (10.8) and (10.11), we obtain that

$$
\begin{aligned}
B(x_1 &+ x_2, y) \\
&= f(x_1 + x_2, y) - [f(x_1 + x_2, 0) + f(0, y)] \\
&= \frac{1}{2}[f(x_1 + x_2, 2y) - f(x_1 + x_2, 0)] - 2f(0, y) \\
&= f(x_1, y) + f(x_2, y) - f(x_1, 0) - f(x_2, 0) - 2f(0, y) \\
&= B(x_1, y) + B(x_2, y), \quad \text{for all } x_1, x_2, y \in X.
\end{aligned}
$$

Similarly, $B(x, y_1 + y_2) = B(x, y_1) + B(x, y_2)$ for all $x, y_1, y_2 \in X$.

Conversely, we assume that there exist two symmetric bi-additive mappings $S_1, S_2 : X \times X \to Y$, and a bi-additive mapping $B : X \times X \to Y$ such that $f(x, y) = S_1(x, x) + B(x, y) + S_2(y, y)$ for all $x, y \in X$. Since B is bi-additive, and S_1, S_2 are symmetric bi-additive,

$$
\begin{aligned}
f(x &+ y, z + w) + f(x - y, z - w) \\
&= S_1(x + y, x + y) + B(x + y, z + w) + S_2(z + w, z + w) \\
&\quad + S_1(x - y, x - y) + B(x - y, z - w) + S_2(z - w, z - w) \\
&= 2[S_1(x, x) + B(x, z) + S_2(z, z)] + 2[S_1(y, y) + B(y, w) + S_2(w, w)] \\
&= 2f(x, z) + 2f(y, w)
\end{aligned}
$$

for all $x, y, z, w \in X$. The proof is now completed. $\qquad\square$

Example 10.4 ([18]). Consider the function $f : \mathbb{R}^2 \times \mathbb{R}^2 \to \mathbb{R}$ given in Example 10.2. By Theorem 10.9, there exist two symmetric bi-additive mappings $S_1, S_2 : \mathbb{R}^2 \times \mathbb{R}^2 \to \mathbb{R}$, and a bi-additive mapping $B : \mathbb{R}^2 \times \mathbb{R}^2 \to \mathbb{R}$ such that

$$f(\mathbf{x}, \mathbf{y}) = S_1(\mathbf{x}, \mathbf{x}) + B(\mathbf{x}, \mathbf{y}) + S_2(\mathbf{y}, \mathbf{y})$$

for all $\mathbf{x}, \mathbf{y} \in \mathbb{R}^2$. In fact

$$S_1(\mathbf{x}, \mathbf{y}) = \frac{a}{2} \left(\mathbf{x}^T A \mathbf{y} + \mathbf{y}^T A \mathbf{x} \right), \qquad S_2(\mathbf{x}, \mathbf{y}) = \frac{c}{2} \left(\mathbf{x}^T A \mathbf{y} + \mathbf{y}^T A \mathbf{x} \right)$$

and

$$B(\mathbf{x}, \mathbf{y}) = \frac{b}{2} \left(\mathbf{x}^T A \mathbf{y} + \mathbf{y}^T A \mathbf{x} \right)$$

for all $\mathbf{x}, \mathbf{y} \in \mathbb{R}^2$.

10.5. Three-variable quadratic functional equation

This section contains the general solution of a three-variable quadratic functional equation

$$f(x+y, z+w, u+v) + f(x-y, z-w, u-v) = 2f(x, z, u) + 2f(y, w, v).$$
(10.12)

The quadratic form $f(x, y, z) = ax^2 + by^2 + cz^2 + dxy + eyz + fzx$ is the solution of the functional equation (10.12).

10.5.1. *Relation between three-variable quadratic functional equation and quadratic functional equation of one variable*

The following theorem shows that the three-variable quadratic functional equation (10.12) induces the quadratic functional equation (4.1).

Theorem 10.10 ([283]). *Let $f : X \times X \times X \to Y$ be a mapping satisfying equation* (10.12), *and let $h : X \to Y$ be a mapping given by*

$$h(x) = f(x, x, x)$$
(10.13)

for all $x \in X$. Then h satisfies equation (4.1).

Proof. From equations (10.12) and (10.13), we obtain

$$h(x+y) + h(x-y) = f(x+y, x+y, x+y) + f(x-y, x-y, x-y)$$
$$= 2f(x, x, x) + 2f(y, y, y)$$
$$= 2h(x) + 2h(y)$$

for all $x, y \in X$. This completes the proof. $\square$

In the following theorem, we will prove that the quadratic functional equation (4.1) induces the three-variable quadratic functional equation (10.12) with an additional condition.

Theorem 10.11 ([283]). *Let $a, b, c, d, e, f \in \mathbb{R}$, and $h : X \to Y$ be a mapping satisfying equation* (4.1). *If $f : X \times X \times X \to Y$ is a mapping given by*

$$f(x, y, z) = ah(x) + bh(y) + ch(z) + \frac{d}{4}[h(x+y) - h(x-y)]$$

$$+ \frac{e}{4}[h(y+z) - h(y-z)] + \frac{f}{4}[h(z+x) - h(z-x)]$$
(10.14)

for all $x, y, z \in X$, then f satisfies equation (10.12). *Also, equation* (10.13) *holds if $a + b + c + d + e + f = 1$.*

Proof. From equations (4.1), and (10.14), we obtain

$$f(x+y, z+w, u+v) + f(x-y, z-w, u-v)$$

$$= ah(x+y) + bh(z+w) + ch(u+v)$$

$$+ \frac{d}{4}[h(x+y+z+w) - h(x+y-z-w)]$$

$$+ \frac{e}{4}[h(z+w+u+v) - h(z+w-u-v)]$$

$$+ \frac{f}{4}[h(u+v+x+y) - h(u+v-x-y)]$$

$$+ ah(x-y) + bh(z-w) + ch(u-v)$$

$$+ \frac{d}{4}[h(x-y+z-w) - h(x-y-z+w)]$$

$$+ \frac{e}{4}[h(z-w+u-v) - h(z-w-u+v)]$$

$$+ \frac{f}{4}[h(u-v+x-y) - h(u-v-x+y)]$$

$$= 2f(x, z, u) + 2f(y, w, v), \quad \text{for all } x, y, z \in X.$$

Moreover, substituting $x = y = 0$, and $y = x$ in equation (4.1), we have $h(0) = 0$, and $h(2x) = 4h(x)$ for all $x \in X$. By equation (10.14), and the above two results, we obtain

$$f(x, x, x) = ah(x) + bh(x) + ch(x) + \frac{d}{4}[h(2x) - h(0)]$$

$$+ \frac{e}{4}[h(2x) - h(0)] + \frac{f}{4}[h(2x) - h(0)]$$

$$= (a + b + c)h(x) + \frac{(d + e + f)}{4}[4h(x)]$$

$$= h(x), \quad \text{if } a + b + c + d + e + f = 1.$$

This completes the proof.   □

Example 10.5 ([283]). Consider the function $h : \mathbb{R}^3 \to \mathbb{R}$ given by $h(\mathbf{x}) = \mathbf{x}^T A \mathbf{x}$, $\forall x \in \mathbb{R}$ where A is a 3×3 real matrix. It is easy to verify that h satisfies equation (4.1). Let $a, b, c, d, e, f \in \mathbb{R}$. Define

$$f(\mathbf{x}, \mathbf{y}, \mathbf{z}) = ah(\mathbf{x}) + bh(\mathbf{y}) + ch(\mathbf{z}) + \frac{d}{f}[h(\mathbf{x} + \mathbf{y}) - h(\mathbf{x} - \mathbf{y})]$$

$$+ \frac{e}{4}[h(\mathbf{y} + \mathbf{z}) - h(\mathbf{y} - \mathbf{z})] + \frac{f}{4}[h(\mathbf{z} + \mathbf{x}) - h(\mathbf{z} - \mathbf{x})]$$

for all $\mathbf{x}, \mathbf{y}, \mathbf{z} \in \mathbb{R}^3$. Then $f : \mathbb{R}^3 \times \mathbb{R}^3 \times \mathbb{R}^3 \to \mathbb{R}$ given by

$$f(\mathbf{x}, \mathbf{y}, \mathbf{z}) = \begin{pmatrix} \mathbf{x} \\ \mathbf{y} \\ \mathbf{z} \end{pmatrix}^T \begin{pmatrix} a & \dfrac{d}{2} & \dfrac{f}{2} \\ \dfrac{d}{2} & b & \dfrac{c}{2} \\ \dfrac{f}{2} & \dfrac{e}{2} & c \end{pmatrix} \begin{pmatrix} A\mathbf{x} \\ A\mathbf{y} \\ A\mathbf{z} \end{pmatrix}$$

for all $\mathbf{x}, \mathbf{y}, \mathbf{z} \in \mathbb{R}^3$. By Theorem 10.11, we see that f satisfies equation (10.12).

Example 10.6 ([283]). Let M_n be the algebra of $n \times n$ real matrices. Consider $h : M_n \to M_n$ given by $h(A) = A^2$, for all $A \in M_n$. Then it is easy to verify that h satisfies equation (4.1). Let $a, b, c, d, e, f \in \mathbb{R}$, and define

$$f(A, B, C) = aA^2 + bB^2 + cC^2 + d\, A \circ B + e\, B \circ C + f\, C \circ A$$

where $A \circ B$, $B \circ C$, $C \circ A$ are the Jordan products $\frac{1}{4}[AB + BA]$, $\frac{1}{4}[BC + CB]$, $\frac{1}{4}[CA + AC]$ of A and B, B and C, C and A, respectively, for all $A, B, C \in M_n$. Then the mapping $f : M_n \times M_n \times M_n \to M_n$ satisfies equations (10.14) and (10.12).

10.5.2. *General solution of three-variable quadratic functional equation*

The following theorem presents the general solution of the three-variable quadratic functional equation (10.12).

Theorem 10.12 ([283]). *Let $f : X \times X \times X \to Y$ be a mapping satisfying (10.12) if and only if there exist three symmetric bi-additive mappings $S_1, S_2, S_3 : X \times X \to Y$, and three bi-additive mappings $B_1, B_2, B_3 : X \times X \to Y$ such that*

$$f(x, y, z) = S_1(x, x) + B_1(x, y) + S_2(y, y) + B_2(y, z) + S_3(z, z) + B_3(z, x)$$

$$(10.15)$$

for all $x, y, z \in X$.

Proof. Assume that f is a solution of equation (10.12). Define $f_1(x) = f(x, 0, 0)$, $f_2(x) = f(0, x, 0)$, and $f_3(x) = f(0, 0, x)$ for all $x \in X$. It is easy to verify that f_1, f_2, and f_3 are quadratic functions. Then there exist three symmetric bi-additive mappings $S_1, S_2, S_3 : X \times X \to Y$ such that $f_1(x) = S_1(x, x)$, $f_2(x) = S_2(x, x)$, and

$f_3(x) = S_3(x, x)$. Define $B_1, B_2, B_3 : X \times X \to Y$ by

$$B_1(x, y) = f(x, y, z) - [f(0, y, z) + f(x, 0, z) - f(0, 0, z)] \qquad (10.16)$$

$$B_2(x, y) = f(x, y, z) - [f(0, y, z) + f(x, 0, z) - f(0, 0, z)] \qquad (10.17)$$

and

$$B_3(x, y) = f(x, y, z) - [f(0, y, z) + f(x, 0, z) - f(0, 0, z)] \qquad (10.18)$$

for all $x, y, z \in X$. From equations (4.1) and (10.16), we have

$$
\begin{aligned}
&B_1(x_1 + x_2, y) \\
&= f(x_1 + x_2, y, z) - f(0, y, z) - f(x_1 + x_2, 0, z) + f(0, 0, z) \\
&= \frac{1}{2}[f(x_1 + x_2, 2y, 2z) + f(x_1 + x_2, 0, 0) - f(x_1 + x_2, 0, 2z) - f(x_1 + x_2, 0, 0)] \\
&\quad - 2f(0, y, z) + 2f(0, 0, z) \\
&= B_1(x_1, y) + B_1(x_2, y)
\end{aligned}
$$

for all $x_1, x_2, y \in X$. Also, one can prove

$$B_1(x, y_1 + y_2) = B_1(x, y_1) + B_1(x, y_2)$$

for all $x, y_1, y_2 \in X$. Therefore B_1 is bi-additive. Similarly, we can prove

$$
\begin{aligned}
B_2(y_1 + y_2, z) &= B_2(y_1, z) + B_2(y_2, z), \\
B_2(y, z_1 + z_2) &= B_2(y, z_1) + B_2(y, z_2), \\
B_3(z_1 + z_2, x) &= B_3(z_1, x) + B_3(z_2, x), \\
B_3(z, x_1 + x_2) &= B_3(z, x_1) + B_3(z, x_2)
\end{aligned}
$$

for all $x_1, x_2, y_1, y_2, z_1, z_2 \in X$.

Conversely, assume that there exist three symmetric bi-additive mappings $S_1, S_2, S_3 : X \times X \to Y$, and three bi-additive mappings $B_1, B_2, B_3 : X \times X \to Y$ such that

$$f(x, y, z) = S_1(x, x) + B_1(x, y) + S_2(y, y) + B_2(y, z) + S_3(z, z) + B_3(z, x)$$

for all $x, y, z \in X$. Since B_i is bi-additive, and S_i is symmetric bi-additive for $i = 1, 2, 3$, we have

$$
\begin{aligned}
&f(x + y, z + w, u + v) + f(x - y, z - w, u - v) \\
&\quad = S_1(x + y, x + y) + B_1(x + y, z + w) + S_2(z + w, z + w) \\
&\qquad + B_2(z + w, u + v) + S_3(u + v, u + v) + B_3(u + v, x + y) \\
&\qquad + S_1(x - y, x - y) + B_1(x - y, z - w) + S_2(z - w, z - w)
\end{aligned}
$$

$$+ B_2(z - w, u - v) + S_3(u - v, u - v) + B_3(u - v, x - y)$$
$$= 2S_1(x, x) + 2B_1(x, z) + 2S_2(z, z) + 2B_2(z, u) + 2S_3(u, u) + 2B_3(u, x)$$
$$+ 2S_1(y, y) + 2B_1(y, w) + 2S_2(w, w) + 2B_2(z, v) + 2S_3(v, v) + 2B_3(v, y)$$
$$= 2f(x, z, u) + 2f(y, w, v), \qquad \text{for all } x, y, z, u, v, w \in X.$$

The proof is now completed. $\square$

Example 10.7 ([283]). Consider the function $f : \mathbb{R}^3 \times \mathbb{R}^3 \times \mathbb{R}^3 \to \mathbb{R}$ given as in Example 10.5. By Theorem 10.12, there exist three symmetric bi-additive mappings $S_1, S_2, S_3 : \mathbb{R}^3 \times \mathbb{R}^3 \to \mathbb{R}$, and three bi-additive mappings $B_1, B_2, B_3 : \mathbb{R}^3 \times \mathbb{R}^3 \to \mathbb{R}$ such that

$$f(\mathbf{x}, \mathbf{y}, \mathbf{z}) = S_1(\mathbf{x}, \mathbf{x}) + B_1(\mathbf{x}, \mathbf{y}) + S_2(\mathbf{y}, \mathbf{y}) + B_2(\mathbf{y}, \mathbf{z}) + S_3(\mathbf{z}, \mathbf{z}) + B_3(\mathbf{z}, \mathbf{x})$$

for all $\mathbf{x}, \mathbf{y}, \mathbf{z} \in \mathbb{R}^3$. In fact

$$S_1(\mathbf{x}, \mathbf{y}) = \frac{a}{4} \left[\mathbf{x}^T A \mathbf{y} + \mathbf{y}^T A \mathbf{x} \right], S_2(\mathbf{x}, \mathbf{y}) = \frac{b}{4} \left[\mathbf{x}^T A \mathbf{y} + \mathbf{y}^T A \mathbf{x} \right],$$

$$S_3(\mathbf{x}, \mathbf{y}) = \frac{c}{4} \left[\mathbf{x}^T A \mathbf{y} + \mathbf{y}^T A \mathbf{x} \right]$$

and

$$B_1(\mathbf{x}, \mathbf{y}) = \frac{d}{4} [\mathbf{x}^T A \mathbf{y} + \mathbf{y}^T A \mathbf{x}], B_2(\mathbf{y}, \mathbf{z}) = \frac{e}{4} [\mathbf{y}^T A \mathbf{z} + \mathbf{z}^T A \mathbf{y}],$$

$$B_3(\mathbf{z}, \mathbf{x}) = \frac{f}{4} [\mathbf{z}^T A \mathbf{x} + \mathbf{x}^T A \mathbf{z}],$$

for all $\mathbf{x}, \mathbf{y}, \mathbf{z} \in \mathbb{R}^3$.

10.6. Two-variable cubic functional equation

This section deals with the general solution of a two-variable cubic functional equation of the form

$$f(2x + y, 2u + v) + f(2x - y, 2u - v)$$
$$= 2f(x + y, u + v) + 2f(x - y, u - v) + 12f(x, u). \tag{10.19}$$

The function $f(x, y) = \alpha x^3 + \beta x^2 y + \gamma x y^2 + \delta y^3$ is a solution of the two-variable cubic functional equation (10.19).

For a mapping $g : X \to Y$, consider the cubic functional equation

$$g(2x + y) + g(2x - y) = 2g(x + y) + 2g(x - y) + 12g(x). \tag{10.20}$$

Definition 10.4. A mapping $f : X \times X \to Y$ is called *additive-quadratic* if f satisfies the system of functional equations

$$f(x + y, z) = f(x, z) + f(y, z), \tag{10.21}$$

and

$$f(x, y + z) + f(x, y - z) = 2f(x, y) + 2f(x, z). \tag{10.22}$$

Definition 10.5. A mapping $f : X \times X \to Y$ is called a *cubic form* if there exist $\alpha, \beta, \gamma, \delta \in \mathbb{R}$ such that $f(x, y) = \alpha x^3 + \beta x^2 y + \gamma x y^2 + \delta y^3$, for all $x, y \in X$.

Definition 10.6. A mapping $g : X \to Y$ is called cubic if and only if there exists a unique function C which is symmetric for each fixed one variable, and additive for each fixed two variables such that $f(x) = C(x, x, x)$ for all $x \in X$.

10.6.1. *Relation between two-variable cubic functional equation (10.19) and cubic functional equation (10.20)*

The following theorem shows that the two-variable cubic functional equation (10.19) induces the cubic functional equation (10.20).

Theorem 10.13 ([290]). *Let $f : X \times X \to Y$ be a mapping satisfying equation (10.19), and let $g : X \to Y$ be a mapping given by*

$$g(x) = f(x, x) \tag{10.23}$$

for all $x \in X$. Then g satisfies equation (10.20).

Proof. From (10.19) and (10.23), we have

$$
\begin{aligned}
g(2x + y) + g(2x - y) &= f(2x + y, 2x + y) + f(2x - y, 2x - y) \\
&= 2f(x + y, x + y) + 2f(x - y, x - y) + 12f(x, x) \\
&= 2g(x + y) + 2g(x - y) + 12g(x)
\end{aligned}
$$

for all $x, y \in X$. The proof is now completed. $\qquad\square$

In the following theorem, we prove that the cubic functional equation (10.20) induces the two-variable cubic functional equation (10.19) with an additional condition.

Theorem 10.14 ([290]). *Let $\alpha, \beta, \gamma, \delta \in \mathbb{R}$, and $g : X \to Y$ be a mapping satisfying equation (10.20). If $f : X \times X \to Y$ is a mapping given by*

$$f(x, y) = \alpha g(x) + \frac{\beta}{24}[g(2x + y) - g(2x - y) - 2g(y)]$$

$$+ \frac{\gamma}{24}[g(x + 2y) + g(x - 2y) - 2g(x)] + \delta g(y) \tag{10.24}$$

for all $x, y \in X$, then f satisfies equation (10.19).

Proof. From equations (10.20) and (10.24), we have

$$f(2x + y, 2u + v) + f(2x - y, 2u - v)$$

$$= \alpha g(2x + y) + \frac{\beta}{24}[g(4x + 2y + 2u + v) - g(4x + 2y - 2u - v) - 2g(2u + v)]$$

$$+ \frac{\gamma}{24}[g(2x + y + 4u + 2v) + g(2x + y - 4u - 2v) - 2g(2x + y)] + \delta g(2u + v)$$

$$+ \alpha g(2x - y) + \frac{\beta}{24}[g(4x - 2y + 2u - v) - g(4x - 2y - 2u + v) - 2g(2u - v)]$$

$$+ \frac{\gamma}{24}[g(2x - y + 4u - 2v) + g(2x - y - 4u + 2v) - 2g(2x - y)] + \delta g(2u - v)$$

$$= 2f(x + y, u + v) + 2f(x - y, u - v) + 12f(x, u)$$

for all $x, y, z, u, v, w \in X$. This completes the proof. $\qquad\square$

The following theorem is useful in finding the general solution of the two-variable cubic functional equation (10.19).

Theorem 10.15 ([290]). *Let $f : X \times X \to Y$ be a mapping satisfying equation (10.19). Then we have the following functional equation*

$$f(x + y + z, u + v + w) + f(x + y - z, u + v - w) + 2f(x, u) + 2f(y, v)$$
$$= 2f(x + y, u + v) + f(x + z, u + w) + f(x - z, u - w)$$
$$+ f(y + z, v + w) + f(y - z, v - w) \qquad (10.25)$$

for all $x, y, z, u, v, w \in X$.

Proof. Substituting $x = y = u = v = 0$ in equation (10.19), we obtain $f(0,0) = 0$. Putting $x = u = 0$ in (10.19), we obtain $f(-y, -v) = -f(y, v)$. Replacing (y, v) by $(0,0)$ and (y, v) by (x, u) in (10.19), we obtain $f(2x, 2u) = 8f(x, u)$, and $f(3x, 3u) = 27f(x, u)$, for all $x, u \in X$, respectively. By induction, we have $f(kx, ku) = k^3 f(x, u)$, for all positive integer k. Now, replacing (x, y) by $(x + y, x - y)$ and (u, v) by $(u + v, u - v)$ in equation (10.19), we obtain

$$f(3x+y, 3u+v)+f(x+3y, u+3v) = 12f(x+y, u+v)+16f(x, u)+16f(y, v). \quad (10.26)$$

Replacing (x, y) by $(x+y, 2y)$ and (u, v) by $(u+v, 2v)$ in equation (10.19), we obtain

$$8f(x + 2y, u + 2v) + 8f(x, u)$$
$$= 12f(x + y, u + v) + 2f(x + 3y, u + 3v) + 2f(x - y, u - v). \qquad (10.27)$$

Interchanging (x, u) with (y, v) in (10.27), adding the resulting equation with (10.27), and using (10.26), we have

$$f(x+2y, u+2v)+f(2x+y, 2u+v) = 6f(x+y, u+v)+3f(x, u)+3f(y, v). \quad (10.28)$$

Equation (10.19) can be written as

$$3f(2x + z, 2u + w) + 3f(2x - z, 2u - w)$$
$$+ 3f(2y + z, 2v + w) + 3f(2y - z, 2v - w)$$
$$= 36f(x, u) + 36f(y, v) + 6f(x + z, u + w) + 6f(x - z, u - w)$$
$$+ 6f(y + z, v + w) + 6f(y - z, v - w). \tag{10.29}$$

Replacing (x, y) by $(2x + z, 2y + z)$, and (u, v) by $(2u + w, 2v + w)$ in (10.28) and (x, y) by $(2x - z, 2y - z)$, and (u, v) by $(2u - w, 2v - w)$ in (10.28) and summing the resulting equations, and then using (10.29), we obtain

$$12f(2x + y, 2u + v) + 2f(2x + y + 3z, 2u + v + 3w)$$
$$+ 2f(2x + y - 3z, 2u + v - 32) + 12f(x + 2y, u + 2v)$$
$$+ 2f(x + 2y + 3z, u + 2v + 3w) + 2f(x + 2y - 3z, u + 2v - 3w)$$
$$= 36f(x, u) + 36f(y, v) + 6f(x + z, u + w) + 6f(x - z, u - w)$$
$$+ 6f(y + z, v + w) + 6f(y - z, v - w)$$
$$+ 48f(x + y + z, u + v + w) + 48f(x + y - z, u + v - w). \tag{10.30}$$

By virtue of (10.28) and (10.19), the left-hand side of (10.29) can be written as

$$3f(2x + z, 2u + w) + 3f(2y - z, 2v - w)$$
$$+ 3f(2x - z, 2u - w) + 3f(2y + z, 2v + w)$$
$$= 2f(2x + y + z, 2u + v + w) + 2f(2x + y - z, 2u + v - w)$$
$$+ 12f(2x + y, 2u + v) + 2f(x + 2y + z, u + 2v + w)$$
$$+ 2f(x + 2y - z, u + 2v - w) + 12f(x + 2y, u + 2v)$$
$$- 96f(x + y, u + v). \tag{10.31}$$

Replacing z by $3z$ and w by $3w$ in (10.31), and then using (10.30), we have

$$3f(2x + 3z, 2u + 3w) + 3f(2y - 3z, 2v - 3w)$$
$$+ 3f(2x - 3z, 2u - 3w) + 3f(2y + 3z, 2v + 3w)$$
$$= 36f(x, u) + 36f(y, v) + 6f(x + z, u + w)$$
$$+ 6f(x - z, u - w) + 6f(y + z, v + w) + 6f(y - z, v - w)$$
$$+ 48f(x + y + z, u + v + w) + 48f(x + y - z, u + v - w)$$
$$- 96f(x + y, u + v). \tag{10.32}$$

Replacing (x, y, u, v) by $(2x+3z, 2x-3z, 2u+3w, 2u-3w)$ and $(2y+3z, 2y-3z, 2v+3w, 2v-3w)$ in (10.28), and applying the resulting equations to the left-hand side of (10.32), we have

$$
\begin{aligned}
3f(2x + 3z, &\, 2u + 3w) + 3f(2y - 3z, 2v - 3w) \\
&+ 3f(2x - 3z, 2u - 3w) + 3f(2y + 3z, 2v + 3w) \\
&= 27[2f(x + z, u + w) + 2f(x - z, u - w) + 12f(x, u)] \\
&\quad + 27[2f(y + z, v + w) + 2f(y - z, v - w)] + 12f(y, v) \\
&\quad - 384f(x, u) - 384f(y, v).
\end{aligned}
\tag{10.33}
$$

From the right-hand side of equations (10.32) and (10.33), we arrive at the equation (10.25). The proof is now completed. $\qquad\square$

10.6.2. *General solution of two-variable cubic functional equation*

The following theorem contains the general solution of the two-variable cubic functional equation (10.19).

Theorem 10.16 ([290]). *Let $f : X \times X \to Y$ be a function. Then f satisfies (10.19) if and only if there exist two mappings $C_1, C_2 : X \times X \times X \to Y$, which are symmetric for each one fixed variable and additive for each two fixed variables, and an additive-quadratic mapping $B : X \times X \to Y$ such that*

$$
f(x, y) = C_1(x, x, x) + B(y, x) + B(x, y) + C_2(y, y, y)
\tag{10.34}
$$

for all $x, y \in X$.

Proof. Assume that f is a solution of equation (10.19). Define $f_1(x) = f(x, 0)$, and $f_2(y) = f(0, y)$, for all $x, y \in X$. It is easy to verify that f_1 and f_2 are cubic functions. Then there exist two mappings $C_1, C_2 : X \times X \times X \to Y$ which are symmetric for each fixed variable, and additive for two fixed variables such that $f_1(x) = C_1(x, x, x)$, and $f_2(y) = C_2(y, y, y)$. Using the above properties, C_1 and C_2 satisfy

$$
\begin{aligned}
C_i(2x + y, &\, 2x + y, 2x + y) \\
&= C_i(2x + y, 2x + y, 2x) + C_i(2x + y, 2x + y, y) \\
&= 8C_i(x, x, x) + 12C_i(x, x, y) + 6C_i(y, y, x) + C_i(y, y, y)
\end{aligned}
\tag{10.35}
$$

for $i = 1, 2$. Define $B : X \times X \to Y$ by

$$
B(x, y) = \frac{1}{6}[f(x + y, x + y) + f(x - y, x - y) - 2f(x, x)]
\tag{10.36}
$$

for all $x, y \in X$. Replacing (y, v) by $(-y, -v)$ in (10.25), we have

$$
\begin{aligned}
f(x - y + z, &\, u - v + w) + f(x - y - z, u - v - w) + 2f(x, u) - 2f(y, v) \\
&= 2f(x - y, u - v) + f(x + z, u + w) + f(x - z, u - w) \\
&\quad - f(y - z, v - w) - f(y + z, v + w).
\end{aligned}
\tag{10.37}
$$

The functional equations (10.25) and (10.37) are useful to show B_1 and B_2 are additive-quadratic mappings. From (10.36) and (10.25), we have that

$$B(x_1, x_2, y) = \frac{1}{6}[f(x_1 + x_2 + y, x_1 + x_2 + y) + f(x_1 + x_2 - y, x_1 + x_2 - y)$$

$$- 2f(x_1 + x_2, x_1 + x_2)]$$

$$= \frac{1}{6}[f(x_1 + y, x_1 + y) + f(x_1 - y, x_1 - y) - 2f(x_1, x_1)]$$

$$+ \frac{1}{6}[f(x_2 + y, x_2 + y) + f(x_2 - y, x_2 - y) - 2f(x_2, x_2)]$$

$$= B(x_1, y) + B(x_2, y).$$

Also,

$$B(x, y_1 + y_2) + B(x, y_1 - y_1)$$

$$= \frac{1}{6}[f(x + y_1 + y_2, x + y_1 + y_2) + f(x - y_1 - y_2, x - y_1 - y_2) - 2f(x, x)]$$

$$+ \frac{1}{6}[f(x + y_1 - y_2, x + y_1 - y_2) + f(x - y_1 + y_2, x - y_1 + y_2) - 2f(x, x)]$$

$$= 2B(x, y_1) + 2B(x, y_2).$$

Hence B is an additive-quadratic function.

Conversely, assume that there exist two mappings $C_1, C_2 : X \times X \times X \to Y$, which are symmetric for each fixed variable and additive for fixed two variables, and an additive-quadratic mapping $B : X \times X \to Y$ such that

$$f(x, y) = C_1(x, x, x) + B(y, x) + B(x, y) + C_2(y, y, y)$$

for all $x, y \in X$. Since B is additive-quadratic, B satisfies (10.21) and (10.22). Hence we have

$$B(x + y, z) = B(x, z) + B(y, z) \tag{10.38}$$

and

$$B(x, y + z) + B(x, y - z) = 2B(x, y) + 2B(x, z). \tag{10.39}$$

Putting $x = y = 0$ in equation (10.38), we obtain

$$B(0, z) = 0. \tag{10.40}$$

Substituting $y = z = 0$ in (10.39), we obtain $B(x, 0) = 0$. Replacing y by x in (10.38), we have $B(2x, z) = 2B(x, z)$. Replacing z by y in (10.39), we arrive at $B(x, 2y) = 4B(x, y)$. Taking $y = 0$ in (10.39), it leads to

$$B(x, -z) = B(x, z). \tag{10.41}$$

Replacing y by $-x$ in (10.38), and using (10.40), we obtain

$$B(-x, z) = -B(x, z). \tag{10.42}$$

Now, replacing y by $y + z$ in (10.40), we have

$$B(x, y + 2z) = 2B(x, y + z) + 2B(x, z) - B(x, y). \tag{10.43}$$

Replacing (x, y) by $(-x, -y)$ in (10.43), and using (10.41) and (10.42), we obtain

$$B(-x, 2z - y) = 2B(-x, z - y) - 2B(x, z) + B(x, y). \tag{10.44}$$

Adding (10.43) with (10.44), we obtain

$$B(x, 2z + y) + B(-x, 2z - y) = 2B(x, z + y) + 2B(-x, z - y). \tag{10.45}$$

Now, using (10.35), we have

$$
\begin{aligned}
f(2x + y, &2u + v) + f(2x - y, 2u - v) \\
&= 8C_1(x, x, x) + 12C_1(x, x, y) + 6C_1(x, y, y) \\
&\quad + C_1(y, y, y) + B(2u, 2x + y) + B(v, 2x + y) \\
&\quad + B(2x, 2u + v) + B(y, 2u + v) + 8C_2(u, u, u) \\
&\quad + 12C_2(u, u, v) + 6C_2(u, v, v) + C_2(u, u, u) \\
&\quad + 8C_1(x, x, x) + 12C_1(x, x, -y) + 6C_1(x, -y, -y) \\
&\quad + C_1(-y, -y, -y) + B(2u, 2x - y) + B(-v, 2x - y) \\
&\quad + B(2x, 2u - v) + B(-y, 2u - v) + 8C_2(u, u, u) \\
&\quad + 12C_2(u, u, -v) + 6C_2(u, -v, -v) + C_2(-v, -v, -v). \tag{10.46}
\end{aligned}
$$

Using (10.39) in (10.45), we obtain

$$
\begin{aligned}
f(2x + y, &2u + v) + f(2x - y, 2u - v) \\
&= 16C_1(x, x, x) + 12C_1(x, y, y) + 16C_2(u, u, u) \\
&\quad + 12C_2(u, v, v) + 16B(u, x) + 4B(u, y) \\
&\quad + 16B(x, u) + 4B(x, v) + 2B(v, x + y) \\
&\quad + 2B(-v, x - y) + 2B(y, u + v) + 2B(-y, u - v). \tag{10.47}
\end{aligned}
$$

Then,

$$
\begin{aligned}
2f(x + y, &u + v) + 2f(x - y, u - v) + 12f(x, u) \\
&= 2C_1(x, x, x) + 6C_1(x, x, y) + 6C_1(x, y, y) \\
&\quad + 2C_1(y, y, y) + 2B(u + v, x + y) + 2B(x + y, u + v)
\end{aligned}
$$

$$+ 2C_2(u, u, u) + 6C_2(u, u, v) + 6C_2(u, v, v)$$
$$+ 2C_2(v, v, v) + 2C_1(x, x, x) + 6C_1(x, x, -y)$$
$$+ 6C_1(x, -y - y) + 2C_1(-y, -y, -y) + 2B(u - v, x - y)$$
$$+ 2B(x - y, u - v) + 2C_2(u, u, u) + 6C_2(u, u, -v)$$
$$+ 6C_2(u, -v, -v) + 2C_2(-v, -v, -v) + 12C_1(x, x, x)$$
$$+ 12B(u, x) + 12B(x, u) + 12C_2(u, u, u). \tag{10.48}$$

Using (10.38) and (10.39) in (10.48), we have

$$2f(x + y, u + v) + 2f(x - y, u - v) + 12f(x, u)$$
$$= 16C_1(x, x, x) + 12C_1(x, y, y) + 16C_2(u, u, u)$$
$$+ 12C_2(u, v, v) + 2B(u, x + y) + 2B(v, x + y) + 2B(x, u + v)$$
$$+ 2B(y, u + v) + 2B(u, x - y) + 2B(-v, x - y) + 2B(x, u - v)$$
$$+ 2B(-y, u - v) + 12B(u, x) + 12B(x, u) \tag{10.49}$$
$$= 16C_1(x, x, x) + 12C_1(x, y, y) + 16C_2(u, u, u)$$
$$+ 12C_2(u, v, v) + 16B(u, x) + 4B(u, y)$$
$$+ 16B(x, u) + 4B(x, v) + 2B(v, x + y)$$
$$+ 2B(-v, x - y) + 2B(y, u + v) + 2B(-y, u - v). \tag{10.50}$$

From (10.47) and (10.49), we see that f satisfies equation (10.19), which completes the proof of the theorem. $\qquad\square$

Chapter 11

The Ulam Stability Problem

11.1. Introduction

So far, we have discussed about various types of functional equations involving one variable, two variables, three variables, and several variables in the previous chapters. In this chapter, we discuss the famous problem posed by the mathematician S.M. Ulam concerning the stability of functional equation, and the subsequent answers given by many mathematicians like D.H. Hyers, Th.M. Rassias, T. Aoki, J.M. Rassias, and P. Gavruta. These results are provided in detailed manner in this chapter.

Very often instead of a functional equation, we consider a functional inequality, and one can ask the following question: *When can one assert that the solutions of the inequality lie near to the solutions of the equation?*

A definition of stability in the case of homomorphisms between groups was suggested by a problem posed by Ulam [336] in 1940: Let $(G_1, \star)$ be a group and let $(G_2, \diamond, d)$ be a metric group with the metric $d(\cdot, \cdot)$. Given $\epsilon > 0$, does there exist a $\delta(\epsilon) > 0$ such that if a mapping $h : G_1 \to G_2$ satisfies the inequality

$$d(h(x \star y), h(x) \diamond h(y)) < \delta$$

for all $x, y \in G_1$, then there is a homomorphism $H : G_1 \to G_2$ with

$$d(h(x), H(x)) < \epsilon$$

for all $x \in G_1$? In other words, if a mapping is almost homomorphism then there is a true homomorphism near it with small error as much as possible. If the answer is affirmative, we would call that the equation $H(x \star y) = H(x) \diamond H(y)$ of homomorphism is stable. The concept of stability for a functional equation arises when we replace the functional equation by an inequality which acts as a perturbation of the equation. Thus the stability question of a functional equation is that how the solution of the pertinent inequality differs from that of the given functional equation? These kinds of questions form the basis of the Ulam stability theory.

11.2. Hyers–Ulam stability

In 1941, Hyers [122] was the first mathematician to present the result concerning the stability of functional equations. He brilliantly answered the question of Ulam for the cases where G and H are assumed to be Banach spaces. The result of Hyers is stated in the following celebrated theorem.

Theorem 11.1 ([122]). *Let E and E' be Banach spaces, and let $f(x)$ be a δ-linear transformation of E into E'. Then the limit*

$$A(x) = \lim_{n \to \infty} \frac{f(2^n x)}{2^n}$$

exists for each $x \in E$, $A(x)$ is a linear transformation, and the inequality

$$\|f(x) - A(x)\| \le \delta$$

is true for all x in E. Moreover, $A(x)$ is the only linear transformation satisfying this inequality.

(*or*)

Let E and E' be Banach spaces. Let $f : E \to E'$ be a function satisfying

$$\|f(x + y) - f(x) - f(y)\| \le \delta \tag{11.1}$$

for some $\delta > 0$ and for all $x, y \in E$. Then there exists a unique additive function $A : E \to E'$ such that

$$\|f(x) - A(x)\| \le \delta \tag{11.2}$$

for all $x \in E$.

Proof. To prove this theorem, we have to show that the following conditions hold:

  (i) $\{\frac{f(2^n x)}{2^n}\}$ is a Cauchy sequence for every fixed $x \in E$.
 (ii) If $A(x) = \lim_{n \to \infty} \frac{f(2^n x)}{2^n}$, then A is additive on E.
(iii) Further A satisfies $\|f(x) - A(x)\| \le \delta$, for all $x \in E$.
(iv) A is unique.

(a) **To show (i):** Letting $y = x$ in (11.1), we obtain

$$\|f(2x) - 2f(x)\| \le \delta \tag{11.3}$$

for every $x \in E$. Replacing x by $2^{k-1}x$ in (11.3), where k is a positive integer greater than or equal to 1, we get

$$\left\| f(2^k x) - 2f(2^{k-1}x) \right\| \le \delta$$

for every $x \in E$, and $k = 1, 2, \ldots, n$ and $n \in \mathbb{N}$. Multiplying both sides of the above inequality by $\frac{1}{2^k}$ and adding the resulting n inequalities, we have

$$\sum_{k=1}^{n} \frac{1}{2^k} \left\| f\left(2^k x\right) - 2f\left(2^{k-1} x\right) \right\| \le \sum_{k=1}^{n} \frac{1}{2^k} \delta \le \delta \left(1 - \frac{1}{2^n}\right). \tag{11.4}$$

Using (11.4), we get

$$\left\| \frac{1}{2^n} f\left(2^n x\right) - f(x) \right\| = \left\| \sum_{k=1}^{n} \frac{1}{2^k} \left(f\left(2^k x\right) - 2f\left(2^{k-1} x\right)\right) \right\|$$

$$\le \sum_{k=1}^{n} \frac{1}{2^k} \left\| f\left(2^k x\right) - 2f\left(2^{k-1} x\right) \right\|$$

$$\le \delta \left(1 - \frac{1}{2^n}\right) \tag{11.5}$$

for all $x \in E$ and $n \in \mathbb{N}$. Using induction, we show that (11.5) holds for all positive integers $n \in \mathbb{N}$, since

$$\left\| \frac{1}{2} f(2x) - f(x) \right\| = \frac{1}{2} \left\| f(2x) - 2f(x) \right\| \le \frac{\delta}{2}.$$

Hence (11.5) is true for $n = 1$. Suppose that (11.5) is true for $n = k$, we get

$$\left\| \frac{1}{2^k} f\left(2^k x\right) - f(x) \right\| \le \delta \left(1 - \frac{1}{2^k}\right).$$

We show that (11.5) is true for $n = k + 1$.

$$\left\| \frac{1}{2^{k+1}} f\left(2^{k+1} x\right) - f(x) \right\| = \frac{1}{2} \left\| \frac{1}{2^k} f\left(2^k y\right) - 2f\left(\frac{y}{2}\right) \right\|$$

$$\le \frac{1}{2} \left(\left\| \frac{1}{2^k} f\left(2^k y\right) - f(y) \right\| + \left\| f(y) - 2f\left(\frac{y}{2}\right) \right\| \right)$$

$$\le \left(1 - \frac{1}{2^{k+1}}\right) \delta.$$

Now, if $n > m > 0$, then $n - m$ is a natural number, and n can be replaced by $n - m$ in (11.5) to get

$$\left\| \frac{f\left(2^{n-m} x\right)}{2^{n-m}} - f(x) \right\| \le \delta \left(1 - \frac{1}{2^{n-m}}\right).$$

Multiplying both sides by $\frac{1}{2^m}$ and simplifying, we get

$$\left\| \frac{f\left(2^{n-m} x\right)}{2^n} - \frac{f(x)}{2^m} \right\| \le \frac{\delta}{2^m} \left(1 - \frac{1}{2^{n-m}}\right)$$

for all $x \in E$. Now, we replace x by $2^m x$ in the above equation, and then we obtain

$$\left\| \frac{f(2^n x)}{2^n} - \frac{f(2^m x)}{2^m} \right\| \leq \delta \left(\frac{1}{2^m} - \frac{1}{2^n} \right).$$

If $m \to \infty$, then $\left(\frac{1}{2^m} - \frac{1}{2^n} \right) \to 0$ and therefore, we have

$$\lim_{m \to \infty} \left\| \frac{f(2^n x)}{2^n} - \frac{f(2^m x)}{2^m} \right\| = 0.$$

Hence $\left\{ \frac{f(2^n x)}{2^n} \right\}$ is a Cauchy sequence in E. Hence the limit of this sequence exists. Define $A : E \to E'$ by

$$A(x) = \lim_{n \to \infty} \frac{f(2^n x)}{2^n}. \tag{11.6}$$

(b) **To show that $A : E \to E'$ defined by (11.6) is additive.** Consider

$$\|A(x + y) - A(x) - A(y)\|$$

$$= \lim_{n \to \infty} \left\| \frac{f(2^n(x + y))}{2^n} - \frac{f(2^n x)}{2^n} - \frac{f(2^n y)}{2^n} \right\|$$

$$= \lim_{n \to \infty} \frac{1}{2^n} \|f(2^n(x + y)) - f(2^n x) - f(2^n y)\|$$

$$\leq \lim_{n \to \infty} \frac{\delta}{2^n}, \qquad \text{using (11.1)}$$

$$= 0.$$

Therefore, $A(x + y) = A(x) + A(y)$, for all $x, y \in E$.

(c) **To show that $\|A(x) - f(x)\| \leq \delta$.** Consider

$$\|A(x) - f(x)\| = \left\| \lim_{n \to \infty} \frac{f(2^n x)}{2^n} - f(x) \right\| \leq \lim_{n \to \infty} \delta \left(1 - \frac{1}{2^n} \right) = \delta.$$

Hence, we obtain $\|A(x) - f(x)\| \leq \delta$, for all $x \in E$.

(d) **To show that A is unique.** Suppose A is not unique. Then there exists another additive function $a : E \to E'$ such that

$$\|a(x) - f(x)\| \leq \delta$$

for all $x \in E$. Now,

$$\|a(x) - A(x)\| = \|a(x) - f(x) + f(x) - A(x)\|$$

$$\leq \|a(x) - f(x)\| + \|f(x) - A(x)\|$$

$$= \delta + \delta$$

$$= 2\delta. \tag{11.7}$$

Since A and a are additive, we have

$$\|A(x) - a(x)\| = \left\| \frac{nA(x)}{n} - \frac{na(x)}{n} \right\|$$

$$= \left\| \frac{A(nx)}{n} - \frac{a(nx)}{n} \right\|$$

$$= \frac{1}{n} \|A(nx) - a(nx)\|$$

$$\leq \frac{2\delta}{n}$$

where $n \in \mathbb{N}$. Taking the limit on both sides, we obtain

$$\lim_{n \to \infty} \|A(x) - a(x)\| \leq \lim_{n \to \infty} \frac{2\delta}{n}$$

$$\leq 0.$$

Hence $A(x) = a(x)$, for all $x \in E$.

Therefore, the additive map A is unique which completes the proof of the theorem.

□

Remark 11.1. Taking the above fact into account, the additive functional equation $f(x + y) = f(x) + f(y)$ is said to have **Hyers–Ulam stability** on (E, E'). In the above theorem, an additive function A is constructed directly from the given function f and it is the most powerful tool to study the stability of several functional equations. This method of proving stability is known as **Direct method** and this method is normally applied to test the Hyers–Ulam–Rassias stability for any type of functional equation.

11.3. Hyers–Ulam–Aoki stability

In 1950, Aoki [8] generalized the Hyers theorem for additive mappings. In generalizing the theorem of Hyers, a transformation $f(x)$ from E into E' is called "approximately linear", when there exist k_0 (≥ 0) and p $(0 \leq p < 1)$ such that

$$\|f(x + y) - f(x) - f(y)\| \leq k_0 (\|x\|^p + \|y\|^p)$$

for any x and y in E.

Let $f(x)$ and $\varphi(x)$ be transformations from E into E'. These are called "near", when there exist k (≥ 0), and p $(0 \leq p < 1)$ such that

$$\|f(x) - \varphi(x)\| \leq k \|x\|^p$$

for any $x \in E$.

Theorem 11.2. ([8]). *If $f(x)$ is an approximately linear transformation from E into E', then there is a linear transformation $\varphi(x)$ near $f(x)$. And such $\varphi(x)$ is unique.*

Proof. By the assumption there are k_0 (≥ 0) and p ($0 \leq p < 1$) such that

$$\left\| \frac{f(2x)}{2} - f(x) \right\| \leq k_0 \|x\|^p. \tag{11.8}$$

We shall now prove that

$$\left\| \frac{f(2^n x)}{2^n} - f(x) \right\| \leq k_0 \|x\|^p \sum_{i=0}^{n-1} 2^{i(p-1)} \tag{11.9}$$

using induction for any integer n.

When $n = 1$, inequality (11.9) clearly holds using (11.8). Assuming the case for n, we shall prove the case for $n + 1$. Replacing x by $2x$ in (11.9), we get

$$\left\| \frac{f(2^{n+1} x)}{2^n} - f(2x) \right\| \leq k_0 \|x\|^p 2^p \sum_{i=0}^{n-1} 2^{i(p-1)}$$

$$\leq 2^p k_0 \|x\|^p \sum_{i=1}^{n} 2^{i(p-1)}.$$

Using (11.8), we have

$$\left\| \frac{f(2^{n+1} x)}{2^{n+1}} - f(x) \right\| \leq \left\| \frac{f(2^{n+1} x)}{2^{n+1}} - \frac{f(2x)}{2} \right\| + \left\| \frac{f(2x)}{2} - f(x) \right\|$$

$$\leq k_0 \|x\|^p \sum_{i=0}^{n} 2^{i(p-1)}.$$

Hence (11.9) is true for the case $n + 1$. Therefore (11.9) holds for any positive integer n.

Since $0 \leq p < 1$, $\sum_{i=0}^{\infty} 2^{i(p-1)}$ converges to $\frac{2}{2-2^p}$. Hence (11.9) becomes,

$$\left\| \frac{f(2^n x)}{2^n} - f(x) \right\| \leq K \|x\|^p, \tag{11.10}$$

where $K = \frac{2k_0}{2-2^p}$.

Let us consider the sequence $\left\{ \frac{f(2^n x)}{2^n} \right\}$. We have

$$\left\| \frac{f(2^m x)}{2^m} - \frac{f(2^n x)}{2^n} \right\| = \frac{1}{2^n} \left\| \frac{f(2^m x)}{2^{m-n}} - f(2^n x) \right\|$$

$$\leq \frac{1}{2^n} K \|2^n x\|^p$$

$$\leq 2^{n(p-1)} K \|x\|^p$$

$$\rightarrow 0 \qquad \text{as } n \rightarrow \infty.$$

Since E' is complete, the sequence $\{\frac{f(2^n x)}{2^n}\}$ converges. Taking $\varphi(x) = \lim_{n \rightarrow \infty} \frac{f(2^n x)}{2^n}$, then $\varphi(x)$ is linear. By the appropriate linearity of $f(x)$

$$\|f(2^n(x+y)) - f(2^n x) - f(2^n y)\| \leq k_0 (\|2^n x\|^p + \|2^n y\|^p)$$
$$\leq 2^{np} k_0 (\|x\|^p + \|y\|^p).$$

Dividing both sides by 2^n and letting $n \rightarrow \infty$, we obtain

$$\varphi(x+y) = \varphi(x) + \varphi(y),$$

which shows that φ is linear. Letting $n \rightarrow \infty$ in (11.10), we obtain

$$\|\varphi(x) - f(x)\| \leq K \|x\|^p \qquad (11.11)$$

which shows that $\varphi(x)$ is near $f(x)$.

To prove the uniqueness of $\varphi(x)$, let $\psi(x)$ be another linear transformation near $f(x)$. Then there exist k' (≥ 0) and p' ($0 \leq p' < 1$) such that

$$\|\psi(x) - f(x)\| \leq k' \|x\|^{p'}. \qquad (11.12)$$

Using (11.11), we have

$$\|\varphi(x) - \psi(x)\| \leq k \|x\|^p + k' \|x\|^{p'}.$$

By the linearity of φ and ψ, we have

$$\|\varphi(x) - \psi(x)\| = \frac{1}{n} \|\varphi(nx) - \psi(nx)\|$$

$$\leq \frac{1}{n} \left(k \|nx\|^p + k' \|nx\|^{p'} \right)$$

$$\leq \frac{1}{n^{1-p}} k \|x\|^p + \frac{1}{n^{1-p'}} k' \|x\|^{p'} \rightarrow 0 \quad \text{as } n \rightarrow \infty.$$

Therefore, $\varphi(x) = \psi(x)$, which completes the proof of the theorem. $\square$

Remark 11.2. The result obtained by Aoki in Theorem 11.2 is called **Hyers–Ulam–Aoki Stability** of additive functional equation.

11.4. Hyers–Ulam–Rassias stability

In the year 1978, Rassias [298] tried to weaken the condition for the Cauchy difference and succeeded in proving what is now known to be the **Hyers–Ulam–Rassias stability** for the additive Cauchy functional equation. This terminology is justified because the theorem of Rassias has strongly influenced mathematicians studying stability problems of functional equation. In fact, Rassias proved the following theorem.

Theorem 11.3 ([298]). *Consider E_1 and E_2 be two Banach spaces, and let $f : E_1 \to E_2$ be a mapping such that $f(tx)$ is continuous in t for each fixed x. Assume that there exist $\theta \geq 0$ and $p \in [0,1)$ such that*

$$\frac{\|f(x+y) - f(x) - f(y)\|}{\|x\|^p + \|y\|^p} \leq \theta, \tag{11.13}$$

for any $x, y \in E_1$. Then there exists a unique linear mapping $T : E_1 \to E_2$ such that

$$\frac{\|f(x) - T(x)\|}{\|x\|^p} \leq \frac{2\theta}{2 - 2^p}, \tag{11.14}$$

for any $x \in E_1$.

Proof. Claim that

$$\frac{1}{\|x\|^p} \left\| \frac{f(2^n x)}{2^n} - f(x) \right\| \leq \theta \sum_{m=0}^{n-1} 2^{m(p-1)} \tag{11.15}$$

for any integer n, and some $\theta \geq 0$. The verification of (11.15) follows by induction on n.

When $n = 1$, (11.15) clearly holds because by the hypothesis we can find θ, that is greater than or equal to zero and p such that $0 \leq p < 1$ with

$$\frac{1}{\|x\|^p} \left\| \frac{f(2x)}{2} - f(x) \right\| \leq \theta. \tag{11.16}$$

Assume now that (11.15) holds for n, and we want to prove it for the case $n + 1$. However, this is true because by (11.15), we obtain

$$\frac{1}{\|2x\|^p} \left\| \frac{f(2^n.2x)}{2^n} - f(2x) \right\| \leq \theta \sum_{m=0}^{n-1} 2^{m(p-1)},$$

therefore

$$\frac{1}{\|x\|^p} \left\| \frac{f(2^{n+1}x)}{2^{n+1}} - \frac{1}{2}f(2x) \right\| \leq \theta \sum_{m=1}^{n} 2^{m(p-1)}.$$

By the triangle inequality, we obtain

$$\left\| \frac{1}{2^{n+1}} f(2^{n+1}x) - f(x) \right\| \leq \left\| \frac{1}{2^{n+1}} f(2^{n+1}x) - \frac{1}{2}f(2x) \right\| + \left\| \frac{1}{2}f(2x) - f(x) \right\|$$

$$\leq \theta \|x\|^p \sum_{m=0}^{n} 2^{m(p-1)}.$$

Thus,

$$\frac{1}{\|x\|^p} \left\| \frac{f\left(2^{n+1}x\right)}{2^{n+1}} - f(x) \right\| \leq \theta \sum_{m=0}^{n} 2^{m(p-1)},$$

and (11.15) is valid for any integer n. It follows then that

$$\frac{1}{\|x\|^p} \left\| \frac{f\left(2^n x\right)}{2^n} - f(x) \right\| \leq \frac{2\theta}{2 - 2^p}, \qquad (11.17)$$

because $\sum_{m=0}^{\infty} 2^{m(p-1)}$ converges to $\frac{2}{2-2^p}$, as $0 \leq p < 1$.

However, for $m > n > 0$,

$$\left\| \frac{1}{2^m} f\left(2^m x\right) - \frac{1}{2^n} f\left(2^n x\right) \right\| = \frac{1}{2^n} \left\| \frac{1}{2^{m-n}} f\left(2^m x\right) - f\left(2^n x\right) \right\|$$

$$< 2^{n(p-1)} \cdot \frac{2\theta}{2 - 2^p} \|x\|^p.$$

Therefore,

$$\lim_{n \to \infty} \left\| \frac{1}{2^m} f\left(2^m x\right) - \frac{1}{2^n} f\left(2^n x\right) \right\| = 0.$$

Since E_2 is a Banach space, which is complete, the sequence $\{\frac{f(2^n x)}{2^n}\}$ converges. Define $T(x) = \lim_{n \to \infty} \frac{1}{2^n} f\left(2^n x\right)$. It follows that

$$\|f\left(2^n(x+y)\right) - f\left(2^n x\right) - f\left(2^n y\right)\| \leq \theta \left(\|2^n x\|^p + \|2^n y\|^p\right)$$

$$= 2^{np}\theta \left(\|x\|^p + \|y\|^p\right).$$

Therefore,

$$\frac{1}{2^n} \|f\left(2^n(x+y)\right) - f\left(2^n x\right) - f\left(2^n y\right)\| \leq 2^{n(p-1)}\theta \left(\|2^n x\|^p + \|2^n y\|^p\right)$$

or

$$\lim_{n \to \infty} \frac{1}{2^n} \|f\left(2^n(x+y)\right) - f\left(2^n x\right) - f\left(2^n y\right)\| \leq \lim_{n \to \infty} 2^{n(p-1)}\theta \left(\|x\|^p + \|y\|^p\right)$$

or

$$\left\| \lim_{n \to \infty} \frac{1}{2^n} f\left(2^n(x+y)\right) - \lim_{n \to \infty} \frac{1}{2^n} f\left(2^n x\right) - \lim_{n \to \infty} \frac{1}{2^n} f\left(2^n y\right) \right\| = 0$$

or

$$\|T(x+y) - T(x) - T(y)\| = 0 \quad \text{for any } x, y \in E_1$$

or

$$T(x+y) = T(x) + T(y) \quad \text{for all } x, y \in E_1.$$

Since $T(x+y) = T(x) + T(y)$ for any $x, y \in E_1$, $T(rx) = rT(x)$ for any rational number r. Next, to prove that T is the unique such linear mapping. Assume that there exist another one, denoted by $g : E_1 \to E_2$ such that $T(x) \neq g(x)$, $x \in E_1$. Then there exist a constant ϵ_1, greater than or equal to zero, and q such that $0 \leq q < 1$ with

$$\frac{\|g(x) - f(x)\|}{\|x\|^q} \leq \epsilon_1.$$

By the triangle inequality, we obtain

$$\|T(x) - g(x)\| \leq \|T(x) - f(x)\| + \|f(x) - g(x)\|$$
$$\leq \epsilon \|x\|^p + \epsilon_1 \|x\|^q.$$

Therefore,

$$\|T(x) - g(x)\| = \left\| \frac{1}{n} T(nx) - \frac{1}{n} g(nx) \right\|$$

$$= \frac{1}{n} \|T(nx) - g(nx)\|$$

$$\leq \frac{1}{n} \left(\epsilon \|nx\|^p + \epsilon_1 \|nx\|^q \right)$$

$$= n^{p-1} \epsilon \|x\|^p + n^{q-1} \epsilon_1 \|x\|^q.$$

Thus, $\lim_{n \to \infty} \|T(x) - g(x)\| = 0$ for all $x \in E_1$ and hence $T(x) = g(x)$ for all $x \in E_1$, which completes the proof. $\square$

11.5. Gajda's counterexample

Rassias noticed that the proof of this theorem also works for $p < 0$ and asked whether such a theorem can also be proved for $p \geq 1$. Gajda answered the question of Rassias for the case $p > 1$ by a slight modification in the expression of $T(x)$ as $T(x) = \lim_{n \to \infty} 2^n f(\frac{x}{2^n})$. His idea to prove the theorem for this case is to replace n by $-n$ in the formula.

It turns out that 1 is the only critical value of p to which Rassias theorem cannot be extended. Gajda [83] showed that theorem is false for $p = 1$ by constructing the following counterexample.

For a fixed $\theta > 0$ and $\mu = \frac{1}{6}\theta$, define a function $f : \mathbb{R} \to \mathbb{R}$ by

$$f(x) = \sum_{n=0}^{\infty} 2^{-n} \phi(2^n x),$$

where the function $\phi : \mathbb{R} \to \mathbb{R}$ given by

$$\phi(x) = \begin{cases} \mu & \text{for } x \in [1, \infty), \\ \mu x & \text{for } x \in (-1, 1), \\ -\mu & \text{for } x \in (-\infty, -1]. \end{cases}$$

Then the function f serves as a counterexample for $p = 1$ as presented in the following theorem.

Theorem 11.4 ([83]). *The function f defined above satisfies*

$$|f(x+y) - f(x) - f(y)| \leq \theta(|x| + |y|) \tag{11.18}$$

for all $x, y \in \mathbb{R}$, while there are no constant $\delta \geq 0$, and no additive function $A : \mathbb{R} \to \mathbb{R}$ satisfying the condition

$$|f(x) - A(x)| \leq \delta|x| \tag{11.19}$$

for all $x, y \in \mathbb{R}$.

Proof. Since f is bounded by means of a uniformly convergent series of continuous functions, f itself is continuous. Moreover,

$$|f(x)| \leq \sum_{n=0}^{\infty} \frac{\mu}{2^n} = 2\mu, \quad x \in \mathbb{R}.$$

We are going to show that f satisfies (11.18). If $x = y = 0$, then (11.18) is trivially true. Next, assume that $0 < |x| + |y| < 1$. Then there exists an $n \in \mathbb{N}$ such that

$$\frac{1}{2^n} \leq |x| + |y| < \frac{1}{2^{n-1}}.$$

Hence, $|2^{n-1}x|$, $|2^{n-1}y|$ and $|2^{n-1}(x+y)| \leq 2^{n-1}(|x| + |y|) < 1$, which imply that for each $n \in \{0, 1, 2, \ldots, N-1\}$ the numbers $2^n x$, $2^n y$ and $2^n(x+y)$ remain in the interval $(-1, 1)$. Since ϕ is linear on this interval, we obtain

$$\phi\left(2^n(x+y)\right) - \phi\left(2^n x\right) - \phi\left(2^n y\right) = 0$$

for $n = 0, 1, 2, \ldots, N-1$. Therefore, we get

$$\frac{|f(x+y) - f(x) - f(y)|}{(|x| + |y|)} \leq \sum_{n=N}^{\infty} \frac{|\phi\left(2^n(x+y)\right) - \phi\left(2^n x\right) - \phi\left(2^n y\right)|}{2^n(|x| + |y|)}$$

$$\leq \sum_{k=0}^{\infty} \frac{3\mu}{2^k 2^N (|x| + |y|)}$$

$$\leq \sum_{k=0}^{\infty} \frac{3\mu}{2^k} = 6\mu = \epsilon.$$

Finally, assume that $|x| + |y| \geq 1$. Then merely by virtue of the boundedness of f, we have

$$\frac{|f(x+y) - f(x) - f(y)|}{|x| + |y|} \leq 6\mu = \epsilon.$$

Thus, we conclude that f satisfies (11.18) for all $x, y \in \mathbb{R}$. Suppose that there exist a $\delta \in [0, \infty)$, and an additive function $T : \mathbb{R} \to \mathbb{R}$ such that (11.19) holds true. Hence, from the continuity of f it follows that T is bounded on some neighborhood of zero. There exists a real constant c such that $T(x) = cx$, $x \in \mathbb{R}$. Hence,

$$|f(x) - cx| \leq \delta|x|, \quad x \in \mathbb{R},$$

which implies that

$$\left| \frac{f(x)}{x} \right| \leq \delta + |c|, \quad x \in \mathbb{R}.$$

On the other hand, we can choose $n \in \mathbb{N}$ so large that $N\mu > \delta + |x|$. Then for any $x \in \left(0, \frac{1}{2^{N-1}}\right)$, we have $2^n x \in (0, 1)$ for each $n \in \{0, 1, 2, \ldots, N-1\}$. Consequently, for such an x, we have

$$\frac{f(x)}{x} \geq \sum_{n=0}^{\infty} \frac{\phi(2^n x)}{2^n x} = \sum_{n=0}^{\infty} = N\mu > \delta + |x|,$$

which is a contradiction. Thus, the function f provides a good example to the effect that Rassias' theorem fails to hold for $p = 1$. This completes the proof of the theorem. $\qquad\square$

11.6. Ulam–Gavruta–Rassias stability

In 1982, Rassias [255–258]) gave a further generalization of the result of Hyers, and proved a theorem using weaker conditions controlled by a product of powers of norms. His theorem is presented, as follows.

Theorem 11.5 ([255, 256]). *Let A be a normed linear space with norm $\| \cdot \|_1$ and B be a Banach space with norm $\| \cdot \|_2$. Assume in addition that $f : A \to B$ is a mapping such that $f(tx)$ is continuous in t for each fixed x. If there exist $a : 0 \leq a < \frac{1}{2}$ and $\delta \geq 0$ such that*

$$\|f(x+y) - [f(x) + f(y)]\|_2 \leq 2\delta \|x\|_1^a \cdot \|y\|_1^a \qquad (11.20)$$

for any $x, y \in A$, *then there exists a linear mapping* $L : A \to B$ *such that*

$$\|f(x) - L(x)\|_2 \le c\|x\|_1^{2a} \tag{11.21}$$

for any $x \in A$, *where* $c = \frac{\delta}{1 - 2^{2a-1}}$.

Uniqueness: *Let* M *be a linear mapping,* $M : A \to B$, *such that*

$$\|f(x) - M(x)\|_2 \le c'\|x\|_1^{2b} \tag{11.22}$$

for any $x \in A$, *where* $b : 0 \le b < \frac{1}{2}$ *and* c' *is a constant. Then*

$$L(x) = M(x) \tag{11.23}$$

for any $x \in A$.

Proof. From (11.20), $y = x$ implies

$$\left\|\frac{f(2x)}{2} - f(x)\right\| \le \delta\|x\|_1^{2a}, \tag{11.24}$$

which completes the proof. □

More generally, the following lemma holds.

Lemma 11.1 ([255, 256]).

$$\left\|\frac{f(2^n x)}{2^n} - f(x)\right\|_2 \le \delta \sum_{i=0}^{n-1} 2^{i(2a-1)}\|x\|_1^{2a} \tag{11.25}$$

for some $\delta \ge 0$ *and for any integer* n.

Proof. To prove Lemma 11.1, we work by induction on n. In fact, the case for $n = 1$ is obvious from (11.24). We assume then that (11.25) holds for $n = k$, and claim that (11.25) is true for $n = k + 1$. Indeed, from (11.25), $n = k$ and $2x = z$, we find

$$\left\|\frac{f(2^k z)}{2^k} - f(z)\right\|_2 \le \delta \sum_{i=0}^{k-1} 2^{i(2a-1)}\|z\|_1^{2a}$$

or $\quad \left\|\frac{f(2^{k+1}x)}{2^{k+1}} - \frac{f(2x)}{2}\right\|_2 \le \delta \sum_{i=0}^{k-1} 2^{(i+1)(2a-1)}\|x\|_1^{2a} \tag{11.26}$

or $\quad \left\|\frac{f(2^{k+1}x)}{2^{k+1}} - \frac{f(2x)}{2}\right\|_2 \le \delta \sum_{i=0}^{k} 2^{i(2a-1)}\|x\|_1^{2a}.$

Therefore, from (11.24) and (11.25), we get

$$\left\| \frac{f\left(2^{k+1}x\right)}{2^{k+1}} - f(x) \right\|_2 \leq \left\| \frac{f\left(2^{k+1}x\right)}{2^{k+1}} - \frac{f(2x)}{2} \right\|_2 + \left\| \frac{f(2x)}{2} - f(x) \right\|_2$$

$$\leq \delta \sum_{i=1}^{k} 2^{i(2a-1)} \|x\|_1^{2a} + \delta \|x\|_1^{2a}$$

$$= \delta \sum_{i=0}^{k} 2^{i(2a-1)} \|x\|_1^{2a}$$

or (11.25) holds for $n = k + 1$, or

$$\left\| \frac{f\left(2^{k+1}x\right)}{2^{k+1}} - f(x) \right\|_2 \leq \delta \sum_{i=0}^{n-1} 2^{i(2a-1)} \|x\|_1^{2a}. \tag{11.27}$$

But

$$\sum_{i=0}^{n-1} 2^{i(2a-1)} < \sum_{i=0}^{\infty} 2^{i(2a-1)} = \frac{1}{1 - 2^{2a-1}} = c_0 \tag{11.28}$$

such that

$$c = \delta c_0. \tag{11.29}$$

Equations (11.24) and (11.27) yield (11.25) which completes the proof of Lemma 11.1.

On the other hand, Lemma 11.1, (11.28) and (11.29) imply

$$\left\| \frac{f\left(2^n x\right)}{2^n} - f(x) \right\|_2 \leq c \|x\|_1^{2a} \tag{11.30}$$

for any $x \in A$ and any integer n, and for some $\delta \geq 0$. This completes the proof. $\square$

Lemma 11.2 ([255, 256]). *The sequence $\left\{ \frac{f(2^n x)}{2^n} \right\}$ converges.*

Proof. To prove Lemma 11.2, we use (11.30) and the completeness of B and prove that the sequence $\left\{ \frac{f(2^n x)}{2^n} \right\}$ is a Cauchy sequence. In fact, if $i > j > 0$, then

$$\left\| \frac{f\left(2^i x\right)}{2^i} - \frac{f\left(2^j x\right)}{2^j} \right\|_2 = \frac{1}{2^j} \left\| \frac{f\left(2^i x\right)}{2^{i-j}} - f\left(2^j x\right) \right\|_2 \tag{11.31}$$

and if we set $2^j x = h$ in (11.31) and employ (11.30), we get

$$\left\| \frac{f\left(2^i x\right)}{2^i} - \frac{f\left(2^j x\right)}{2^j} \right\|_2 = \frac{1}{2^j} \left\| \frac{f\left(2^{i-j} h\right)}{2^{i-j}} - f(h) \right\|_2$$

or

$$\lim_{j \to \infty} \left\| \frac{f\left(2^i x\right)}{2^i} - \frac{f\left(2^j x\right)}{2^j} \right\|_2 = 0 \tag{11.32}$$

because $a : 0 \le a < \frac{1}{2}$.

From (11.31), and the completeness of B, we complete the required proof of Lemma 11.2. Therefore, we set

$$L(x) = \lim_{n \to \infty} \frac{f\left(2^n x\right)}{2^n}. \tag{11.33}$$

It is clear now from (11.20) and (11.33) that

$$\|f\left(2^n x + 2^n y\right) - [f\left(2^n x\right) + f\left(2^n y\right)]\|_2 \le 2\delta \left\|2^n x\right\|_1^a \left\|2^n y\right\|_1^a$$

or

$$\frac{1}{2^n} \|f\left(2^n x + 2^n y\right) - [f\left(2^n x\right) + f\left(2^n y\right)]\|_2 \le 2\delta 2^{2na-n} \|x\|_1^a \|y\|_1^a$$

or

$$\left\| \lim_{n \to \infty} \left\{ \frac{f\left(2^n (x+y)\right)}{2^n} \right\} - \lim_{n \to \infty} \frac{f\left(2^n x\right)}{2^n} - \lim_{n \to \infty} \frac{f\left(2^n y\right)}{2^n} \right\|_2 = 0$$

or

$$\|L(x+y) - L(x) - L(y)\|_2 = 0$$

for any $x, y \in A$, or

$$L(x+y) = L(x) + L(y) \tag{11.34}$$

for any $x, y \in A$. From (11.34), we get

$$L(qx) = qL(x) \tag{11.35}$$

for any $q \in \mathbb{Q}$. The proof is now completed. $\qquad \square$

Lemma 11.3 ([255, 256]). *Let B^* be the space of continuous linear functionals and consider the mapping*

$$T : t \to d(L(tx)) \quad or \quad T : \mathbb{R} \to \mathbb{R} \tag{11.36}$$

such that

$$T(t) = d(L(tx)), \tag{11.37}$$

where $d \in B^*$, $t \in \mathbb{R}$ and $x \in A$, x is fixed. Then T is a continuous mapping.

Proof. To prove Lemma 11.3, we work as follows. Let

$$T_n(t) = d\left(f\left(2^n t x\right)/2^n\right) \tag{11.38}$$

such that

$$T(t) = \lim_{n \to \infty} |T_n(t)|, \tag{11.39}$$

where $x \in A$, x is fixed and $t \in \mathbb{R}$, $d \in B^*$. Then $T_n(t)$ are continuous, and therefore T is measurable as the pointwise limit of continuous mappings T_n. Moreover, T is homomorphism with respect to addition '+', that is,

$$T(x + y) = T(x) + T(y) \tag{11.40}$$

for any $x, y \in \mathbb{R}$. From (11.40) and the measurability of T, the required result for T is implied and therefore we complete the proof of the lemma.   □

Remark 11.3. The above type of stability involving a product of powers of norms is called **Ulam–Gavruta–Rassias Stability** by Bouikhalene, Elquorachi [33], Nakmahachalasint [210], Park, Najati [230], Pietrzyk [250] and Sibaha *et al.* [321].

11.7. Generalized Hyers–Ulam–Rassias stability

In the year 1994, Gavruta [84] replaced the unbounded Cauchy difference in Theorem 11.4 of Rassias, and provided a further generalization by using a general control function. The following theorem provides Gavruta's result.

Theorem 11.6 ([84]). *Let $(G, +)$ be an abelian group and $(X, \|\cdot\|)$ be a Banach space. Let $\varphi : G \times G \to [0, \infty)$ be a mapping satisfying*

$$\tilde{\varphi}(x, y) = \sum_{k=0}^{\infty} 2^{-k} \varphi\left(2^k x, 2^k y\right) < \infty \tag{11.41}$$

for all $x, y \in G$. Let $f : G \to X$ be such that

$$\|f(x + y) - f(x) - f(y)\| \le \varphi(x, y) \tag{11.42}$$

for all $x, y \in G$. Then there exists a unique mapping $T : G \to X$ such that

$$T(x + y) = T(x) + T(y) \tag{11.43}$$

for all $x, y \in G$, and

$$\|f(x) - T(x)\| \le \frac{1}{2}\tilde{\varphi}(x, x) \tag{11.44}$$

for all $x \in G$.

Proof. Substituting $x = y$ in (11.42) implies

$$\|f(2x) - 2f(x)\| \leq \varphi(x, x).$$

Thus

$$\left\|2^{-1}f(2x) - f(x)\right\| \leq \frac{1}{2}\varphi(x, x) \qquad (11.45)$$

for all $x \in G$. Replacing x by $2x$, inequality (11.45) gives

$$\left\|2^{-1}f\left(2^2x\right) - f(2x)\right\| \leq \frac{1}{2}\varphi(2x, 2x) \qquad (11.46)$$

for all $x \in G$. From (11.45) and (11.46), it follows that

$$\left\|2^{-2}f\left(2^2x\right) - f(x)\right\| \leq 2\left\|2^{-2}f\left(2^2x\right) - 2^{-1}f(2x)\right\| + \left\|2^{-1}f(2x) - f(x)\right\|$$
$$\leq 2^{-1}\left[\frac{1}{2}\varphi(2x, 2x)\right] + \frac{1}{2}\varphi(x, x).$$

Hence

$$\left\|2^{-2}f\left(2^2x\right) - f(x)\right\| \leq \frac{1}{2}\left[\varphi(x, x) + \frac{1}{2}\varphi(2x, 2x)\right] \qquad (11.47)$$

for all $x \in G$. Replacing x by $2x$, inequality (11.47) becomes

$$\left\|2^{-2}f\left(2^3x\right) - f(2x)\right\| \leq \frac{1}{2}\left[\varphi(2x, 2x) + \frac{1}{2}\varphi\left(2^2x, 2^2x\right)\right],$$

and therefore

$$\left\|2^{-3}f\left(2^3x\right) - f(x)\right\| \leq \left\|2^{-3}f\left(2^3x\right) - 2^{-1}f(2x)\right\| + \left\|2^{-1}f(2x) - f(x)\right\|$$
$$\leq 2^{-1}\left(\frac{1}{2}\left[\varphi(2x, 2x) + \frac{1}{2}\varphi\left(2^2x, 2^2x\right)\right]\right) + \frac{1}{2}\varphi(x, x).$$

Thus

$$\left\|2^{-3}f\left(2^3x\right) - f(x)\right\| \leq \frac{1}{2}\left[\varphi(x, x) + \frac{1}{2}\varphi(2x, 2x) + \frac{1}{2^2}\varphi\left(2^2x, 2^2x\right)\right] \qquad (11.48)$$

for all $x \in G$.

Applying an induction argument to n, we obtain

$$\left\|2^{-n}f\left(2^{n}x\right) - f(x)\right\| \leq \frac{1}{2}\sum_{k=0}^{n-1}2^{-k}\varphi\left(2^{k}x, 2^{k}x\right) \tag{11.49}$$

for all $x \in G$. Indeed,

$$\left\|2^{-(n+1)}f(2^{n+1}x) - f(x)\right\| \leq \left\|2^{-(n+1)}f(2^{n+1}x) - 2^{-1}f(2x)\right\| + \left\|2^{-1}f(2x) - f(x)\right\|,$$

and with (11.49) and (11.45), we obtain

$$\left\|2^{-(n+1)}f\left(2^{n+1}x\right) - f(x)\right\| \leq 2^{-1}\frac{1}{2}\sum_{k=0}^{n-1}2^{-k}\varphi\left(2^{k+1}x, 2^{k+1}x\right) + \frac{1}{2}\varphi(x, x)$$

$$= \frac{1}{2}\sum_{k=0}^{n}2^{-k}\varphi\left(2^{k}x, 2^{k}x\right).$$

We claim that the sequence $\{2^{-n}f(2^{n}x)\}$ is a Cauchy sequence. Indeed for $n > m$, we have

$$\left\|2^{-n}f\left(2^{n}x\right) - 2^{-m}f\left(2^{m}x\right)\right\| = 2^{-m}\left\|2^{-(n-m)}f(2^{n-m}2^{m}x) - f(2^{m}x)\right\|$$

$$\leq 2^{-m}\frac{1}{2}\sum_{k=0}^{n-m-1}2^{-k}\varphi\left(2^{k+m}x, 2^{k+m}x\right)$$

$$= \frac{1}{2}\sum_{p=m}^{n-1}2^{-p}\varphi\left(2^{p}x, 2^{p}x\right).$$

Taking the limit as $m \to \infty$, we obtain

$$\lim_{m\to\infty}\left\|2^{-n}f\left(2^{n}x\right) - 2^{-m}f\left(2^{m}x\right)\right\| = 0.$$

Since X is a Banach space, it follows that the sequence $\{2^{-n}f(2^{n}x)\}$ converges. Denote

$$T(x) = \lim_{n\to\infty}\frac{f\left(2^{n}x\right)}{2^{n}}.$$

We claim that T satisfies (11.43).

From (11.42), we have

$$\|f\left(2^{n}x + 2^{n}y\right) - f\left(2^{n}x\right) - f\left(2^{n}y\right)\| \leq \varphi\left(2^{n}x, 2^{n}y\right)$$

for all $x, y \in G$. Therefore

$$\left\|2^{-n}f\left(2^{n}x + 2^{n}y\right) - 2^{-n}f\left(2^{n}x\right) - 2^{-n}f\left(2^{n}y\right)\right\| \leq 2^{-n}\varphi\left(2^{n}x, 2^{n}y\right). \tag{11.50}$$

From (11.41), it follows that

$$\lim_{n \to \infty} 2^{-n} \varphi \left(2^n x, 2^n y \right) = 0.$$

Then (11.50) implies

$$\|T(x + y) - T(x) - T(y)\| = 0.$$

To prove (11.44), taking the limit in (11.49) as $n \to \infty$, we obtain

$$\|T(x) - f(x)\| \leq \frac{1}{2} \tilde{\varphi}(x, x) \quad \text{for all } x \in G.$$

It remains to show that T is uniquely defined. Let $F : G \to X$ be another such mapping with

$$F(x + y) = F(x) + F(y)$$

and (11.44) is satisfied.

Then

$$\|T(x) - F(x)\| = \left\| 2^{-n} T\left(2^n x\right) - 2^{-n} F\left(2^n x\right) \right\|$$
$$\leq \left\| 2^{-n} T\left(2^n x\right) - 2^{-n} f\left(2^n x\right) \right\| + \left\| 2^{-n} f\left(2^n x\right) - 2^{-n} F\left(2^n x\right) \right\|$$
$$= 2^{-n} \tilde{\varphi}\left(2^n x, 2^n x\right)$$
$$= 2^{-n} \sum_{k=0}^{\infty} 2^{-k} \varphi \left(2^{k+n} x, 2^{k+n} x \right)$$
$$- \sum_{p=n}^{\infty} 2^{-p} \varphi \left(2^p x, 2^p x \right).$$

Thus

$$\|T(x) - F(x)\| \leq \sum_{p=n}^{\infty} 2^{-p} \varphi \left(2^p x, 2^p x \right) \quad \text{for all } x \in G. \tag{11.51}$$

Taking the limit in (11.51) as $n \to \infty$, we obtain $T(x) = F(x)$, for all $x \in G$, which completes the proof of theorem. $\qquad \square$

Remark 11.4. The above stability is called **Generalized Hyers–Ulam–Rassias Stability.**

11.8. Rassias stability controlled by the mixed product-sum of powers of norms

In 2008, Rassias *et al.* [284] discussed the stability of quadratic functional equation

$$f(mx + y) + f(mx - y) = 2f(x + y) + 2f(x - y) + 2\left(m^2 - 2\right) f(x) - 2f(y) \tag{11.52}$$

for any arbitrary but fixed real constant m with $m \neq 0$; $m \neq \pm 1$; $m \neq \pm\sqrt{2}$ using mixed product-sum of powers of norms. He proved the following theorem.

Theorem 11.7 ([284]). *Let $f : E \to F$ be a mapping which satisfies the inequality*

$$\|f(mx+y) + f(mx-y) - 2f(x+y) - 2f(x-y) - 2(m^2-2)f(x) + 2f(y)\|_F$$

$$\leq \epsilon\{\|x\|_E^p \|y\|_E^p + (\|x\|_E^{2p} + \|y\|_E^{2p})\} \tag{11.53}$$

for all $x, y \in E$ with $x \perp y$, where ϵ and p are constants with $\epsilon, p > 0$ and either $m > 1$; $p < 1$ or $m < 1$; $p > 1$ with $m \neq 0$; $m \neq \pm 1$; $m \neq \pm\sqrt{2}$ and $-1 \neq |m|^{p-1} < 1$. Then the limit

$$Q(x) = \lim_{n \to \infty} \frac{f(m^n x)}{m^{2n}} \tag{11.54}$$

exists for all $x \in E$ and $Q : E \to F$ is the unique orthogonally Euler–Lagrange quadratic mapping such that

$$\|f(x) - Q(x)\|_F \leq \frac{\epsilon}{2|m^2 - m^{2p}|} \|x\|_E^{2p} \tag{11.55}$$

for all $x \in E$, where $(E, \perp)$ is an orthogonality normed space with norm $\|\cdot\|_E$ and $(F, \|\cdot\|_F)$ is a Banach space.

Proof. Replacing (x, y) with $(0, 0)$ in (11.53), we obtain $2|2 - m^2|\|f(0)\| = 0$ or $f(0) = 0$ if $m^2 \neq 2$. Again substituting (x, y) by $(x, 0)$ in (11.53), we get

$$\left\|f(mx) - m^2 f(x)\right\|_F \leq \frac{1}{2}\epsilon \|x\|_E^{2p}$$

or

$$\left\|\frac{f(mx)}{m^2} - f(x)\right\|_F \leq \frac{1}{2}\frac{\epsilon}{m^2} \|x\|_E^{2p} \ (m \neq 0) \tag{11.56}$$

for all $x \in E$. Now, replacing x by mx and dividing by m^2 in (11.57), and then adding the resulting inequality with (11.56), we obtain

$$\left\|\frac{f(m^2 x)}{m^4} - f(x)\right\|_F \leq \frac{1}{2}\frac{\epsilon}{m^2}\left(1 + \frac{m^{2p}}{m^2}\right)\|x\|_E^{2p} \tag{11.57}$$

for all $x \in E$. Using induction on n, we obtain that

$$\left\|\frac{f(m^n x)}{m^{2n}} - f(x)\right\|_F \leq \frac{1}{2}\frac{\epsilon}{m^2} \sum_{k=0}^{n-1} \frac{m^{2pk}}{m^{2k}} \|x\|_E^{2p}$$

$$\leq \frac{1}{2}\frac{\epsilon}{m^2} \sum_{k=0}^{\infty} \frac{m^{2pk}}{m^{2k}} \|x\|_E^{2p} \tag{11.58}$$

for all $x \in E$. In order to prove the convergence of the sequence $\{\frac{f(m^n)x}{m^{2n}}\}$, replace x by $m^\ell x$ and divide by $m^{2\ell}$ in (11.58); then for any $n, \ell > 0$, we obtain

$$\left\| \frac{f\left(m^{n+\ell}x\right)}{m^{2(\ell+n)}} - \frac{f\left(m^\ell x\right)}{m^{2\ell}} \right\|_F = \frac{1}{m^{2\ell}} \left\| \frac{f\left(m^{n+\ell}x\right)}{m^{2n}} - f\left(m^\ell x\right) \right\|_F$$

$$\leq \frac{1}{2} \frac{\epsilon}{m^2} \frac{1}{m^{2\ell(1-p)}} \sum_{k=0}^{\infty} \frac{m^{2pk}}{m^{2k}} \|x\|_E^{2p}. \tag{11.59}$$

Since $m^{2(1-p)} < 1$, the right-hand side of (11.59) tends to 0 as $\ell \to \infty$ for all $x \in E$. Thus $\{\frac{f(m^n)x}{m^{2n}}\}$ is a Cauchy sequence. Since F is complete, there exists a mapping $Q : E \to F$ such that

$$Q(x) = \lim_{n \to \infty} \frac{f\left(m^n x\right)}{m^{2n}} \quad \text{for all } x \in E.$$

By letting $n \to \infty$ in (11.58), we arrive at the formula (11.55) for all $x \in E$. To prove Q satisfies (11.53), replace (x, y) by $(m^n x, m^n y)$ in (11.53) and divide by m^{2n}, then it follows that

$$\frac{1}{m^{2n}} \| f\left(m^n(mx+y)\right) + f\left(m^n(mx-y)\right) - 2f\left(m^n(x+y)\right) - 2f\left(m^n(x-y)\right)$$

$$- 2\left(m^2-2\right) f\left(m^n x\right) + 2f\left(m^n y\right) \|_F$$

$$\leq \frac{\epsilon}{m^{2n}} \{ \|m^n x\|_E^p + (\|m^n x\|_E^{2p} + \|m^n y\|_E^{2p}) \}.$$

Taking limit as $n \to \infty$ in the above inequality, we get

$$\| Q(mx+y) + Q(mx-y) - 2Q(x+y) - 2Q(x-y)$$
$$- 2\left(m^2-2\right) Q(x) + 2Q(y) \|_F \leq 0$$

which gives

$$Q(mx+y) + Q(mx-y) = 2Q(x+y) + 2Q(x-y) + 2(m^2-2)Q(x) - 2Q(y)$$

for all $x, y \in E$ with $x \perp y$. Therefore $Q : E \to F$ is an orthogonally Euler–Lagrange quadratic mapping which satisfies (11.53). To prove the uniqueness of Q, let Q' be another orthogonally Euler–Lagrange quadratic mapping satisfying (11.53), and the inequality (11.56), we have

$$\| Q(x) - Q'(x) \|_F \leq \frac{1}{m^{2n}} \{ \|Q(m^n x) - f(m^n x)\|_F + \|f(m^n x) - Q'(m^n x)\|_F \}$$

$$\leq \frac{\epsilon}{m^2} \sum_{j=0}^{\infty} \frac{1}{m^{2(k+n)(1-p)}} \|x\|_E^{2p}$$

$$\to 0 \quad \text{as } n \to \infty$$

for all $x \in E$. Therefore Q is unique. This completes the proof of the theorem. $\square$

11.9. Hyers–Ulam stability of Cauchy additive functional equation on restricted domains

The Hyers–Ulam stability of functional equations in several variables can be easily derived from a simple result on stability of a functional equation in single variable.

Let us start with the following theorem, concerning stability of the equation $\Psi \circ f \circ a = f$. For a function, a mapping on a non-empty set K into K in the sequel, we write $a^0(x) = x$ for $x \in K$ and $a^n = a \circ a^{n-1}$ for $n \in \mathbb{N}$, where $\mathbb{N}$ denotes the set of positive integers.

The following theorem is useful for proving Hyers–Ulam stability of Cauchy additive functional equation on restricted domains.

Theorem 11.8 ([82]). *Assume that (Y, d) is a complete metric space, K is a non-empty set, $f : K \to Y$, $\Psi : Y \to Y$, $a : K \to K$, $h : K \to [0, \infty)$, $\lambda \in [0, \infty)$, $d(\psi \circ f \circ a(x), f(x)) \le h(x)$ for $x \in K$,*

$$d(\Psi(x), \Psi(y)) \le \lambda d(x, y) \quad \text{for } x, y \in Y, \tag{11.60}$$

and

$$H(x) = \sum_{i=0}^{\infty} \lambda^i h\left(a^i(x)\right) < \infty \quad \text{for } x \in K. \tag{11.61}$$

Then, for every $x \in K$, the limit $F(x) = \lim_{n \to \infty} \Psi^n \circ f \circ a^n(x)$ exists, and $F : K \to Y$ is a unique function such that $\Psi \circ F \circ a = F$, and $d(f(x), F(x)) \le H(x)$, for $x \in K$.

Proof. Note that

$$d\left(\Psi^n \circ f \circ a^n(x)\right) \le \sum_{i=1}^{n} d\left(\Psi^i \circ f \circ a^i(x), \Psi^{i-1} \circ f \circ a^{i-1}(x)\right)$$

$$\le \sum_{i=1}^{n} \lambda^{i-1} d\left(\Psi \circ f \circ a^i(x), f \circ a^{i-1}(x)\right)$$

$$\le \sum_{i=0}^{n-1} \lambda^i h\left(a^i(x)\right)$$

for $x \in K$ and $n \in \mathbb{N}$. Hence, for every $x \in K$, and $m \in \mathbb{N}$, $k \in \mathbb{N} \cup \{0\}$,

$$d\left(\Psi^{m+k} \circ f \circ a^{m+k}(x), \Psi^k \circ f \circ a^k(x)\right) \le \lambda^k d\left(\Psi^m \circ f \circ a^{m+k}(x), f \circ a^k(x)\right)$$

$$\le \lambda^k \sum_{i=0}^{m-1} \lambda^i h\left(a^{k+i}(x)\right)$$

$$\le \sum_{i=k}^{m+k-1} \lambda^i h\left(a^i(x)\right).$$

Since $\sum_{i=k}^{m+k-1} \lambda^i h(a^i(x)) \to 0$ with $k \to \infty$, $F(x) = \lim_{n \to \infty} \Psi^n \circ f \circ a^n(x)$ exists and $d(F(x), f(x)) \leq H(x)$ for every $x \in K$. Moreover, Ψ is continuous (in view of (11.60)), whence for every $x \in K$, we have

$$F(x) = \lim_{n \to \infty} \Psi^{n+1} \circ f \circ a^{n+1}(x) = \Psi \left(\lim_{n \to \infty} \Psi^n \circ f \circ a^n(a(x)) \right) = \Psi \circ F \circ a(x).$$

It remains to show the uniqueness of F. We suppose that $G : K \to Y$, $d(f(x), G(x)) \leq H(x)$ for $x \in K$, and $\Psi \circ G \circ a = G$. By induction it is easy to show that $\Psi^n \circ G \circ a^n = G$, and $\Psi^n \circ F \circ a^n = F$ for $n \in \mathbb{N}$. Hence, for $x \in K$,

$$d(F(x), G(x)) = d\left(\Psi^n \circ F \circ a^n(x), \Psi^n \circ G \circ a^n(x)\right)$$

$$\leq \lambda^n d\left(F \circ a^n(x), G \circ a^n(x)\right)$$

$$= 2 \sum_{i=n}^{\infty} \lambda^i h\left(a^i(x)\right).$$

Since, for every $x \in K$, $\sum_{i=n}^{\infty} \lambda^i h\left(a^i(x)\right) \to 0$ with $n \to \infty$, this completes the proof. $\qquad\square$

Let us recall that a groupoid $(G, +)$ (that is, a non-empty set G endowed with a binary operation $+ : G^2 \to G$) is uniquely divisible by 2 provided, for each $x \in X$, there is a unique $y \in X$ with $x = 2y$ (we write $x = x + x$ for $x \in G$): such an element y will be denoted by $\frac{x}{2}$ or $\frac{1}{2}x$ in the sequel. In what follows, we use the notion: $2^0 x = x$, $2^n x = 2(2^{n-1}x)$, and (only for groupoids uniquely divisible by 2) $2^{-n}x = \frac{1}{2}(2^{-n+1}x)$ for $x \in G$, $n \in \mathbb{N}$.

A groupoid $(G, +)$ is square symmetric provided the operation $+$ is square symmetric, i.e., $2(x + y) = 2x + 2y$ for $x, y \in G$; it is easy to show by induction that, for each $n \in \mathbb{N}$ (and also for all integers n, if the groupoid is uniquely divisible by 2),

$$2^n(x + y) = 2^n x + 2^n y, \quad \text{for } x, y \in G. \tag{11.62}$$

Clearly, every commutative semigroup is a square symmetric groupoid. Next, let X be a linear space over a field $\mathbb{K}$, $a, b \in \mathbb{K}$, $z \in X$, and define a binary operation $\star : X^2 \to X$ by: $x \star y = ax + by + z$ for $x, y \in X$. Then it is easy to check that $(X, \star)$ provides a simple example of a square symmetric groupoid.

Finally, we say that $(G, +, d)$ is a complete metric groupoid provided $(G, +)$ is a groupoid, (G, d) is a complete metric space, and the operation $+$ is continuous with respect to the metric d.

In the following theorem, $(X, +)$, and $(Y, +)$ are square symmetric groupoids, $(Y, +, d)$ is a complete metric groupoid, $K \subset X$, and $\chi : X^2 \to [0, \infty)$. The stability of the Cauchy functional equation

$$F(x + y) = F(x) + F(y), \tag{11.63}$$

for $x, y \in K$ with $x + y \in K$, is presented in the following theorem.

Theorem 11.9 ([37]). *Suppose that X is uniquely divisible by 2, $\frac{1}{2}K \subset K$ (i.e., $\frac{1}{2} \in K$ for $a \in K$), and there exist $\xi, \eta \in (0, \infty)$ such that $\xi\eta < 1$,*

$$\chi\left(\frac{x}{2}, \frac{y}{2}\right) \leq \eta\chi(x, y) \tag{11.64}$$

for $x, y \in K$,

$$d(2x, 2y) \leq \xi d(x, y) \tag{11.65}$$

for $x, y \in Y$. Let $\varphi : K \to Y$ satisfy

$$d(\varphi(x + y), \varphi(x) + \varphi(y)) \leq \chi(x, y) \tag{11.66}$$

for $x, y \in K$ with $x + y \in K$. Then there is a unique solution $F : K \to Y$ of (11.63) with

$$d(\varphi(x), F(x)) \leq \frac{\eta\chi(x, x)}{1 - \xi\eta}, \quad \text{for } x \in K.$$

Proof. From (11.66), we obtain $d\left(\varphi(x), 2\varphi\left(\frac{x}{2}\right)\right) \leq \chi\left(\frac{x}{2}, \frac{x}{2}\right)$ for $x \in K$. Hence, by Theorem 11.8 (with $f = \varphi$, $\Psi(z) = 2z$, $\lambda = \xi$, $h(x) = \chi\left(\frac{x}{2}, \frac{x}{2}\right)$, and $a(x) = \frac{x}{2}$), for every $x \in K$, the limit $F(x)$ exists and $d(f(x0, F(x)) \leq H(x)$. Take $x, y \in K$ with $x + y \in K$. Using (11.62) and (11.66), we have

$$d\left(2^n\varphi\left(2^{-n}(x + y)\right), 2^n\varphi\left(2^{-n}x\right) + 2^n\varphi\left(2^{-n}y\right)\right) \leq (\xi\eta)^n\chi(x, y)$$

for $n \in \mathbb{N}$, and subsequently letting $n \to \infty$, we obtain $F(x + y) = F(x) + F(y)$.

Thus, we have shown that equation (11.63) holds. Suppose $F_0 : K \to Y$ is also a solution of (11.63), and $d(f(x), F_0(x)) \leq H(x)$ for every $x \in K$. Then $\Psi \circ F_0 \circ a = F_0$, whence, by Theorem 11.8, $F = F_0$, which implies the uniqueness of F. This completes the proof of the theorem. $\square$

Theorem 11.10 ([37]). *Suppose that Y is uniquely divisible by 2, $2K \subset K$ (i.e., $2a \in K$ for $a \in K$), and there exist $\xi, \eta \in [0, \infty)$ such that $\xi\eta < 1$,*

$$d\left(\frac{1}{2}x, \frac{1}{2}y\right) \leq \xi d(x, y) \tag{11.67}$$

for $x, y \in Y$,

$$\chi(2x, 2y) \leq \eta\chi(x, y) \tag{11.68}$$

for $x, y \in K$. Let $\varphi : K \to Y$ satisfy (11.66). Then there is a unique solution $G : K \to Y$ of (11.63) with

$$d(\varphi(x), G(x)) \leq \frac{\xi^2\chi(x, x)}{1 - \xi\eta}, \quad \text{for } x \in K.$$

Proof. From (11.66) and (11.67), we obtain

$$d\left(\frac{1}{2}\varphi(2x), \varphi(x)\right) \leq \xi d(\varphi(2x), 2\varphi(x)) \leq \xi\chi(x,x)$$

for $x \in K$. Now, we can use Theorem 11.8 analogously as in the proof of Theorem 11.9 (with $\lambda = \xi$, $f = 2\varphi$, $\Psi(z) = \frac{1}{2}z$, $h(x) = \xi\chi(x,x)$ and $a(x) = 2x$). Then, with $G = \frac{1}{2}F$, for every $x \in K$, we get

$$d(\varphi(x), G(x)) \leq \xi d(f(x), F(x)) \leq \xi^2\chi(x,x) \sum_{i=0}^{\infty} (\xi\eta)^i.$$

Next, by (11.62) and (11.66), for every $x, y \in K$ with $x + y \in K$, we have

$$d\left(2^{-n}\varphi\left(2^n(x+y)\right), 2^{-n}\varphi\left(2^n x\right) + 2^{-n}\varphi\left(2^n y\right)\right) \leq (\xi\eta)^n \chi(x,y)$$

for $n \in \mathbb{N}$, which implies that F is a solution of (11.63) and so is G. The proof of uniqueness is analogous as in the proof of Theorem 11.9. This completes the proof of the theorem. □

11.10. Application of Ulam stability theory

In the year 1978, according to Gruber [114] stability problems are of particular interest in probability theory, and in the case of functional equations of different types. Furthermore, we quote that the Ulam stability results can be applied to mathematical statistics, stochastic analysis, algebra, geometry as well as psychology, and sociology.

Several mathematicians have made remarkable interesting applications of the Hyers–Ulam–Rassias stability theory to various mathematical problems. Stability theory is applied in the study of nonlinear analysis, especially in fixed point theory. In nonlinear analysis, it is well known that finding the expression of the asymptotic derivative of a nonlinear operator can be a difficult problem. In this sense, how the Hyers–Ulam–Rassias stability theory can be used to evaluate the asymptotic derivative of some nonlinear operators is explained.

Jung [145] investigated the Hyers–Ulam stability for Jensen's equation on a restricted domain, and he applied the result to the study of an interesting asymptotic property of additive mappings.

The stability properties of different functional equations can have applications to unrelated fields. For instance, Zhou [348] used a stability property of the functional equation $f(x+y) + f(x-y) = 2f(x)$ to prove a conjecture of Ditzian about the relationship between the smoothness of a mapping, and the degree of its approximation by the associated Bernstein polynomials.

Park and Bae [248] obtained the general solution, and the stability of the functional equation

$$f(x + y + z, u + v + w) + f(x + y - z, u + v + w) + 2f(x, u, -w) + 2f(y, v, -w)$$
$$= f(x + y, u + w) + f(x + y, v + w) + f(x + z, u + w)$$
$$+ f(x - z, u + v + w) + f(y + z, v + w) + f(y - z, u + v - w). \qquad (11.69)$$

The function $f(x, y) = x^3 + ax + b - y^2$ having level curves as elliptic curves is a solution of the above functional equation (11.69). The mapping $f(x, y) = x^3 + ax + b - y^2$ is useful in studying cryptography and other applications. In the same paper, they have presented an application of the stability of the equation (11.69) by showing that the canonical height function of an elliptic curve over $\mathbb{Q}$ is a solution of equation (11.69).

Chapter 12

Stability of Functional Equations in Various Spaces

12.1. Introduction

In the previous chapter, the stabilities pertinent to Hyers, Th.M. Rassias, Aoki, J.M. Rassias, and Gavruta in Banach spaces are discussed. In this chapter, we present latest results associated with Hyers–Ulam stability of functional equations in various abstract spaces like paranormed spaces, non-Archimedean spaces, random normed spaces, intuitionistic fuzzy normed spaces, quasi-Banach spaces, fuzzy Banach spaces, and etc.

12.2. Stability in paranormed spaces

We present here the definition of **paranormed space** and **Fréchet space** which will be useful in investigating the stability of functional equations.

Definition 12.1 ([242]). Let X be a vector space. A paranorm $P : X \to [0, \infty)$ is a function on X such that

(1) $P(0) = 0$;
(2) $P(-x) = P(x)$;
(3) $P(x + y) \leq P(x) + P(y)$ (triangle inequality);
(4) if $\{t_n\}$ is a sequence of scalars with $t_n \to t$ and $\{x_n\} \subset X$ with $P(x_n - x) \to 0$, then $P(t_n x_n - tx) \to 0$ (continuity of multiplication).
The pair (X, P) is called a **paranormed space** if P is a paranorm on X. The paranorm is called **total** if in addition, we have
(5) $P(x) = 0$ implies $x = 0$.

A **Fréchet space** is a total and complete paranormed space.
Throughout this section, (X, P) is a Fréchet space, and that $(Y, \|.\|)$ is a Banach space.
Note that $P(2x) \leq 2P(x)$ for all $x \in Y$.

12.2.1. Hyers–Ulam stability of the Cauchy additive functional equation

The Hyers–Ulam stability of the Cauchy additive functional equation in paranormed spaces is presented in the following theorems.

Theorem 12.1 ([242]). *Let r, θ be positive real numbers with $r > 1$, and let $f : Y \to X$ be an odd mapping such that*

$$P(f(x + y) - f(x) - f(y)) \leq \theta \left(\|x\|^r + \|y\|^r \right) \tag{12.1}$$

for all $x, y \in Y$. Then there exists a unique Cauchy additive mapping $A : Y \to X$ such that

$$P(f(x) - A(x)) \leq \frac{2\theta}{2^r - 2} \|x\|^r \tag{12.2}$$

for all $x \in Y$.

Proof. Letting $y = x$ in (12.1), we obtain

$$P(f(2x) - 2f(x)) \leq 2\theta \|x\|^r$$

for all $x \in Y$. So

$$P\left(f(x) - 2f\left(\frac{x}{2}\right) \right) \leq \frac{2}{2^r} \theta \|x\|^r$$

for all $x \in Y$. Hence

$$P\left(2^l f\left(\frac{x}{2^l}\right) - 2^m f\left(\frac{x}{2^m}\right) \right) \leq \sum_{j=l}^{m-1} P\left(2^j f\left(\frac{x}{2^j}\right) - 2^{j+1} f\left(\frac{x}{2^{j+1}}\right) \right)$$

$$\leq \frac{2}{2^r} \sum_{j=l}^{m-1} \frac{2^j}{2^{rj}} \theta \|x\|^r \tag{12.3}$$

for all non-negative integers m and l with $m > l$, and all $x \in Y$. It follows from (12.3) that the sequence $\left\{ 2^n f\left(\frac{x}{2^n}\right) \right\}$ is a Cauchy sequence for all $x \in Y$. Since X is complete, the sequence $\left\{ 2^n f\left(\frac{x}{2^n}\right) \right\}$ converges. So one can define the mapping $A : Y \to X$ by

$$A(x) := \lim_{n \to \infty} 2^n f\left(\frac{x}{2^n}\right)$$

for all $x \in Y$. Moreover, letting $l = 0$ and passing the limit $m \to \infty$ in (12.3), we obtain (12.2). It follows from (12.1) that

$$P(A(x + y) - A(x) - A(y)) = \lim_{n \to \infty} P\left(2^n \left(f\left(\frac{x+y}{2^n}\right) - f\left(\frac{x}{2^n}\right) - f\left(\frac{y}{2^n}\right) \right) \right)$$

$$\leq \lim_{n \to \infty} 2^n P\left(f\left(\frac{x+y}{2^n}\right) - f\left(\frac{x}{2^n}\right) - f\left(\frac{y}{2^n}\right) \right)$$

$$\leq \lim_{n \to \infty} \frac{2^n \theta}{2^{nr}} \left(\|x\|^r + \|y\|^r \right) = 0$$

for all $x, y \in Y$. Hence $A(x + y) = A(x) + A(y)$ for all $x, y \in Y$ and so the mapping $A : Y \to X$ is Cauchy additive.

Now, let $T : Y \to X$ be another Cauchy additive mapping satisfying (12.2). Then we have

$$P(A(x) - T(x)) = P\left(2^n \left(A\left(\frac{x}{2^n}\right) - T\left(\frac{x}{2^n}\right) \right) \right)$$

$$\leq 2^n \left(P\left(A\left(\frac{x}{2^n}\right) - f\left(\frac{x}{2^n}\right) \right) + P\left(T\left(\frac{x}{2^n}\right) - f\left(\frac{x}{2^n}\right) \right) \right)$$

$$\leq \frac{4 \cdot 2^n}{(2^r - 2)\, 2^{nr}} \theta \|x\|^r,$$

which tends to zero as $n \to \infty$ for all $x \in Y$. So we can conclude that $A(x) = T(x)$ for all $x \in Y$. This proves the uniqueness of A. Thus the mapping $A : Y \to X$ is a unique Cauchy additive mapping satisfying (12.2). This completes the proof of the theorem. $\qquad \square$

Theorem 12.2 ([242]). *Let r be a positive real number with $r < 1$, and let $f : X \to Y$ be an odd mapping such that*

$$\|f(x + y) - f(x) - f(y)\| \leq P(x)^r + P(y)^r \tag{12.4}$$

for all $x, y \in X$. Then there exists a unique Cauchy additive mapping $A : X \to Y$ such that

$$\|f(x) - A(x)\| \leq \frac{2}{2 - 2^r} P(x)^r \tag{12.5}$$

for all $x \in X$.

Proof. Letting $y = x$ in (12.4), we obtain

$$\|2f(x) - f(2x)\| \leq 2P(x)^r,$$

and so

$$\left\| f(x) - \frac{1}{2} f(2x) \right\| \leq P(x)^r$$

for all $x \in X$. Hence

$$\left\| \frac{1}{2^l} f\left(2^l x\right) - \frac{1}{2^m} f\left(2^m x\right) \right\| \leq \sum_{j=l}^{m-1} \left\| \frac{1}{2^j} f\left(2^j x\right) - \frac{1}{2^{j+1}} f\left(2^{j+1} x\right) \right\|$$

$$\leq \sum_{j=l}^{m-1} \frac{2^{rj}}{2^j} P(x)^r \tag{12.6}$$

for all non-negative integers m and l with $m > l$ and all $x \in X$. It follows from (12.6) that the sequence $\left\{ \frac{1}{2^n} f\left(2^n x\right) \right\}$ is a Cauchy sequence for all $x \in X$. Since Y is complete, the sequence $\left\{ \frac{1}{2^n} f\left(2^n x\right) \right\}$ converges. So one can define the mapping $A : X \to Y$ by

$$A(x) = \lim_{n \to \infty} \frac{1}{2^n} f\left(2^n x\right)$$

for all $x \in X$. Moreover, letting $l = 0$, and passing the limit $m \to \infty$ in (12.6), we obtain (12.5). It follows from (12.4) that

$$\|A(x+y) - A(x) - A(y)\| = \lim_{n \to \infty} \frac{1}{2^n} \left\| f\left(2^n(x+y)\right) - f\left(2^n x\right) - f\left(2^n y\right) \right\|$$

$$\leq \lim_{n \to \infty} \frac{2^{nr}}{2^n} \left(P(x)^r + P(y)^r \right) = 0$$

for all $x, y \in X$. Thus $A(x+y) = A(x) + A(y)$ for all $x, y \in X$, and so the mapping $A : X \to Y$ is Cauchy additive.

Now, let $T : X \to Y$ be another Cauchy additive mapping satisfying (12.5). Then we have

$$\|A(x) - T(x)\| = \frac{1}{2^n} \left\| A\left(2^n x\right) - T\left(2^n x\right) \right\|$$

$$\leq \frac{1}{2^n} \left(\left\| A\left(2^n x\right) - f\left(2^n x\right) \right\| + \left\| T\left(2^n x\right) - f\left(2^n x\right) \right\| \right)$$

$$\leq \frac{4 \cdot 2^{nr}}{(2 - 2^r) \, 2^n} P(x)^r,$$

which tends to zero as $n \to \infty$ for all $x \in X$. So we can conclude that $A(x) = T(x)$ for all $x \in X$. This proves the uniqueness of A. Thus the mapping $A : X \to Y$ is a unique Cauchy additive mapping satisfying (12.5). The proof is now completed. $\square$

12.2.2. *Hyers–Ulam stability of the quadratic functional equation in paranormed spaces*

The Hyers–Ulam stability of the quadratic functional equation in paranormed spaces is presented in the following theorems.

Theorem 12.3 ([242]). *Let r, θ be positive real numbers with $r > 2$, and let $f : Y \to X$ be a mapping satisfying $f(0) = 0$, and*

$$P(f(x+y) + f(x-y) - 2f(x) - 2f(y)) \leq \theta \left(\|x\|^r + \|y\|^r \right) \tag{12.7}$$

for all $x, y \in Y$. Then there exists a unique quadratic mapping $Q_2 : Y \to X$ such that

$$P(f(x) - Q_2(x)) \leq \frac{2\theta}{2^r - 4} \|x\|^r \tag{12.8}$$

for all $x \in Y$.

Proof. Letting $y = x$ in (12.7), we obtain

$$P(f(2x) - 4f(x)) \leq 2\theta\|x\|^r$$

for all $x \in Y$. So

$$P\left(f(x) - 4f\left(\frac{x}{2}\right)\right) \leq \frac{2}{2^r}\theta\|x\|^r$$

for all $x \in Y$. Hence

$$P\left(4^l f\left(\frac{x}{2^l}\right) - 4^m f\left(\frac{x}{2^m}\right)\right) \leq \sum_{j=l}^{m-1} P\left(4^j f\left(\frac{x}{2^j}\right) - 4^{j+1} f\left(\frac{x}{2^{j+1}}\right)\right)$$

$$\leq \frac{2}{2^r} \sum_{j=l}^{m-1} \frac{4^j}{2^{rj}}\theta\|x\|^r \tag{12.9}$$

for all non-negative integers m and l with $m > l$, and all $x \in Y$. It follows from (12.9) that the sequence $\left\{4^n f\left(\frac{x}{2^n}\right)\right\}$ is a Cauchy sequence for all $x \in Y$. Since X is complete, the sequence $\left\{4^n f\left(\frac{x}{2^n}\right)\right\}$ converges. So one can define the mapping $Q_2 : Y \to X$ by

$$Q_2(x) := \lim_{n\to\infty} 4^n f\left(\frac{x}{2^n}\right)$$

for all $x \in Y$. Moreover, letting $l = 0$, and passing the limit $m \to \infty$ in (12.9), we obtain (12.8). It follows from (12.7) that

$$P(Q_2(x + y) + Q_2(x - y) - 2Q_2(x) - 2Q_2(y))$$

$$= \lim_{n\to\infty} P\left(4^n\left(f\left(\frac{x+y}{2^n}\right) + f\left(\frac{x-y}{2^n}\right) - 2f\left(\frac{x}{2^n}\right) - 2f\left(\frac{y}{2^n}\right)\right)\right)$$

$$\leq 4^n P\left(f\left(\frac{x+y}{2^n}\right) + f\left(\frac{x-y}{2^n}\right) - 2f\left(\frac{x}{2^n}\right) - 2f\left(\frac{y}{2^n}\right)\right)$$

$$\leq \lim_{n\to\infty} \frac{4^n\theta}{2^{nr}}\left(\|x\|^r + \|y\|^r\right) = 0$$

for all $x, y \in Y$. Hence $Q_2(x + y) + Q_2(x - y) = 2Q_2(x) + 2Q_2(y)$ for all $x, y \in Y$ and so the mapping $Q_2 : Y \to X$ is quadratic.

Now, let $T : Y \to X$ be another quadratic mapping satisfying (12.8). Then we have

$$P(Q_2(x) - T(x)) = P\left(4^n\left(Q_2\left(\frac{x}{2^n}\right) - T\left(\frac{x}{2^n}\right)\right)\right)$$

$$\leq 4^n\left(P\left(Q_2\left(\frac{x}{2^n}\right) - f\left(\frac{x}{2^n}\right)\right) + P\left(T\left(\frac{x}{2^n}\right) - f\left(\frac{x}{2^n}\right)\right)\right)$$

$$\leq \frac{4 \cdot 4^n}{(2^r - 4)\,2^{nr}}\theta\|x\|^r,$$

which tends to zero as $n \to \infty$ for all $x \in Y$. So we conclude that $Q_2(x) = T(x)$ for all $x \in Y$. This proves the uniqueness of Q_2. Thus the mapping $Q_2 : Y \to X$ is a unique quadratic mapping satisfying (12.8). This completes the proof of the theorem. □

Theorem 12.4 ([242]). *Let r be a positive real number with $r < 2$, and let $f : X \to Y$ be a mapping satisfying $f(0) = 0$, and*

$$\|f(x+y) + f(x-y) - 2f(x) - 2f(y)\| \leq P(x)^r + P(y)^r \tag{12.10}$$

for all $x, y \in X$. Then there exists a unique quadratic mapping $Q_2 : X \to Y$ such that

$$\|f(x) - Q_2(x)\| \leq \frac{2}{4 - 2^r} P(x)^r \tag{12.11}$$

for all $x \in X$.

Proof. Letting $y = x$ in (12.10), we obtain

$$\|4f(x) - f(2x)\| \leq 2P(x)^r,$$

and so

$$\left\| f(x) - \frac{1}{4}f(2x) \right\| \leq \frac{1}{2}P(x)^r$$

for all $x \in X$. Hence

$$\left\| \frac{1}{4^l}f\left(2^l x\right) - \frac{1}{4^m}f\left(2^m x\right) \right\| \leq \sum_{j=1}^{m-1} \left\| \frac{1}{4^j}f\left(2^j x\right) - \frac{1}{4^{j+1}}f\left(2^{j+1}x\right) \right\|$$

$$\leq \frac{1}{2} \sum_{j=l}^{m-1} \frac{2^{rj}}{4^j} P(x)^r \tag{12.12}$$

for all non-negative integers m and l with $m > l$, and all $x \in X$. It follows from (12.12) that the sequence $\left\{ \frac{1}{4^n}f(2^n x) \right\}$ is a Cauchy sequence for all $x \in X$. Since Y is complete, the sequence $\left\{ \frac{1}{4^n}f(2^n x) \right\}$ converges. So one can define the mapping $Q_2 : X \to Y$ by

$$Q_2(x) := \lim_{n \to \infty} \frac{1}{4^n}f\left(2^n x\right)$$

for all $x \in X$. Moreover, letting $l = 0$ and passing the limit $m \to \infty$ in (12.12), we obtain (12.11). It follows from (12.10) that

$$\|Q_2(x+y) + Q_2(x-y) - 2Q_2(x) - 2Q_2(y)\|$$

$$= \lim_{n \to \infty} \frac{1}{4^n} \left\| f\left(2^n(x+y)\right) + f\left(2^n(x-y)\right) - 2f\left(2^n x\right) - 2f\left(2^n y\right) \right\|$$

$$\leq \lim_{n \to \infty} \frac{2^{nr}}{4^n} \left(P(x)^r + P(y)^r\right) = 0$$

for all $x, y \in X$. Thus $Q_2(x+y) + Q_2(x-y) = 2Q_2(x) + 2Q_2(y)$ for all $x, y \in X$ and so the mapping $Q_2 : X \to Y$ is quadratic.

Now, let $T : X \to Y$ be another quadratic mapping satisfying (12.11). Then we have

$$\|Q_2(x) - T(x)\| = \frac{1}{4^n} \left\| Q_2\left(2^n x\right) - T\left(2^n x\right) \right\|$$

$$\leq \frac{1}{4^n} \left(\left\| Q_2\left(2^n x\right) - f\left(2^n x\right) \right\| + \left\| T\left(2^n x\right) - f\left(2^n x\right) \right\| \right)$$

$$\leq \frac{4 \cdot 2^{nr}}{(4 - 2^r) 4^n} P(x)^r,$$

which tends to zero as $n \to \infty$ for all $x \in X$. So we can conclude that $Q_2(x) = T(x)$ for all $x \in X$. This proves the uniqueness of Q_2. Thus the mapping $Q_2 : X \to Y$ is a unique quadratic mapping satisfying (12.11). The proof is now completed. $\square$

12.3. Stability in non-Archimedean spaces

Definition 12.2. By a non-Archimedean field, we mean a field K equipped with a function (valuation) $|\cdot|$ from K into $[0, \infty)$ such that $|r| = 0$ if and only if $r = 0$, $|rs| = |r||s|$ and $|r + s| \leq \max\{|r|, |s|\}$ for all $r, s \in K$.

Clearly $|1| = |-1| = 1$ and $|n| \leq 1$ for all $n \in \mathbb{N}$.

Let X be a vector space over a scalar field $\mathbb{K}$ with a non-Archimedean non-trivial valuation $|\cdot|$. A function $\|\cdot\| : X \to \mathbb{R}$ is a *non-Archimedean norm (valuation)* if it satisfies the following conditions:

(i) $\|x\| = 0$ if and only if $x = 0$;
(ii) $\|rx\| = |r| \, \|x\|$ ($r \in \mathbb{K}, x \in X$);
(iii) the strong triangle inequality (ultrametric); namely,

$$\|x + y\| \leq \max\{\|x\|, \|y\|\} \qquad (x, y \in X).$$

Then $(X, \|\cdot\|)$ is called a non-Archimedean space. Due to the fact that

$$\|x_n - x_m\| \leq \max\{\|x_{j+1} - x_j\| : m \leq j \leq n - 1\} \quad (n > m)$$

a sequence $\{x_n\}$ is Cauchy if and only if $\{x_{n+1} - x_n\}$ converges to zero in a non-Archimedean space. By a complete non-Archimedean space, we mean that every Cauchy sequence is convergent in the space.

12.3.1. *Generalized Hyers–Ulam stability of the Cauchy additive functional equation in non-Archimedean spaces*

By assuming H as an additive semigroup and X as a complete non-Archimedean space, the generalized Hyers–Ulam stability of the Cauchy functional equation in non-Archimedean spaces is presented in [198], which is presented in the following theorem.

Theorem 12.5 ([198]). *Let $\varphi : H \times H \to [0, \infty)$ be a function such that*

$$\lim_{n \to \infty} \frac{\varphi(2^n x, 2^n y)}{|2|^n} = 0 \qquad (x, y \in H), \tag{12.13}$$

and let for each $x \in H$ the limit

$$\lim_{n \to \infty} \max \left\{ \frac{\varphi(2^j x, 2^j x)}{|2|^j} : 0 \leq j < n \right\}, \tag{12.14}$$

denoted by $\tilde{\varphi}(x)$, exist. Suppose that $f : H \to X$ is a mapping satisfying

$$\|f(x + y) - f(x) - f(y)\| \leq \varphi(x, y) \qquad (x, y \in H). \tag{12.15}$$

Then there exists an additive mapping $T : H \to X$ such that

$$\|f(x) - T(x)\| \leq \frac{1}{|2|} \tilde{\varphi}(x) \qquad (x \in H). \tag{12.16}$$

Moreover, if

$$\lim_{k \to \infty} \lim_{n \to \infty} \max \left\{ \frac{\varphi(2^j x, 2^j x)}{|2|^j} : k \leq j < n + k \right\} = 0,$$

then T is the unique additive mapping satisfying (12.16).

Proof. Putting $y = x$ in (12.15), we obtain

$$\|f(2x) - 2f(x)\| \leq \varphi(x, x) \qquad (x \in H). \tag{12.17}$$

Let $x \in H$. Replacing x by $2^{n-1} x$ in (12.17), we obtain

$$\left\| \frac{f(2^n x)}{2^n} - \frac{f(2^{n-1} x)}{2^{n-1}} \right\| \leq \frac{\varphi(2^{n-1} x, 2^{n-1} x)}{|2|^n} \qquad (x \in H). \tag{12.18}$$

It follows from (12.18) and (12.13) that the sequence $\{\frac{f(2^n x)}{2^n}\}$ is Cauchy. Since X is complete, we conclude that $\{\frac{f(2^n x)}{2^n}\}$ is convergent in X. Set $T(x) := \lim_{n \to \infty} \frac{f(2^n x)}{2^n}$.

Using induction, one can show that

$$\left\|\frac{f\left(2^n x\right)}{2^n} - f(x)\right\| \leq \frac{1}{|2|} \max\left\{\frac{\varphi\left(2^k x, 2^k x\right)}{|2|^k} : 0 \leq k < n\right\} \qquad (12.19)$$

for all $n \in \mathbb{N}$ and all $x \in H$. Letting $n \to \infty$ in (12.19), and using (12.14) one obtains (12.16). Replacing x and y by $2^n x$ and $2^n y$, respectively, in (12.15), we have

$$\left\|\frac{f\left(2^n(x+y)\right)}{2^n} - \frac{f\left(2^n x\right)}{2^n} - \frac{2^n y}{2^n}\right\| \leq \frac{\varphi\left(2^n x, 2^n y\right)}{|2|^n} \qquad (x, y \in H).$$

Taking the limit as $n \to \infty$ and using (12.13), we obtain $T(x + y) = T(x) + T(y)$ $(x, y \in H)$.

If T' is another additive mapping satisfying (12.16), then

$$\|T(x) - T'(x)\| = \lim_{k\to\infty} |2|^{-k} \left\|T\left(2^k x\right) - T'\left(2^k x\right)\right\|$$

$$\leq \lim_{k\to\infty} |2|^{-k} \max\left\{\left\|T\left(2^k x\right) - f\left(2^k x\right)\right\|, \left\|f\left(2^k x\right) - T'\left(2^k x\right)\right\|\right\}$$

$$\leq \frac{1}{|2|} \lim_{k\to\infty} \lim_{n\to\infty} \max\left\{\frac{\varphi\left(2^j x, 2^j x\right)}{|2|^j} : k \leq j < n + k\right\}$$

$$= 0 \quad (x \in H).$$

Therefore $T = T'$. This completes the proof of the theorem. $\qquad\square$

Corollary 12.1 ([198]). *Let $\rho : [0, \infty) \to [0, \infty)$ be a function satisfying*

$$\rho(|2|t) \leq \rho(|2|)\rho(t) \ (t \geq 0), \quad \rho(|2|) < |2|.$$

Let $\delta > 0$, let H be a normed space, and let $f : H \to X$ fulfill the inequality

$$\|f(x + y) - f(x) - f(y)\| \leq \delta(\rho(\|x\|) + \rho(\|y\|)) \quad (x, y \in H).$$

Then there exists a unique additive mapping $T : H \to X$ such that

$$\|f(x) - T(x)\| \leq \frac{2}{|2|}\delta\rho(\|x\|) \quad (x \in H). \qquad (12.20)$$

Proof. Defining $\varphi : H \times H \to [0, \infty)$ by $\varphi(x, y) := \delta(\rho(\|x\|) + \rho(\|y\|))$, we have

$$\lim_{n\to\infty} \frac{\varphi\left(2^n x, 2^n y\right)}{|2|^n} \leq \lim_{n\to\infty} \left(\frac{\rho(|2|)}{|2|}\right)^n \varphi(x, y) = 0 \quad (x, y \in H),$$

$$\tilde{\varphi}(x) = \lim_{n\to\infty} \max\left\{\frac{\varphi\left(2^j x, 2^j x\right)}{|2|^j} : 0 \leq j < n\right\} = \varphi(x, x),$$

and

$$\lim_{k\to\infty}\lim_{n\to\infty}\max\left\{\frac{\varphi\left(2^j x, 2^j x\right)}{|2|^j} : k \le j < n+k\right\} = \lim_{k\to\infty}\frac{\varphi\left(2^k x, 2^k x\right)}{|2|^k} = 0.$$

Applying Theorem 12.5, we conclude the required result. This completes the proof.
□

12.3.2. *Generalized Hyers–Ulam stability of quadratic functional equation in non-Archimedean spaces*

By assuming G as an additive group and X as a complete non-Archimedean space, the generalized Hyers–Ulam stability of the quadratic functional equation in non-Archimedean spaces is provided in [198], which is presented in the following theorem.

Theorem 12.6 ([198]). *Let $\psi : G \times G \to [0, \infty)$ be a function such that*

$$\lim_{n\to\infty}\frac{\psi\left(2^n x, 2^n y\right)}{|4|^n} = 0 \quad (x, y \in G),$$

and let for each $x \in G$ the limit

$$\lim_{n\to\infty}\max\left\{\frac{\psi\left(2^j x, 2^j x\right)}{|4|^j} : 0 \le j < n\right\},$$

denoted by $\tilde{\psi}(x)$, exist. Suppose that $f : G \to X$ is a mapping satisfying $f(0) = 0$, and

$$\|f(x+y) + f(x-y) - 2f(x) - 2f(y)\| \le \psi(x, x) \quad (x \in G). \tag{12.21}$$

Then there exists a quadratic mapping $Q : G \to X$ such that

$$\|f(x) - Q(x)\| \le \frac{1}{|4|}\tilde{\psi}(x) \quad (x \in G). \tag{12.22}$$

Moreover, if

$$\lim_{k\to\infty}\lim_{n\to\infty}\max\left\{\frac{\psi\left(2^j x, 2^j x\right)}{|4|^j} : k \le j < n+k\right\} = 0,$$

then Q is the unique quadratic mapping satisfying (12.22).

Proof. Putting $y = x$ in (12.21), we obtain

$$\|f(2x) - 4f(x)\| \le \psi(x, x) \quad (x \in G). \tag{12.23}$$

Replacing x by $2^{n-1}x$ in (12.23), we obtain

$$\left\|\frac{f\left(2^n x\right)}{4^n} - \frac{f\left(2^{n-1}x\right)}{4^{n-1}}\right\| \le \frac{\psi\left(2^{n-1}x, 2^{n-1}x\right)}{|4|^n} \quad (x \in G).$$

Hence the sequence $\{\frac{f(2^n x)}{4^n}\}$ is Cauchy. Using the same method as in the proof of Theorem 12.5, we conclude that $Q(x) := \lim_{n\to\infty} \frac{f(2^n x)}{4^n}$ defines a quadratic mapping satisfying (12.22). This completes the proof of the theorem.

Corollary 12.2 ([198]). *Let* $\tau : [0, \infty) \to [0, \infty)$ *be a function satisfying*

$$\tau(|2|t) \leq \tau(|2|)\tau(t) \ (t \geq 0), \quad \tau(|2|) < |2|.$$

Let $\delta > 0$, *let* G *be a normed space and let* $f : G \to X$ *fulfill* $f(0) = 0$, *and the inequality*

$$\|f(x+y) + f(x-y) - 2f(x) - 2f(y)\| \leq \delta(\tau\|x\|\tau\|y\|) \quad (x, y \in G).$$

Then there exists a unique quadratic mapping $Q : G \to X$ *such that*

$$\|f(x) - Q(x)\| \leq \frac{\delta}{|4|}\tau(\|x\|)^2 \quad (x \in G).$$

Proof. Defining $\psi : G \times G \to [0, \infty)$ by $\psi(x, y) := \delta(\tau\|x\|\tau\|y\|)$ we have

$$\lim_{n\to\infty} \frac{\psi(2^n x, 2^n y)}{|4|^n} \leq \lim_{n\to\infty} \left(\frac{\tau(|2|)}{|2|}\right)^{2n} \psi(x, y) = 0 \quad (x, y \in G),$$

$$\tilde{\psi}(x) = \lim_{n\to\infty} \max\left\{\frac{\psi(2^j x, 2^j x)}{|4|^j} : 0 \leq j < n\right\} = \psi(x, x),$$

and

$$\lim_{k\to\infty}\lim_{n\to\infty} \max\left\{\frac{\psi(2^j x, 2^j x)}{|4|^j} : k \leq j < n+k\right\} = \lim_{k\to\infty} \frac{\psi(2^k x, 2^k x)}{|4|^k} = 0.$$

Applying Theorem 12.6, we conclude the required result. The proof is completed. □

12.4. Stability in generalized quasi-Banach spaces and in generalized p-Banach spaces

We recall some basic facts concerning quasi-Banach spaces, and some preliminary results.

Definition 12.3 ([226]). Let X be a linear space. A quasi-norm is a real-valued function on X satisfying the following:

(1) $\|x\| \geq 0$ for all $x \in X$, and $\|x\| = 0$ if and only if $x = 0$.
(2) $\|\lambda x\| = |\lambda| \cdot \|x\|$ for all $\lambda \in \mathbb{R}$ and all $x \in X$.
(3) There is a constant $K \geq 1$ such that $\|x + y\| \leq K(\|x\| + \|y\|)$ for all $x, y \in X$.

The pair $(X, \| \cdot \|)$ is called a quasi-normed space if $\| \cdot \|$ is a quasi-norm on X. A quasi-norm $\| \cdot \|$ is called a p-norm $0 < p \leq 1$ if

$$\|x + y\|^p \leq \|x\|^p + \|y\|^p$$

for all $x, y \in X$. In this case, a quasi-Banach space is called a p-Banach space.

Definition 12.4 ([226]). Let X be a linear space. A generalized quasi-norm is a real-valued function on X satisfying the following:

(1) $\|x\| \geq 0$ for all $x \in X$, and $\|x\| = 0$ if and only if $x = 0$.
(2) $\|\lambda x\| = |\lambda| \cdot \|x\|$ for all $\lambda \in \mathbb{R}$ and all $x \in X$.
(3) There is a constant $K \geq 1$ such that $\left\| \sum_{j=1}^{\infty} x_j \right\| \leq \sum_{j=1}^{\infty} K \|x_j\|$ for all $x_1, x_2, \cdots \in X$ with $\sum_{j=1}^{\infty} x_j \in X$.

The pair $(X, \| \cdot \|)$ is called a generalized quasi-normed space if $\| \cdot \|$ is a generalized quasi-norm on X. The smallest possible K is called the modulus of concavity of $\| \cdot \|$. A generalized quasi-Banach space is a complete generalized quasi-normed space. A generalized quasi-norm $\| \cdot \|$ is called a p-norm $(0 < p \leq 1)$ if

$$\|x + y\|^p \leq \|x\|^p + \|y\|^p$$

for all $x, y \in X$. In this case, a generalized quasi-Banach space is called a generalized p-Banach space.

12.4.1. *Stability of quadratic functional equation in quasi-Banach spaces*

In the following theorems, let X be a generalized quasi-normed vector space with generalized quasi-norm $\| \cdot \|$, and Y be a generalized quasi-Banach space with generalized quasi-norm $\| \cdot \|$. Let K be the modulus of concavity of $\| \cdot \|$. The stability of the quadratic functional equation in quasi-Banach spaces is presented in the following theorems.

Theorem 12.7 ([226]). *Let $f : X \to Y$ be a mapping satisfying $f(0) = 0$ for which there exists a function $\varphi : X \times X \to [0, \infty)$ such that*

$$\tilde{\varphi}(x, y) := \sum_{j=1}^{\infty} 4^j \varphi \left(\frac{x}{2^j}, \frac{y}{2^j} \right) < \infty, \tag{12.24}$$

and

$$\|f(x + y) + f(x - y) - 2f(x) - 2f(y)\| \leq \varphi(x, y) \tag{12.25}$$

for all $x, y \in X$. Then there exists a unique quadratic mapping $Q : X \to Y$ such that

$$\|f(x) - Q(x)\| \leq \frac{K}{4} \tilde{\varphi}(x, x) \tag{12.26}$$

for all $x \in X$.

Proof. Letting $y = x$ in (12.25), we obtain

$$\|f(2x) - 4f(x)\| \leq \varphi(x, x) \tag{12.27}$$

for all $x \in X$. So

$$\left\| 4^l f\left(\frac{x}{2^l}\right) - 4^m f\left(\frac{x}{2^m}\right) \right\| \leq K \sum_{j=l}^{m-1} 4^j \varphi\left(\frac{x}{2^{j+1}}, \frac{x}{2^{j+1}}\right) \tag{12.28}$$

for all non-negative integers m and l with $m > l$, and all $x \in X$. It follows from (12.24) and (12.28) that the sequence $\{4^n f(\frac{x}{2^n})\}$ is a Cauchy sequence for all $x \in X$. Since Y is complete, the sequence $\{4^n f(\frac{x}{2^n})\}$ converges. So one can define the mapping $Q : X \to Y$ by

$$Q(x) := \lim_{n \to \infty} 4^n f\left(\frac{x}{2^n}\right)$$

for all $x \in X$.

By (12.25) and (12.24), we have

$$\|Q(x + y) + Q(x - y) - 2Q(x) - 2Q(y)\|$$

$$= \lim_{n \to \infty} 4^n \left\| f\left(\frac{x+y}{2^n}\right) + f\left(\frac{x-y}{2^n}\right) - 2f\left(\frac{x}{2^n}\right) - 2f\left(\frac{y}{2^n}\right) \right\|$$

$$\leq \lim_{n \to \infty} 4^n \varphi\left(\frac{x}{2^n}, \frac{y}{2^n}\right) = 0$$

for all $x, y \in X$. So

$$Q(x + y) + Q(x - y) = 2Q(x) + 2Q(y)$$

for all $x, y \in X$. Moreover, letting $l = 0$, and passing the limit $m \to \infty$ in (12.28), we obtain (12.26).

Now, let $Q' : X \to Y$ be another quadratic mapping satisfying (12.26). Then we have

$$\|Q(x) - Q'(x)\| = 4^n \left\| Q\left(\frac{x}{2^n}\right) - Q'\left(\frac{x}{2^n}\right) \right\|$$

$$\leq 4^n K \left(\left\| Q\left(\frac{x}{2^n}\right) - f\left(\frac{x}{2^n}\right) \right\| + \left\| Q'\left(\frac{x}{2^n}\right) - f\left(\frac{x}{2^n}\right) \right\| \right)$$

$$\leq \frac{2K}{4} \cdot 4^n \tilde{\varphi}\left(\frac{x}{2^n}, \frac{x}{2^n}\right),$$

which tends to zero as $n \to \infty$ for all $x \in X$. So we can conclude that $Q(x) = Q'(x)$ for all $x \in X$. The proof is completed. $\square$

Corollary 12.3 ([226]). *Let $r > 2$, and θ be positive real numbers, and let $f : X \to Y$ be a mapping such that*

$$\|f(x + y) + f(x - y) - 2f(x) - 2f(y)\| \leq \theta \left(\|x\|^r + \|y\|^r\right)$$

for all $x, y \in X$. Then there exists a unique quadratic mapping $Q : X \to Y$ such that

$$\|f(x) - Q(x)\| \leq \frac{2K\theta}{2^r - 4}\|x\|^r$$

for all $x \in X$.

Proof. Define $\varphi(x, y) = \theta \left(\|x\|^r + \|y\|^r\right)$, and apply Theorem 12.7, and the proof is completed. $\qquad\square$

Theorem 12.8 ([226]). *Let $f : X \to Y$ be a mapping satisfying $f(0) = 0$ for which there exists a function $\varphi : X \times X \to [0, \infty)$ such that*

$$\tilde{\varphi}(x, y) := \sum_{j=0}^{\infty} \frac{1}{4^j} \varphi\left(2^j x, 2^j y\right) < \infty, \tag{12.29}$$

$$\|f(x + y) + f(x - y) - 2f(x) - 2f(y)\| \leq \varphi(x, y) \tag{12.30}$$

for all $x, y \in X$. Then there exists a unique quadratic mapping $Q : X \to Y$ such that

$$\|f(x) - Q(x)\| \leq \frac{K}{4} \tilde{\varphi}(x, x) \tag{12.31}$$

for all $x \in X$.

Proof. It follows from (12.27) that

$$\left\|f(x) - \frac{1}{4}f(2x)\right\| \leq \frac{1}{4}\varphi(x, x)$$

for all $x \in X$. Hence

$$\left\|\frac{1}{4^l}f\left(2^l x\right) - \frac{1}{4^m}f\left(2^m x\right)\right\| \leq K \sum_{j=l}^{m-1} \frac{1}{4^{j+1}} \varphi\left(2^j x, 2^j x\right) \tag{12.32}$$

for all non-negative integers m and l with $m > l$ and all $x \in X$. It follows from (12.29) and (12.32) that the sequence $\left\{\frac{1}{4^n} f\left(2^n x\right)\right\}$ is a Cauchy sequence for all $x \in X$. Since Y is complete, the sequence $\left\{\frac{1}{4^n} f\left(2^n x\right)\right\}$ converges. So one can define the mapping $Q : X \to Y$ by

$$Q(x) := \lim_{n \to \infty} \frac{1}{4^n} f\left(2^n x\right)$$

for all $x \in X$.

By (12.30) and (12.29), we have

$$\|Q(x+y) + Q(x-y) - 2Q(x) - 2Q(y)\|$$

$$= \lim_{n\to\infty} \frac{1}{4^n} \|f\left(2^n(x+y)\right) + f\left(2^n(x-y)\right) - 2f\left(2^n x\right) - 2f\left(2^n y\right)\|$$

$$\leq \lim_{n\to\infty} \frac{1}{4^n} \varphi\left(2^n x, 2^n y\right) = 0$$

for all $x, y \in X$. So

$$Q(x+y) + Q(x-y) = 2Q(x) + 2Q(y)$$

for all $x, y \in X$. Moreover, letting $l = 0$, and passing the limit $m \to \infty$ in (12.32), we obtain (12.31). The rest of the proof is similar to that of Theorem 12.7. The proof is now completed. $\square$

Corollary 12.4 ([226]). *Let $r < 2$ and θ be positive real numbers, and let $f : X \to Y$ be a mapping such that for all $x, y \in X$. Then there exists a unique quadratic mapping $Q : X \to Y$ such that*

$$\|f(x) - Q(x)\| \leq \frac{2K\theta}{4 - 2^r} \|x\|^r$$

for all $x \in X$.

Proof. Define $\varphi(x, y) = \theta\left(\|x\|^r + \|y\|^r\right)$, and apply Theorem 12.8, and this completes the proof. $\square$

12.4.2. *Stability of quadratic functional equation in generalized p-Banach space*

In the following theorems, X is a generalized quasi-normed vector space with generalized quasi-norm $\| \cdot \|$, and Y is a generalized p-Banach space with generalized quasi-norm $\| \cdot \|$.

Theorem 12.9 ([226]). *Let $r > 2$ and θ be positive real numbers, and let $f : X \to Y$ be a mapping such that*

$$\|f(x+y) + f(x-y) - 2f(x) - 2f(y)\| \leq \theta\left(\|x\|^r + \|y\|^r\right) \tag{12.33}$$

for all $x, y \in X$. Then there exists a unique quadratic mapping $Q : X \to Y$ such that

$$\|f(x) - Q(x)\| \leq \frac{2\theta}{(2^{pr} - 4^p)^{\frac{1}{p}}} \|x\|^r \tag{12.34}$$

for all $x \in X$.

Proof. Letting $y = x$ in (12.33), we obtain

$$\|f(2x) - 4f(x)\| \leq 2\theta \|x\|^r \tag{12.35}$$

for all $x \in X$. So

$$\left\| f(x) - 4f\left(\frac{x}{2}\right) \right\| \leq \frac{2}{2^r} \theta \|x\|^r$$

for all $x \in X$. Since Y is a generalized p-Banach space,

$$\left\| 4^l f\left(\frac{x}{2^l}\right) - 4^m f\left(\frac{x}{2^m}\right) \right\|^p \leq \sum_{j=l}^{m-1} \left\| 4^j f\left(\frac{x}{2^j}\right) - 4^{j+1} f\left(\frac{x}{2^{j+1}}\right) \right\|$$

$$\leq \sum_{j=l}^{m-1} \frac{4^{pj}}{2^{prj}} \cdot \frac{2^p}{2^{pr}} \theta^p \|x\|^{pr} \tag{12.36}$$

for all non-negative integers m and l with $m > l$, and all $x \in X$. So the sequence $\left\{ 4^n f\left(\frac{x}{2^n}\right) \right\}$ is a Cauchy sequence for all $x \in X$. Since Y is complete, the sequence $\left\{ 4^n f\left(\frac{x}{2^n}\right) \right\}$ converges. So one can define the mapping $Q : X \to Y$ by

$$Q(x) := \lim_{n \to \infty} 4^n f\left(\frac{x}{2^n}\right)$$

for all $x \in X$. By (12.33), we have

$$\|Q(x+y) + Q(x-y) - 2Q(x) - 2Q(y)\|$$

$$= \lim_{n \to \infty} 4^n \left\| f\left(\frac{x+y}{2^n}\right) + f\left(\frac{x-y}{2^n}\right) - 2f\left(\frac{x}{2^n}\right) - 2f\left(\frac{y}{2^n}\right) \right\|$$

$$\leq \lim_{n \to \infty} \frac{4^n}{2^{rn}} \theta \left(\|x\|^r + \|y\|^r \right) = 0$$

for all $x, y \in X$. So

$$Q(x+y) + Q(x-y) = 2Q(x) + 2Q(y)$$

for all $x, y \in X$. Moreover, letting $l = 0$, and passing the limit $m \to \infty$ in (12.36), we obtain (12.34).

Now, let $Q' : X \to Y$ be another quadratic mapping satisfying (12.34). Then we have

$$\|Q(x) - Q'(x)\|^p = 4^{pn} \left\| Q\left(\frac{x}{2^n}\right) - Q'\left(\frac{x}{2^n}\right) \right\|^p$$

$$\leq 4^{pn} \left(\left\| Q\left(\frac{x}{2^n}\right) - f\left(\frac{x}{2^n}\right) \right\|^p + \left\| Q'\left(\frac{x}{2^n}\right) - f\left(\frac{x}{2^n}\right) \right\|^p \right)$$

$$\leq 2 \cdot \frac{4^{pn}}{2^{prn}} \cdot \frac{2^p \theta^p}{2^{pr} - 4^p} \|x\|^{pr},$$

which tends to zero as $n \to \infty$ for all $x \in X$. So we can conclude that $Q(x) = Q'(x)$ for all $x \in X$. This completes the proof. $\qquad\square$

Theorem 12.10 ([226]). *Let $r < 2$ and θ be positive real numbers, and let $f : X \to Y$ be a mapping such that*

$$\|f(x+y) + f(x-y) - 2f(x) - 2f(y)\| \leq \theta \left(\|x\|^r + \|y\|^r\right) \tag{12.37}$$

for all $x, y \in X$. Then there exists a unique quadratic mapping $Q : X \to Y$ such that

$$\|f(x) - Q(x)\| \leq \frac{2\theta}{(4^p - 2^{pr})^{\frac{1}{p}}} \|x\|^r \tag{12.38}$$

for all $x \in X$.

Proof. Letting $y = x$ in (12.37), we obtain

$$\|f(2x) - 4f(x)\| \leq 2\theta \|x\|^r \tag{12.39}$$

for all $x \in X$. So

$$\left\| f(x) - \frac{1}{4} f(2x) \right\| \leq \frac{1}{2} \theta \|x\|^r$$

for all $x \in X$. Since Y is a generalized p-Banach space,

$$\left\| \frac{1}{4^l} f\left(2^l x\right) - \frac{1}{4^m} f\left(2^m x\right) \right\|^p \leq \sum_{j=l}^{m-1} \left\| \frac{1}{4^j} f\left(2^j x\right) - \frac{1}{4^{j+1}} f\left(2^{j+1} x\right) \right\|^p$$

$$\leq \sum_{j=l}^{m-1} \frac{2^{prj}}{4^{pj}} \cdot \frac{\theta^p}{2^p} \|x\|^{pr} \tag{12.40}$$

for all non-negative integers m and l with $m > l$, and all $x \in X$. So the sequence $\left\{ \frac{1}{4^n} f\left(2^n x\right) \right\}$ is a Cauchy sequence for all $x \in X$. Since Y is complete, the sequence $\left\{ \frac{1}{4^n} f\left(2^n x\right) \right\}$ converges. So one can define the mapping $Q : X \to Y$ by

$$Q(x) := \lim_{n \to \infty} \frac{1}{4^n} f\left(2^n x\right)$$

for all $x \in X$. By (12.37), we have

$$\|Q(x+y) + Q(x-y) - 2Q(x) - 2Q(y)\|$$
$$= \lim_{n \to \infty} \frac{1}{4^n} \|f\left(2^n (x+y)\right) + f\left(2^n (x-y)\right) - 2f\left(2^n x\right) - 2f\left(2^n y\right)\|$$
$$\leq \lim_{n \to \infty} \frac{2^{rn}}{4^n} \theta \left(\|x\|^r + \|y\|^r\right) = 0$$

for all $x, y \in X$. So

$$Q(x+y) + Q(x-y) = 2Q(x) + 2Q(y)$$

for all $x, y \in X$. Moreover, letting $l = 0$, and passing the limit $m \to \infty$ in (12.40), we obtain (12.38).

Now, let $Q' : X \to Y$ be another quadratic mapping satisfying (12.38). Then we have

$$\|Q(x) - Q'(x)\|^p = \frac{1}{4^{pn}} \|Q(2^n x) - Q'(2^n x)\|^p$$

$$\leq \frac{1}{4^{pn}} (\|Q(2^n x) - f(2^n x)\|^p + \|Q^{prime}(2^n x) - f(2^n x)\|^p)$$

$$\leq 2 \cdot \frac{2^{prn}}{4^{pn}} \cdot \frac{2^p \theta^p}{4^p - 2^{pr}} \|x\|^{pr},$$

which tends to zero as $n \to \infty$ for all $x \in X$. So we can conclude that $Q(x) = Q'(x)$ for all $x \in X$. The proof is now completed. $\qquad\square$

12.5. Stability in fuzzy Banach spaces

We present here the basic definition of fuzzy normed vector space and fuzzy Banach space.

Definition 12.5 ([175]). Let X be a real vector space. A function $N : X \times \mathbb{R} \to [0, 1]$ is called a fuzzy norm on X if, for all $x, y \in X$, and all $s, t \in \mathbb{R}$:

(N_1) $N(x, t) = 0$ for $t \leq 0$,
(N_2) $x = 0$ if and only if $N(x, t) = 1$ for all $t > 0$,
(N_3) $N(cx, t) = N(x, \frac{t}{|c|})$ if $c \neq 0$,
(N_4) $N(x + y, s + t) \geq \min\{N(x, s), N(y, t)\}$,
(N_5) $N(x, \cdot)$ is a non-decreasing function of $\mathbb{R}$, and $\lim_{n \to \infty} N(x, t) = 1$,
(N_6) for $x \neq 0$, $N(x, \cdot)$ is continuous on $\mathbb{R}$.

The pair (X, N) is called a *fuzzy normed vector space*.

Definition 12.6 ([175]). Let (X, N) be a fuzzy normed vector space. A sequence $\{x_n\}$ in X is said to be *convergent* or *converges* if there exists a $x \in X$ such that $\lim_{n \to \infty} N(x_n - x, t) = 1$ for all $t > 0$. In this case, x is called the limit of the sequence $\{x_n\}$, and we denote it by $N\text{-}\lim_{n \to \infty} x_n = x$.

Definition 12.7 ([175]). Let (X, N) be a fuzzy normed vector space. A sequence $\{x_n\}$ in X is called Cauchy if for each $\epsilon > 0$ and each $t > 0$ there exists an $n_0 \in \mathbb{N}$ such that, for all $n \geq n_0$ and all $p > 0$, we have $N(x_{n+p} - x_n, t) > 1 - \epsilon$.

It is well known that every convergent sequence in a fuzzy normed vector space is Cauchy. If each Cauchy sequence is convergent, then the fuzzy norm is said to be *complete*, and the fuzzy normed vector space is called a *fuzzy Banach space*.

We say that a mapping $f : X \to Y$ between fuzzy normed vector spaces X and Y is continuous at a point $x_0 \in X$ if, for each sequence $\{x_n\}$ converging to x_0 in X, the sequence $|f(x_n)|$ converges to $f(x_0)$. If $f : X \to Y$ at each $x \in X$, then $f : X \to Y$ is said to be *continuous* on X.

12.5.1. *Generalized Hyers–Ulam stability of the quadratic functional equation in fuzzy Banach spaces*

The generalized Hyers–Ulam stability of the quadratic functional equation in fuzzy Banach spaces is presented in [175] with the assumption that X to be a vector space and (Y, N) to be a fuzzy Banach space, which is presented in the following theorem.

Theorem 12.11 ([175]). *Let $\varphi : X \times X \to [0, \infty)$ be a function such that*

$$\tilde{\varphi}(x, y) := \sum_{n=0}^{\infty} 4^{-n} \varphi\left(2^n x, 2^n y\right) < \infty \tag{12.41}$$

for all $x, y \in X$. Let $f : X \to Y$ be a mapping with $f(0) = 0$ such that

$$\lim_{t \to \infty} N(f(x + y) + f(x - y) - 2f(x) - 2f(y), t\varphi(x, y)) = 1 \tag{12.42}$$

uniformly on $X \times X$. Then $Q(x) = N\text{-}\lim_{n \to \infty} \frac{f(2^n x)}{4^n}$ exists for each $x \in X$, and defines a quadratic mapping $Q : X \to Y$ such that if for some $\delta > 0$, $\alpha > 0$

$$N(f(x + y) + f(x - y) - 2f(x) - 2f(y), \delta\varphi(x, y)) \geq \alpha \tag{12.43}$$

for all $x, y \in X$, then

$$N\left(f(x) - Q(x), \frac{\delta}{4}\tilde{\varphi}(x, x)\right) \geq \alpha \tag{12.44}$$

for all $x \in X$. Furthermore, the quadratic mapping $Q : X \to Y$ is a unique mapping such that

$$\lim_{t \to \infty} N(f(x) - Q(x), t\tilde{\varphi}(x, x)) = 1 \tag{12.45}$$

uniformly on X.

Proof. For a given $\epsilon > 0$, by (12.42), we can find some $t_0 > 0$ such that

$$N(f(x + y) + f(x - y) - 2f(x) - 2f(y), t\varphi(x, y)) \geq 1 - \epsilon \tag{12.46}$$

for all $y \geq t_0$. By induction on n, we show that

$$N\left(4^n f(x) - f\left(2^n x\right), t\sum_{k=0}^{n-1} 4^{n-k-1} \varphi\left(2^k x, 2^k x\right)\right) \geq 1 - \epsilon \tag{12.47}$$

for all $t \geq t_0$, all $x \in X$, and all $n \in \mathbb{N}$. Letting $y = x$ in (12.46), we obtain

$$N(4f(x) - f(2x), t\varphi(x, x)) \geq 1 - \epsilon \tag{12.48}$$

for all $x \in X$ and all $t \geq t_0$. So we obtain (12.47) for $n = 1$. Assume that (12.47) holds for $n \in \mathbb{N}$. Then

$$N(4^{n+1}f(x) - f(2^{n+1}x), t \sum_{k=0}^{n} 4^{n-k}\varphi(2^k x, 2^k x))$$

$$\geq \min\left\{ N\left(4^{n+1}f(x) - 4f(2^n x), t_0 \sum_{k=0}^{n-1} 4^{n-k}\varphi(2^k x, 2^k x) \right), \right.$$

$$\left. N(4f(2^n x) - f(2^{n+1}x), t_0\varphi(2^n x, 2^n x)) \right\}$$

$$\geq \min\{1 - \epsilon, 1 - \epsilon\} = 1 - \epsilon. \tag{12.49}$$

This completes the induction argument. Letting $t = t_0$, and replacing n and x by p and $2^n x$ in (12.47), respectively, we obtain

$$N\left(\frac{f(2^n x)}{4^n} - \frac{f(2^{n+p}x)}{4^{n+p}}, \frac{t_0}{4^{n+p}} \sum_{k=0}^{p-1} 4^{p-k-1}\varphi\left(2^{n+k}x, 2^{n+k}x\right) \right) \geq 1 - \epsilon \quad (12.50)$$

for all integers $n \geq 0$, $p > 0$.

It follows from (12.41), and the equality

$$\sum_{k=0}^{p-1} 4^{-n-k-1}\varphi\left(2^{n+k}x, 2^{n+k}x\right) = \frac{1}{4} \sum_{k=n}^{n++p-1} 4^{-k}\varphi\left(2^k x, 2^k x\right) \tag{12.51}$$

that for a given $\delta > 0$ there is an $n_0 \in \mathbb{N}$ such that

$$\frac{t_0}{2} \sum_{k=n}^{n+p-1} 4^{-k}\varphi\left(2^k x, 2^k x\right) < \delta \tag{12.52}$$

for all $n \geq n_0$, and $p > 0$. Now we deduce from (12.50) that

$$N\left(\frac{f(2^n x)}{4^n} - \frac{f(2^{n+p}x)}{4^{n+p}}, \delta \right)$$

$$\geq N\left(\frac{f(2^n x)}{4^n} - \frac{f(2^{n+p}x)}{4^{n+p}}, \frac{t_0}{4^{n+p}} \sum_{k=0}^{p-1} 4^{p-k-1}\varphi\left(2^{n+k}x, 2^{n+k}x\right) \right)$$

$$\geq 1 - \epsilon \tag{12.53}$$

for all $n \geq n_0$, and all $p > 0$. Thus the sequence $\{\frac{f(2^n x)}{4^n}\}$ is Cauchy in Y. Since Y is a fuzzy Banach space, the sequence $\{\frac{f(2^n x)}{4^n}\}$ converges to some $Q(x) \in Y$. So we can define a mapping $Q : X \to Y$ by $Q(x) := N\text{-}\lim_{n \to \infty} \frac{f(2^n x)}{4^n}$; namely, for each $t > 0$ and $x \in X$, $\lim_{n \to \infty} N(\frac{f(2^n x)}{4^n} - Q(x), t) = 1$.

Let $x, y \in X$. Fix $t > 0$ and $0 < \epsilon < 1$. Since $\lim_{n \to \infty} 4^{-n} \varphi(2^n x, 2^n y) = 0$, there is an $n_1 > n_0$ such that $t_0 \varphi(2^n x, 2^n y) < \frac{4^n t}{4}$ for all $n \geq n_1$. Hence for all $n \geq n_1$, we have

$$N(Q(x + y) + Q(x - y) - 2Q(x) - 2Q(y), t)$$
$$\geq \min \left\{ N\left(Q(x+y) - 4^{-n} f\left(2^n x + 2^n y\right), \frac{t}{8} \right), \right.$$
$$N\left(Q(x-y) - 4^{-n} f\left(2^n x - 2^n y\right), \frac{t}{8} \right),$$
$$N\left(2Q(x) - 4^{-n} \cdot 2f\left(2^n x\right), \frac{t}{4} \right), N\left(2Q(y) - 4^{-n} \cdot 2f\left(2^n y\right), \frac{t}{4} \right),$$
$$\left. N\left(f\left(2^n(x+y)\right) - f\left(2^n(x-y)\right) - 2f\left(2^n x\right) - 2f\left(2^n y\right), \frac{4^n t}{4} \right) \right\}.$$

$$(12.54)$$

The first four terms on the right-hand side of the above inequality tend to 1 as $n \to \infty$, and the fifth term is greater than

$$N\left(f\left(2^n(x+y)\right) + f\left(2^n(x-y)\right) - 2f\left(2^n x\right) - 2f\left(2^n y\right), t_0 \varphi\left(2^n x, 2^n y\right) \right), \quad (12.55)$$

which is greater than or equal to $1 - \epsilon$. Thus

$$N(Q(x+y) + Q(x-y) - 2Q(x) - 2Q(y), t) \geq 1 - \epsilon \qquad (12.56)$$

for all $t > 0$. Since $N(Q(x+y) + Q(x-y) - 2Q(x) - 2Q(y), t) = 1$ for all $t > 0$, by (N_2), $Q(x+y) + Q(x-y) - 2Q(x) - 2Q(y) = 0$ for all $x \in X$. Thus the mapping $Q : X \to Y$ is quadratic, that is, $Q(x+y) + Q(x-y) = 2Q(x) + 2Q(y)$ for all $x, y \in X$.

Now, let, for some positive δ, and α, (12.43) hold. Let

$$\varphi_n(x, y) = \sum_{k=0}^{n-1} 4^{-k-1} \varphi\left(2^n x, 2^n y\right) \qquad (12.57)$$

for all $x, y \in X$. Let $x \in X$. By the same reasoning as in the beginning of the proof, one can deduce from (12.43) that

$$N\left(4^n f(x) - f\left(2^n x\right), \delta \sum_{k=0}^{n-1} 4^{n-k-1} \varphi\left(2^k x, 2^k x\right) \right) \geq \alpha \qquad (12.58)$$

for all positive integers n. Let $t > 0$. We have

$$N(f(x) - Q(x), \delta \varphi_n(x, x) + t)$$
$$\geq \min \left\{ N\left(f(x) - \frac{f(2^n x)}{4^n}, \delta \varphi_n(x, x) \right), N\left(\frac{f(2^n x)}{4^n} - Q(x), t \right) \right\}. \quad (12.59)$$

Combining (12.58) and (12.59) and the fact that $\lim_{n\to\infty} N\left(\frac{f(2^n x)}{4^n} - Q(x), t\right) = 1$, we observe that

$$N(f(x) - Q(x), \delta\varphi_n(x, x) + t) \geq \alpha \qquad (12.60)$$

for large enough $n \in \mathbb{N}$. Thanks to the continuity of the function $N(f(x) - Q(x), \cdot)$, we see that $N\left(f(x) - Q(x), \frac{\delta}{4}\tilde\varphi(x, x) + t\right) \geq \alpha$. Letting $t \to 0$, we conclude that

$$N\left(f(x) - Q(x), \frac{\delta}{4}\tilde\varphi(x, x)\right) \geq \alpha. \qquad (12.61)$$

To end the proof, it remains to prove the uniqueness assertion. Let T be another quadratic mapping satisfying (12.45). Fix $c > 0$. Given that $\epsilon > 0$, by (12.45) for Q and T, we can find some $t_0 > 0$ such that

$$N\left(f(x) - Q(x), \frac{t}{2}\tilde\varphi(x, x)\right) \geq 1 - \epsilon,$$

$$N\left(f(x) - T(x), \frac{t}{2}\tilde\varphi(x, x)\right) \geq 1 - \epsilon \qquad (12.62)$$

for all $x \in X$ and all $t \geq 2t_0$. Fix some $x \in X$ and find some integer n_0 such that

$$t_0 \sum_{k=n}^{\infty} 4^{-k}\varphi\left(2^k x, 2^k x\right) < \frac{c}{2} \qquad (12.63)$$

for all $n \geq n_0$. Since

$$\sum_{k=n}^{\infty} 4^{-k}\varphi\left(2^k x, 2^k x\right) = \frac{1}{4^n}\sum_{k=n}^{\infty} 4^{-(k-n)}\varphi\left(2^{k-n}\left(2^n x\right), 2^{k-n}\left(2^n x\right)\right)$$

$$= \frac{1}{4^n}\sum_{m=0}^{\infty} 4^{-m}\varphi\left(2^m\left(2^n x\right), 2^m\left(2^n x\right)\right)$$

$$= \frac{1}{4^n}\tilde\varphi\left(2^n x, 2^n x\right), \qquad (12.64)$$

we have

$$N(Q(x) - T(x), c)$$

$$\geq \min\left\{N\left(\frac{f\left(2^n x\right)}{4^n} - Q(x), \frac{c}{2}\right), N\left(T(x) - \frac{f\left(2^n x\right)}{4^n}, \frac{c}{2}\right)\right\}$$

$$\geq \min\left\{N\left(f\left(2^n x\right) - Q\left(2^n x\right), 4^n t_0 \sum_{k=n}^{\infty} 4^{-k}\varphi\left(2^k x, 2^k x\right)\right),\right.$$

$$\left. N\left(T\left(2^n x\right) - f\left(2^n x\right), 4^n t_0 \sum_{k=n}^{\infty} 4^{-k}\varphi\left(2^k x, 2^k x\right)\right)\right\}$$

$$\geq 1 - \epsilon. \qquad (12.65)$$

It follows that $N(Q(x) - T(x), c) = 1$ for all $c > 0$. Thus $Q(x) = T(x)$ for all $x \in X$. The proof is now completed. $\qquad\square$

Corollary 12.5 ([175]). *Let $\theta \geq 0$, and let p be a real number with $0 < p < 2$. Let $f : X \to Y$ be a mapping with $f(0) = 0$ such that*

$$\lim_{t \to \infty} N\left(f(x+y) + f(x-y) - 2f(x) - 2f(y), t\theta\left(\|x\|^p + \|y\|^p\right)\right) = 1 \qquad (12.66)$$

uniformly on $X \times X$. Then $Q(x) := N\text{-}\lim_{n \to \infty} \frac{f(2^n x)}{4^n}$ exists for each $x \in X$, and defines a quadratic mapping $Q; X \to Y$ such that if for some $\delta > 0$, $\alpha > 0$

$$N\left(f(x+y) + f(x-y) - 2f(x) - 2f(y), \delta\theta\left(\|x\|^p + \|y\|^p\right)\right) \geq \alpha \qquad (12.67)$$

for all $x, y \in X$, then

$$N\left(f(x) - Q(x), \frac{2\delta\theta}{4 - 2^p}\|x\|^p\right) \geq \alpha \qquad (12.68)$$

for all $x \in X$.

Furthermore, the quadratic mapping $Q : X \to Y$ is a unique mapping such that

$$\lim_{t \to \infty} N\left(f(x) - Q(x), \frac{8}{4 - 2^p}t\theta\|x\|^p\right) = 1 \qquad (12.69)$$

uniformly on X.

Proof. Define $\varphi(x, y) = \theta\left(\|x\|^p + \|y\|^p\right)$, and apply Theorem 12.11 to get the desired result. $\qquad\square$

Similarly, we can obtain the following theorem, and the details of the proof is omitted.

Theorem 12.12 ([175]). *Let $\varphi : X \times X \to [0, \infty)$ be a function such that*

$$\tilde{\varphi}(x, y) = \sum_{n=1}^{\infty} 4^n \varphi\left(\frac{x}{2^n}, \frac{x}{2^n}\right) < \infty \qquad (12.70)$$

for all $x, y \in X$. Let $f : X \to Y$ be a mapping satisfying (12.42), and $f(0) = 0$. Then $Q(x) = N\text{-}\lim_{n \to \infty} 4^n f(\frac{x}{2^n})$ exists for each $x \in X$, and defines a quadratic mapping $Q : X \to Y$ such that if for some $\delta > 0$, $\alpha > 0$

$$N(f(x+y) + f(x-y) - 2f(x) - 2f(y), \delta\varphi(x,y)) \geq \alpha \qquad (12.71)$$

for all $x, y \in X$, then

$$N\left(f(x) - Q(x), \frac{\delta}{4}\tilde{\varphi}(x, x)\right) \geq \alpha \qquad (12.72)$$

for all $x \in X$. Furthermore, the quadratic mapping $Q : X \to Y$ is a unique mapping such that

$$\lim_{t \to \infty} N(f(x) - Q(x), t\tilde{\varphi}(x, x)) = 1 \tag{12.73}$$

uniformly on X.

Corollary 12.6 ([175]). *Let $\theta \geq 0$, and let p be a real number with $p > 2$. Let $f : X \to Y$ be a mapping satisfying (12.66), and $f(0) = 0$. Then $Q(x) := N\text{-}\lim_{n \to \infty} 4^n f\left(\frac{x}{2^n}\right)$ exists for each $x \in X$, and defines a quadratic mapping $Q : X \to Y$ such that if for some $\delta > 0$, $\alpha > 0$*

$$N\left(f(x+y) + f(x-y) - 2f(x) - 2f(y), \delta\theta\left(\|x\|^p + \|y\|^p\right)\right) \geq \alpha \tag{12.74}$$

for all $x, y \in X$, then

$$N\left(f(x) - Q(x), \frac{2\delta\theta}{2^p - 4}\|x\|^p\right) \geq \alpha \tag{12.75}$$

for all $x \in X$.

     Furthermore, the quadratic mapping $Q : X \to Y$ is a unique mapping such that

$$\lim_{t \to \infty} N\left(f(x) - Q(x), \frac{8}{2^p - 4}t\theta\|x\|^p\right) = 1 \tag{12.76}$$

uniformly on X.

Proof. Define $\varphi(x, y) := \theta\left(\|x\|^p + \|y\|^p\right)$, and apply Theorem 12.12 to get the desired result. $\square$

12.6. Stability in Menger probabilistic normed spaces

In this section, we present stability results for functional equation in Menger probabilistic normed spaces.

Definition 12.8 ([67]). A function $F : \mathbb{R} \to [0, 1]$ is called a distribution function if it is non-decreasing and left-continuous, with $\sup_{t \in \mathbb{R}} F(t) = 1$, and $\inf_{t \in \mathbb{R}} F(t) = 0$.

     The class of all distribution functions F with $F(0) = 0$ is denoted by D_+. Now ϵ_0 is the element of D_+ defined by

$$\epsilon_0 = \begin{cases} 1, & t > 0, \\ 0, & t \leq 0. \end{cases}$$

Definition 12.9 ([67]). A binary operation $\star : [0, 1] \times [0, 1]$ is said to be a t-norm if it satisfies the following conditions:

(1) $\star$ is commutative and associative;
(2) $\star$ is continuous;
(3) $a \star 1 = a$ for all $a \in [0, 1]$;
(4) $a \star b \leq c \star d$ whenever $a \leq c$ and $b \leq d$ for each $a, b, c, d \in [0, 1]$.

Definition 12.10 ([67]). Let X be a real vector space, F a mapping from X to D_+ (for any $x \in X$, $F(x)$ is denoted by F_x), and $\star$ a t-norm. The triple $(X, F, \star)$ is called a Menger Probabilistic Normed space (briefly Menger PN-space) if the following conditions are satisfied:

(1) $F_x(0) = 0$, for all $x \in X$;
(2) $F_x = \epsilon_0$ if and only if $x = \theta$;
(3) $F_{\alpha x}(t) = F_x(\frac{t}{|\alpha|})$ for all $\alpha \in \mathbb{R}$, $\alpha \neq 0$ and $x \in X$;
(4) $F_{x+y}(t_1 + t_2) \geq F_x(t_1) \star F_x(t_2)$, for all $x, y \in X$ and $t_1, t_2 > 0$.

Definition 12.11 ([67]). Let $(X, F, \star)$ be a Menger PN-space, and let $\{x_n\}$ be a sequence in X. Then $\{x_n\}$ is said to be convergent if there exists $x \in X$ such that

$$\lim_{n \to \infty} F_{x_n - x}(t) = 1$$

for all $t > 0$. In this case x is called the limit of $\{x_n\}$.

Definition 12.12 ([67]). The sequence $\{x_n\}$ in Menger PN-space $(X, F, \star)$ is called Cauchy if for each $\epsilon > 0$ and $\delta > 0$, there exists some n_0 such that $F_{x_n - x_m}(\delta) > 1 - \epsilon$ for all $m, n \geq n_0$.

Clearly, every convergent sequence in Menger PN-space is Cauchy. If each Cauchy sequence is convergent sequence in a Menger PN-space $(X, F, \star)$, then $(X, F, \star)$ is called Menger Probabilistic Banach space (briefly, Menger PB-space).

Definition 12.13 ([67]). Let $(X, F, \star)$ be a Menger PN-space, and $(Y, G, \star)$ be a Menger PB-space. A mapping $f : X \to Y$ is said to be P-approximately quadratic if

$$G_{f(x++y)+f(x-y)-2f(x)-2f(y)}(t + s) \geq F_x(t) \star F_y(s), \tag{12.77}$$

for all $x, y \in X$, $t, s \in [0, \infty)$.

12.6.1. *Hyers–Ulam–Rassias stability of the P-approximately quadratic functional equation*

The following theorem gives Hyers–Ulam–Rassias stability of the P-approximately quadratic functional equation.

Theorem 12.13 ([67]). *Let $f : X \to Y$ be a P-approximately quadratic functional equation. Then there exists a unique quadratic mapping $Q : X \to Y$ such that*

$$G_{Q(x)-f(x)}(t) \geq F_x(t) \tag{12.78}$$

for all $x \in X$, $t > 0$.

Proof. Put $x = y$, and $s = t$ in (12.77) to obtain

$$G_{f(2x)-4f(x)}(2t) \geq F_x(t). \tag{12.79}$$

Replacing x by $2^n x$ in (12.79), we see that

$$G_{f(2^{n+1}x)-4f(2^n x)}(2t) \geq F_{2^n x}(t).$$

It follows that

$$G_{f(2^{n+1}x)-4f(2^n x)}\left(2^{n+1}t\right) \geq F_x(t).$$

Whence

$$G_{\frac{f(2^{n+1}x)}{4^{n+1}} - \frac{f(2^n x)}{4^n}}\left(2^{-n-1}t\right) \geq F_x(t).$$

If $n > m > 0$, then

$$G_{\frac{f(2^n x)}{4^n} - \frac{f(2^m x)}{4^m}}\left(\sum_{k=m+1}^{n} 2^{-k-1}t\right) \geq G_{\sum_{k=m+1}^{n}\left(\frac{f(2^k x)}{4^k}\right) - \frac{f(2^{k-1}x)}{4^{k-1}}}\left(\sum_{k=m+1}^{n} 2^{-k-1}t\right)$$

$$\geq \prod_{k=m+1}^{n} G_{\frac{f(2^k x)}{4^k} - \frac{f(2^{k-1}x)}{4^{k-1}}}\left(2^{-k-1}t\right)$$

$$\geq F_x(t). \tag{12.80}$$

Let $c > 0$, and ϵ be given. Since

$$\lim_{n\to\infty} F_x(t) = 1,$$

there is some $t_0 > 0$ such that $F_x(t_0) \geq 1 - \epsilon$. Fix some $t > t_0$. The convergence of the series $\sum_{n=1}^{\infty} 2^{-n-1}t$ shows that there exists some $n_0 \geq 0$ such that for each $n > m \geq n_0$, the inequality $\sum_{k=m+1}^{n} 2^{-k-1}t < c$ holds. It follows that

$$G_{\frac{f(2^n x)}{4^n} - \frac{f(2^m x)}{4^m}}(c) \geq G_{\frac{f(2^n x)}{4^n} - \frac{f(2^m x)}{4^m}}\left(\sum_{k=m+1}^{n} 2^{-k-1}t_0\right) \geq F_x(t_0) \geq 1 - \epsilon.$$

Hence $\{\frac{f(2^n x)}{4^n}\}$ is a Cauchy sequence in $(Y, G, \star)$. Since $(Y, G, \star)$ is a Menger PB-space, this sequence converges to some $Q(x) \in Y$. Hence, we can define a mapping $Q : X \to Y$ such that $\lim_{n\to\infty} G_{\frac{f(2^n x)}{4^n} - Q(x)} = 1$. Moreover, if we put $m = 0$ in (12.80), we observe that

$$G_{\frac{f(2^n x)}{4^n} - f(x)}\left(\sum_{k=1}^{n} 2^{-k-1}t\right) \geq F_x(t).$$

Therefore,

$$G_{\frac{f(2^n x)}{4^n} - f(x)}(t) \geq F_x\left(\frac{t}{\sum_{k=1}^{n} 2^{-k-1}}\right). \tag{12.81}$$

Next, we will show that Q is quadratic. Let $x, y \in X$; then we have

$$G_{Q(x+y)+Q(x-y)-2Q(x)-2Q(y)}(t)$$

$$\geq G_{Q(x+y)-\frac{f(2^n(x+y))}{4^n}}\left(\frac{t}{5}\right) \star G_{Q(x-y)-\frac{f(2^n(x-y))}{4^n}}\left(\frac{t}{5}\right)$$

$$\star G_{2\frac{f(2^n x)}{4^n}-2Q(x)}\left(\frac{t}{5}\right) \star G_{2\frac{f(2^n y)}{4^n}-2Q(y)}\left(\frac{t}{5}\right)$$

$$\star G_{\frac{f(2^n(x+y))}{4^n}+\frac{f(2^n(x-y))}{4^n}-2\frac{f(2^n x)}{4^n}-2\frac{f(2^n y)}{4^n}}\left(\frac{t}{5}\right).$$

The first four terms on the right-hand side of the above inequality tend to 1 as $n \to \infty$, $t \to \infty$ and the fifth term, by (12.77), is greater than or equal to $F_{2^n x}\left(\frac{4^n t}{10}\right) \star F_{2^n y}\left(\frac{4^n t}{10}\right) = F_x\left(\frac{2^n t}{10}\right) \star F_y\left(\frac{2^n t}{10}\right)$, which tends to 1 as $n \to \infty$. Therefore $Q(x+y)+Q(x-y) = 2Q(x)+2Q(y)$. Next we approximate the difference between f and Q. For every $x \in X$ and $t > 0$, by (12.81) for large enough n, we have

$$G_{Q(x)-f(x)}(t) \geq G_{Q(x)-\frac{f(2^n x)}{4^n}}\left(\frac{t}{2}\right) \star G_{\frac{f(2^n x)}{4^n}-f(x)}\left(\frac{t}{2}\right) \geq F_x(t).$$

Let Q' be another quadratic function from X to Y which satisfies (12.78). We have

$$G_{Q(x)-Q'(x)}(t) \geq G_{Q(x)-f(x)}\left(\frac{t}{2}\right) \star G_{f(x)-Q'(x)}\left(\frac{t}{2}\right) \geq F_x(t)$$

for each $t > 0$. Therefore $Q = Q'$. This completes the proof. $\square$

12.7. Stability in intuitionistic fuzzy normed spaces

In this section, we present stability results for functional equation in intuitionistic fuzzy normed spaces.

Definition 12.14 ([4]). A binary operation $\star : [0,1] \times [0,1]$ is said to be a *continuous t-norm* if it satisfies the following conditions:

(a) $\star$ is commutative and associative;
(b) $\star$ is continuous;
(c) $a \star 1 = a$ for all $a \in [0,1]$;
(d) $a \star b \leq c \star d$ whenever $a \leq c$ and $b \leq d$ for each $a, b, c, d \in [0,1]$.

Definition 12.15 ([4]). A binary operation $\Diamond : [0,1] \times [0,1] \to [0,1]$ is said to be a *continuous t-conorm* if it satisfies the following conditions:

(a') $\Diamond$ is associative and commutative;
(b') $\Diamond$ is continuous;
(c') $a \Diamond 0 = a$ for all $a \in [0,1]$;
(d') $a \Diamond b \leq c \Diamond d$ whenever $a \leq c$ and $b \leq d$ for each $a, b, c, d \in [0,1]$.

Definition 12.16 ([4]). The five-tuple $(X, \mu, v, \star, \Diamond)$ is said to be an *intuitionistic fuzzy normed space* (for short, **IFNS**) if X is a vector space. $\star$ is a continuous t-norm, $\Diamond$ is a continuous t-conorm, and μ, v are fuzzy sets on $X \times (0, \infty)$ satisfying the following conditions for each $x, y \in X$ and $s, t > 0$:

(i) $\mu(x, t) + v(x, t) \leq 1$;

(ii) $\mu(x, t) > 0$;

(iii) $\mu(x, t) = 1$ if and only if $x = 0$;

(iv) $\mu(\alpha x, t) = \mu(x, \frac{t}{|\alpha|})$ for each $\alpha \neq 0$;

(v) $\mu(x, t) \star \mu(y, s) \leq \mu(x + y, t + s)$;

(vi) $\mu(x, \cdot) : (0, \infty) \to [0, 1]$ is continuous;

(vii) $\lim_{n \to \infty} \mu(x, t) = 1$ and $\lim_{t \to 0} \mu(x, t) = 0$;

(viii) $v(x, t) < 1$;

(ix) $v(x, t) = 0$ if and only if $x = 0$;

(x) $v(\alpha x, t) = v(x, \frac{t}{|\alpha|})$ for each $\alpha \neq 0$;

(xi) $v(x, t) \Diamond v(y, s) \geq v(x + y, t + s)$;

(xii) $v(x, \cdot) : (0, \infty) \to [0, 1]$ is continuous;

(xiii) $\lim_{t \to \infty} v(x, t) = 0$ and $\lim_{t \to 0} v(x, t) = 1$.

In this case (μ, v) is called an *intuitionistic fuzzy norm*.

Example 12.1 ([4]). Let $(X, \| \cdot \|)$ be a normed space, $a \star b = ab$ and $a \Diamond b = \min\{a + b, 1\}$ for all $a, b \in [0, 1]$. For all $x \in X$ and every $t > 0$, consider

$$\mu(x, t) = \begin{cases} \dfrac{t}{t + \|x\|} & \text{if } t > 0, \\ 0 & \text{if } t \leq 0, \end{cases}$$

and

$$v(x, t) = \begin{cases} \dfrac{\|x\|}{t + \|x\|} & \text{if } t > 0, \\ 0 & \text{if } t \leq 0. \end{cases}$$

Then $(X, \mu, v, \star, \Diamond)$ is an IFNS.

Let $(X, \mu, v, \star, \Diamond)$ be an IFNS. A sequence $x = (x_k)$ is said to be *intuitionistic fuzzy convergent* to $L \in X$ if, for every $\epsilon > 0$, there exists $k_0 \in \mathbb{N}$ such that $\mu(x_k - L, t) > 1 - \epsilon$ and $v(x_k - L, t) < \epsilon$ for all $k \geq k_0$. In this case, we write (μ, v)-$\lim x_k = L$ or $x_k \xrightarrow{(\mu, v)} L$ as $k \to \infty$.

Let $(X, \mu, v, \star, \Diamond)$ be an IFNS. A $x = (x_k)$ is said to be *intuitionistic fuzzy Cauchy sequence* if, for every $\epsilon > 0$ and $t > 0$, there exists $k_0 \in \mathbb{N}$ such that $\mu(x_k - x_l, t) > 1 - \epsilon$ and $v(x_k - x_l, t) < \epsilon$ for all $k, l \geq k_0$.

An IFNS $(X, \mu, v, \star, \Diamond)$ is said to be *complete* if every intuitionistic fuzzy Cauchy sequence is intuitionistic fuzzy convergent in $(X, \mu, v, \star, \Diamond)$. In this case (X, μ, v) is called intuitionistic fuzzy Banach space.

12.7.1. *Hyers–Ulam–Rassias stability of Cauchy additive functional equation in INFS*

The following theorems provide the Hyers–Ulam–Rassias stability of Cauchy additive functional equation in INFS.

Theorem 12.14 ([4]). *Let X be a linear space, and (Y, μ, v) be an intuitionistic fuzzy Banach space. Let $\varphi : X \times X \to [0, \infty)$ be a control function such that*

$$\tilde{\varphi}(x, y) = \sum_{n=0}^{\infty} 2^{-n} \varphi(2^n x, 2^n y) < \infty \tag{12.82}$$

for all $x, y \in X$. Let $f : X \to Y$ be a uniformly approximately additive function with respect to φ in the sense that

$$\lim_{t \to \infty} \mu(f(x + y) - f(x) - f(y), t\varphi(x, y)) = 1,$$

and

$$\lim_{t \to \infty} v(f(x + y) - f(x) - f(y), t\varphi(x, y)) = 0 \tag{12.83}$$

uniformly in $X \times X$. Then $T(x) = (\mu, v)\text{-}\lim_{n \to \infty} \frac{f(2^n x)}{2^n}$ for each $x \in X$ exists and defines an additive mapping $T : X \to Y$ such that if for some $\delta > 0$, $\alpha > 0$ and all $x, y \in X$,

$$\mu(f(x + y) - f(x) - f(y), \delta\varphi(x, y)) > \alpha,$$

and

$$v(f(x + y) - f(x) - f(y), \delta\varphi(x, y)) < 1 - \alpha, \tag{12.84}$$

then

$$\mu\left(T(x) - f(x), \frac{\delta}{2}\tilde{\varphi}(x, x)\right) > \alpha,$$

and

$$v\left(T(x) - f(x), \frac{\delta}{2}\tilde{\varphi}(x, x)\right) < 1 - \alpha.$$

Proof. Given $\epsilon > 0$. By (12.83), we can find some $t_0 > 0$ such that

$$\mu(f(x+y) - f(x) - f(y), t\varphi(x,y)) \geq 1 - \epsilon,$$

and

$$v(f(x+y) - f(x) - f(y), t\varphi(x,y)) \leq \epsilon \tag{12.85}$$

for all $x, y \in X$, and all $t \geq t_0$. By induction on n, we shall show that

$$\mu\left(f\left(2^n x\right) - 2^n f(x), t \sum_{k=0}^{n-1} 2^{n-k-1}\varphi\left(2^k x, 2^k x\right)\right) \geq 1 - \epsilon,$$

$$v\left(f\left(2^n x\right) - 2^n f(x), t \sum_{k=0}^{n-1} 2^{n-k-1}\varphi\left(2^k x, 2^k x\right)\right) \leq \epsilon, \tag{12.86}$$

for all $x \in X$, $t \geq t_0$, and all positive integers n. Putting $y = x$ in (12.85), we obtain (12.86) for $n = 1$. Let (12.86) hold for some positive integer n. Then

$$\mu\left(f\left(2^{n+1}x\right) - 2^{n+1}f(x), t \sum_{k=0}^{n} 2^{n-k}\varphi\left(2^k x, 2^k x\right)\right)$$

$$\geq \mu\left(f\left(2^{n+1}x\right) - 2f\left(2^n x\right), t_0\varphi\left(2^n x, 2^n x\right)\right)$$

$$\star \mu\left(2f\left(2^n x\right) - 2^{n+1}f(x), t_0 \sum_{k=0}^{n-1} 2^{n-k}\varphi\left(2^k x, 2^k x\right)\right)$$

$$\geq (1 - \epsilon) \star (1 - \epsilon) = 1 - \epsilon,$$

and

$$v\left(f\left(2^{n+1}x\right) - 2^{n+1}f(x), t \sum_{k=0}^{n} 2^{n-k}\varphi\left(2^k x, 2^k x\right)\right)$$

$$\leq v\left(f\left(2^{n+1}x\right) - 2f\left(2^n x\right), t_0\varphi\left(2^n x, 2^n x\right)\right)$$

$$\Diamond v\left(2f\left(2^n x\right) - 2^{n+1}f(x), t_0 \sum_{k=0}^{n-1} 2^{n-k}\varphi\left(2^k x, 2^k x\right)\right) \leq \epsilon \Diamond \epsilon = \epsilon.$$

This completes the induction argument. Let $t = t_0$ and put $n = p$; then by replacing x with $2^n x$ in (12.86), we obtain

$$\mu\left(\frac{f\left(2^{n+p}x\right)}{2^{n+p}} - \frac{f\left(2^n x\right)}{2^n}, \frac{t_0}{2^{n+p}} \sum_{k=0}^{p-1} 2^{p-k-1}\varphi\left(2^{n+k}x, 2^{n+k}x\right)\right) \geq 1 - \epsilon,$$

and

$$v\left(\frac{f\left(2^{n+p}x\right)}{2^{n+p}} - \frac{f\left(2^n x\right)}{2^n}, \frac{t_0}{2^{n+p}}\sum_{k=0}^{p-1}2^{p-k-1}\varphi\left(2^{n+k}x, 2^{n+k}x\right)\right) \le \epsilon, \qquad (12.87)$$

for all integers $n \ge 0$ and $p > 0$. The convergence of (12.82), and

$$\sum_{k=0}^{p-1}2^{-n-k-1}\varphi\left(2^{n+k}x, 2^{n+k}x\right) = \frac{1}{2}\sum_{k=n}^{n+p-1}2^{-k}\varphi\left(2^k x, 2^k x\right)$$

imply that for given $\delta > 0$ there is $n_0 \in \mathbb{N}$ such that

$$\frac{t_0}{2}\sum_{k=n}^{n+p-1}2^{-k}\varphi\left(2^k x, 2^k x\right) < \delta,$$

for all $n \ge n_0$ and all $p > 0$. Now we deduce that from (12.87) that

$$\mu\left(\frac{\left(2^{n+p}x\right)}{2^{n+p}} - \frac{f\left(2^n x\right)}{2^n}, \delta\right)$$

$$\ge \mu\left(\frac{f\left(2^{n+p}x\right)}{2^{n+p}} - \frac{f\left(2^n x\right)}{2^n}, \frac{t_0}{2^{n+p}}\sum_{k=0}^{p-1}2^{p-k-1}\varphi\left(2^{n+k}x, 2^{n+k}x\right)\right)$$

$$\ge 1 - \epsilon,$$

and

$$v\left(\frac{\left(2^{n+p}x\right)}{2^{n+p}} - \frac{f\left(2^n x\right)}{2^n}, \delta\right)$$

$$\le v\left(\frac{f\left(2^{n+p}x\right)}{2^{n+p}} - \frac{f\left(2^n x\right)}{2^n}, \frac{t_0}{2^{n+p}}\sum_{k=0}^{p-1}2^{p-k-1}\varphi\left(2^{n+k}x, 2^{n+k}x\right)\right)$$

$$\le \epsilon,$$

for all $n \ge n_0$ and all $p > 0$. Hence $\{\frac{f(2^n x)}{2^n}\}$ is a Cauchy sequence in Y. Since Y is an intuitionistic fuzzy Banach space, $\{\frac{f(2^n x)}{2^n}\}$ converges to some $T(x) \in Y$. Hence, we can define a mapping $T : X \to Y$ such that $T(x) = (\mu, v)\text{-}\lim_{n\to\infty}\frac{f(2^n x)}{2^n}$, namely, for each $t > 0$, and $x \in X$,

$$\mu\left(T(x) - \frac{f\left(2^n x\right)}{2^n}, t\right) = 1 \quad \text{and} \quad v\left(T(x) - \frac{f\left(2^n x\right)}{2^n}, t\right) = 0.$$

Now, let $x, y \in X$. Choose any fix value of $t > 0$, and $\epsilon \in (0, 1)$. Since $\lim_{n\to\infty}2^{-n}\varphi(2^n x, 2^n y) = 0$, there exists $n_1 > n_0$ such that $t_0\varphi(2^n x, 2^n y) < \frac{2^n t}{4}$

for all $n \geq n_1$. Hence for each $n \geq n_1$, we have

$$\mu(T(x+y) - T(x) - T(y), t)$$

$$\geq \mu\left(T(x+y) - \frac{f\left(2^n(x+y)\right)}{2^n}, \frac{t}{4}\right)$$

$$\star \mu\left(T(x) - \frac{f\left(2^n x\right)}{2^n}, \frac{t}{4}\right)$$

$$\star \mu\left(T(y) - \frac{f\left(2^n y\right)}{2^n}, \frac{t}{4}\right)$$

$$\star \mu\left(f\left(2^n(x+y)\right) - f\left(2^n x\right) - f\left(2^n y\right), \frac{2^n t}{4}\right), \qquad (12.88)$$

and also

$$\mu\left(f\left(2^n(x+y)\right) - f\left(2^n x\right) - f\left(2^n y\right), \frac{2^n t}{4}\right)$$

$$\geq \mu\left(f\left(2^n(x+y)\right) - f\left(2^n x\right) - f\left(2^n y\right), t\varphi\left(2^n x, 2^n y\right)\right). \qquad (12.89)$$

Letting $n \to \infty$ in (12.88) and using (12.85), (12.89), we obtain

$$\mu(T(x+y) - T(x) - T(y), t) \geq 1 - \epsilon$$

for all $t > 0$ and $\epsilon \in (0, 1)$. Similarly, we obtain

$$v(T(x+y) - T(x) - T(y), t) \leq \epsilon$$

for all $t > 0$ and $\epsilon \in (0, 1)$. It follows that

$$\mu(T(x+y) - T(x) - T(y), t) = 1,$$

and

$$v(T(x+y) - T(x) - T(y), t) = 0,$$

for all $t > 0$. Therefore $T(x+y) = T(x) + T(y)$.

Lastly, suppose that for some positive δ and α, (12.84) holds, and

$$\varphi_n(x, y) = \sum_{k=0}^{n-1} 2^{-k-1} \varphi\left(2^k x, 2^k y\right),$$

for all $x, y \in X$. By a similar argument as in the beginning of the proof, one can deduce from (12.84)

$$\mu\left(f\left(2^n x\right) - 2^n f(x), \delta \sum_{k=0}^{n-1} 2^{n-k-1} \varphi\left(2^k x, 2^k x\right)\right) \geq \alpha,$$

and

$$v\left(f\left(2^n x\right) - 2^n f(x), \delta \sum_{k=0}^{n-1} 2^{n-k-1} \varphi\left(2^k x, 2^k x\right)\right) \leq 1 - \alpha, \qquad (12.90)$$

for all positive integers n. For $s > 0$, we have

$$\mu(f(x) - T(x), \delta \varphi_n(x, x) + s)$$

$$\geq \mu\left(f(x) - \frac{f\left(2^n x\right)}{2^n}, \delta \varphi_{(}x, x)\right)$$

$$\star \mu\left(\frac{f\left(2^n x\right)}{2^n} - T(x), s\right),$$

and

$$v(f(x) - T(x), \delta \varphi_n(x, x) + s)$$

$$\leq v\left(f(x) - \frac{f\left(2^n x\right)}{2^n}, \delta \varphi_{(}x, x)\right)$$

$$\Diamond v\left(\frac{f\left(2^n x\right)}{2^n} - T(x), s\right). \qquad (12.91)$$

Combining (12.90) with (12.91), and using the fact that

$$\lim_{n \to \infty} \mu\left(\frac{f\left(2^n x\right)}{2^n} - T(x), s\right) = 1,$$

and

$$\lim_{n \to \infty} v\left(\frac{f\left(2^n x\right)}{2^n} - T(x), s\right) = 0,$$

we obtain

$$\mu(f(x) - T(x), \delta \varphi_n(x, x) + s) \geq \alpha,$$

and

$$v(f(x) - T(x), \delta \varphi_n(x, x) + s) \leq 1 - \alpha,$$

for sufficiently large n. From the (upper semi) continuity of real functions $\mu(f(x) - T(x), \cdot)$, and $v(f(x) - T(x), \cdot)$, we see that

$$\mu\left(f(x) - T(x), \frac{\delta}{2} \tilde{\varphi}(x, x) + s\right) \geq \alpha,$$

and

$$v\left(f(x) - T(x), \frac{\delta}{2} \tilde{\varphi}(x, x) + s\right) \leq 1 - \alpha.$$

Taking the limit $s \to \infty$, we obtain

$$\mu\left(f(x) - T(x), \frac{\delta}{2}\tilde{\varphi}(x,x)\right) \geq \alpha,$$

and

$$v\left(f(x) - T(x), \frac{\delta}{2}\tilde{\varphi}(x,x)\right) \leq 1 - \alpha.$$

The proof is now completed. $\qquad\qquad\qquad\qquad\qquad\qquad\qquad\qquad\qquad$ $\square$

Theorem 12.15 ([4]). *Let X be a linear space and (Y,μ,v) be an intuitionistic fuzzy Banach space. Let $\varphi : X \times X \to [0,\infty)$ be a control function satisfying (12.82). Let $f : X \to Y$ be an uniformly approximately additive function with respect to φ. Then there is a unique additive mapping $T : X \to Y$ such that*

$$\lim_{n\to\infty} \mu(f(x) - T(x), t\tilde{\varphi}(x,x)) = 1,$$

and $\qquad\qquad\qquad\qquad\qquad\qquad\qquad\qquad\qquad\qquad\qquad\qquad$ (12.92)

$$\lim_{n\to\infty} v(f(x) - T(x), t\tilde{\varphi}(x,x)) = 0$$

uniformly in X.

Proof. The existence of uniform limit (12.92) immediately follows from Theorem 12.14. It remains to prove the uniqueness assertion. Let S be another additive mapping satisfying (12.92). Choose $c > 0$. Given $\epsilon > 0$, there is some $t_0 > 0$ such that from (12.92), we have

$$\mu\left(f(x) - T(x), \frac{t}{2}\tilde{\varphi}(x,x)\right) \geq 1 - \epsilon,$$

$$\mu\left(f(x) - S(x), \frac{t}{2}\tilde{\varphi}(x,x)\right) \geq 1 - \epsilon,$$

and

$$v\left(f(x) - T(x), \frac{t}{2}\tilde{\varphi}(x,x)\right) \leq \epsilon,$$

$$v\left(f(x) - S(x), \frac{t}{2}\tilde{\varphi}(x,x)\right) \leq \epsilon,$$

for all $x \in X$, and all $t \geq t_0$. For some $x \in X$, we can find some integer n_0 such that

$$t_0 \sum_{k=n}^{\infty} 2^{-k}\varphi\left(2^k x, 2^k x\right) < \frac{c}{2},$$

for all $n \geq n_0$. Since

$$\sum_{k=n}^{\infty} 2^{-k} \varphi\left(2^k x, 2^k x\right) = \frac{1}{2^n} \sum_{k=n}^{\infty} 2^{-(k-n)} \varphi(2^{(k-n)} 2^n x, 2^{(k-n)} 2^n x)$$

$$= \frac{1}{2^n} \tilde{\varphi}(2^n x, 2^n x),$$

we have

$$\mu(s(x) - T(x), c) \geq \mu\left(\frac{f(2^n x)}{2^n} - T(x), \frac{c}{2}\right) \star \mu\left(S(x) - \frac{f(2^n x)}{2^n}, \frac{c}{2}\right)$$

$$= \mu\left(f(2^n x) - T(2^n x), 2^{n-1} c\right) \star \mu\left(S(2^n x) - f(2^n x), 2^{n-1} c\right)$$

$$= \mu\left(f(2^n x) - T(2^n x), t_0 \tilde{\varphi}(2^n x, 2^n x)\right)$$

$$\star \mu\left(S(2^n x) - f(2^n x), t_0 \tilde{\varphi}(2^n x, 2^n x)\right) \geq 1 - \epsilon,$$

and similarly

$$V(S(x) - T(x), c) \leq v\left(\frac{f(2^n x)}{2^n} - T(x), \frac{c}{2}\right) \Diamond v\left(S(x) - \frac{f(2^n x)}{2^n}, \frac{c}{2}\right)$$

$$= v\left(f(2^n x) - T(2^n x), t_0 \tilde{\varphi}(2^n x, 2^n x)\right)$$

$$\Diamond v\left(S(2^n x) - f(2^n x), t_0 \tilde{\varphi}(2^n x, 2^n x)\right) \leq \epsilon.$$

It follows that

$$\mu(S(x) - T(x), c) = 1, \quad \text{and} \quad v(S(x) - T(x), c) = 0$$

for all $c > 0$. Hence $T(x) = S(x)$ for all $x \in X$. This completes the proof. $\square$

Corollary 12.7 ([4]). *Let X be a normed linear space, and (Y, μ, v) be an intuitionistic fuzzy Banach space. Let $f : X \to Y$ be a function such that for all $\theta \geq 0$, $0 \leq q \leq 1$*

$$\lim_{t \to \infty} \mu\left(f(x + y) - f(x) - f(y), t\theta\left(\|x\|^q + \|y\|^q\right)\right) = 1,$$

and

$$\lim_{t \to \infty} v\left(f(x + y) - f(x) - f(y), t\theta\left(\|x\|^q + \|y\|^q\right)\right) = 0,$$

uniformly in $X \times X$. Then there exists a unique additive mapping $T : X \to Y$ such that

$$\lim_{t \to \infty} \mu\left(T(x) - f(x), \frac{2\theta t \|x\|^q}{1 - 2^{q-1}}\right) = 1,$$

and

$$\lim_{t \to \infty} v\left(T(x) - f(x), \frac{2\theta t \|x\|^q}{1 - 2^{q-1}}\right) = 0,$$

uniformly in X.

Proof. The proof follows immediately from Theorems 12.14, and 12.15 by considering the control function $\varphi(x,y) = \theta\left(\|x\|^q + \|y\|^q\right)$ for some $\theta > 0$. The proof is now completed. $\qquad\square$

12.8. Stability in matrix normed spaces

In this section, first we present the definition of matrix normed spaces, then we discuss the Hyers–Ulam stability of Cauchy additive and quadratic functional equations in matrix normed spaces in various theorems. We will use the following notations:

$M_n(X)$ is the set of all $n \times n$-matrices in X;

$e_j \in M_{1,n}(\mathbb{C})$ is that jth component is 1, and the other components are zero;

$E_{ij} \in M_n(\mathbb{C})$ is that (i,j)-component is 1, and the other components are zero;

$E_{ij} \otimes x \in M_n(X)$ is that (i,j)-component is x, and the other components are zero.
For $x \in M_n(X)$, $y \in M_k(X)$,

$$x \oplus y = \begin{pmatrix} x & 0 \\ 0 & y \end{pmatrix}.$$

Definition 12.17. Let $(X, \|\cdot\|)$ be a normed space. Note that $(X, \|\cdot\|_n)$ is a matrix normed space if and only if $(M_n(X), \|\cdot\|_n)$ is a normed space for each positive integer n and $\|AxB\|_k \le \|A\| \|B\| \|x\|_n$ holds for $A \in M_{k,n}(\mathbb{C})$, $x = (x_{ij}) \in M_n(X)$ and $B \in M_{n,k}(\mathbb{C})$.

Definition 12.18. $(X, \|\cdot\|_n)$ is a matrix Banach space if and only if x is a Banach space and $(X, \|\cdot\|_n)$ is matrix normed space.

Definition 12.19. A matrix normed space $(X, \|\cdot\|_n)$ is called and L^∞-*matrix normed space* if $\|x \oplus y\|_{n+k} = \max\{\|x\|_n, \|y\|_k\}$ holds for all $x \in M_n(X)$ and all $y \in M_k(X)$.

Let E, F be vector spaces. For a given mapping $h : E \to F$ and a given positive integer n, define $h_n : M_n(E) \to M_n(F)$ by

$$h_n\left([x_{ij}]\right) = [h\left(x_{ij}\right)]$$

for all $[x_{ij}] \in M_n(E)$.

12.8.1. *Hyers–Ulam stability of the Cauchy additive functional equation in matrix normed spaces*

In the following theorems, let $(X, \|\cdot\|_n)$ be a matrix normed space and $(Y, \|\cdot\|_n)$ be a matrix Banach space.

The Hyers–Ulam stability of the Cauchy additive functional equation in matrix normed spaces is presented in the following theorem.

Lemma 12.1 ([176]). *Let* $(X, \| \cdot \|_n)$ *be a matrix normed space. Then we have the following conditions:*

(i) $\|E_{kl} \otimes x\|_n = \|x\|$ *for* $x \in X$.

(ii) $\|x_{kl}\| \leq \|[x_{ij}]\|_n \leq \sum_{i,j=1}^n \|x_{ij}\|$ *for* $[x_{ij}] \in M_n(X)$.

(iii) $\lim_{n \to \infty} x_n = x$ *if and only if* $\lim_{n \to \infty} x_{nij} = x_{ij}$ *for* $x_n = [x_{nij}]$, $x = [x_{ij}] \in M_k(X)$.

Proof.

(i) Since $E_{kl} \otimes x = \dot{e}_k x e_l$ and $\|\dot{e}_k\| = \|e_l\| = 1$, $\|E_{kl} \otimes x\|_n \leq \|x\|$. Since $e_k (E_{kl} \otimes x) \dot{e}_l = x$, $\|x\| \leq \|E_{kl} \otimes x\|_n$. So, $\|E_{kl} \otimes x\|_n = \|x\|$.

(ii) Since $e_k x \dot{e}_l = x_{kl}$ and $\|e_k\| = \|\dot{e}_l\| = 1$, we have

$$\|x_{kl}\| \leq \|[x_{ij}]\|_n.$$

Since $[x_{ij}] = \sum_{i,j=1}^n E_{ij} \otimes x_{ij}$, we obtain

$$\|[x_{ij}]\|_n = \left\| \sum_{i,j=1}^n E_{ij} \otimes x_{ij} \right\|_n \leq \sum_{i,j=1}^n \|E_{ij} \otimes x_{ij}\|_n = \sum_{i,j=1}^n \|x_{ij}\|.$$

(iii) By (ii), we have

$$\|x_{nkl} - x_{kl}\| \leq \|[x_{nij} - x_{ij}]\|_n = \|[x_{nij}] - [x_{ij}]\|_n \leq \sum_{i,j=1}^n \|x_{nij} - x_{ij}\|.$$

Hence, we get the desired result. This completes the proof of the lemma. $\qquad \square$

For a mapping $f : X \to Y$, define $Df : X \times X \to Y$, and $Df_n : M_n(X \times X) \to M_n(Y)$ by

$$Df(a, b) = f(a + b) - f(a) - f(b),$$

and

$$Df_n ([x_{ij}], [y_{ij}]) = f_n ([x_{ij} + y_{ij}]) - f_n ([x_{ij}]) - f_n ([y_{ij}])$$

for all $a, b \in X$, and all $x = [x_{ij}]$, $y = [y_{ij}] \in M_n(X)$.

Theorem 12.16 ([176]). *Let* $f : X \to Y$ *be a mapping and let* $\phi : X \times X \to [0, \infty)$ *be a function such that*

$$\Phi(a, b) = \frac{1}{2} \sum_{l=0}^{\infty} \frac{1}{2^l} \phi \left(2^l a, 2^l b \right) < +\infty, \qquad (12.93)$$

$$\|Df_n ([x_{ij}], [y_{ij}])\|_n \leq \sum_{i,j=1}^n \phi(x_{ij}, y_{ij}) \qquad (12.94)$$

for all $a, b \in X$, and all $x = [x_{ij}]$, $y = [y_{ij}] \in M_n(X)$. Then there exists a unique additive mapping $A : X \to Y$ such that

$$\|f_n([x_{ij}]) - A_n([x_{ij}])\|_n \leq \sum_{i,j=1}^{n} \Phi(x_{ij}, x_{ij}) \tag{12.95}$$

for all $x = [x_{ij}] \in M_n(X)$.

Proof. Assume $n = 1$ in (12.94), then we have

$$\|f(a+b) - f(a) - f(b)\| \leq \phi(a, b)$$

for all $a, b \in X$. Then there exists a unique additive mapping $A : X \to Y$ such that

$$\|f(a) - A(a)\| \leq \Phi(a, a)$$

for all $a \in X$. Define the mapping $A : X \to Y$ by

$$A(a) = \lim_{l \to \infty} \frac{1}{2^l} f\left(2^l a\right)$$

for all $a \in X$. In view of Lemma 12.1, we have

$$\|f_n([x_{ij}]) - A_n([x_{ij}])\|_n \leq \sum_{i,j=1}^{n} \|f(x_{ij}) - A(x_{ij})\| \leq \sum_{i,j=1}^{n} \Phi(x_{ij}, x_{ij})$$

for all $x = [x_{ij}] \in M_n(X)$. Thus, $A : X \to Y$ is a unique additive mapping satisfying (12.95), as desired. The proof of the theorem is completed. $\square$

Corollary 12.8 ([176]). *Let r, θ be positive real numbers with $r < 1$. Let $f : X \to Y$ be a mapping such that*

$$\|Df_n([x_{ij}], [y_{ij}])\|_n \leq \sum_{i,j=1}^{n} \theta\left(\|x_{ij}\|^r + \|y_{ij}\|^r\right) \tag{12.96}$$

for all $x = [x_{ij}]$, $y = [y_{ij}] \in M_n(X)$. Then there exists a unique additive mapping $A : X \to Y$ such that

$$\|f_n([x_{ij}]) - A_n([x_{ij}])\|_n \leq \sum_{i,j=1}^{n} \frac{2\theta}{2 - 2^r} \|x_{ij}\|^r$$

for all $x = [x_{ij}] \in M_n(X)$.

Proof. Letting $\phi(a, b) = \theta\left(\|a\|^r + \|b\|^r\right)$ in Theorem 12.16, we obtain the result, and the proof is completed. $\square$

Theorem 12.17 ([176]). *Let $f : X \to Y$ be a mapping and let $\phi : X \times X \to [0, \infty)$ be a function satisfying (12.92), and*

$$\Phi(a, b) := \frac{1}{2} \sum_{l=1}^{\infty} 2^l \phi\left(\frac{a}{2^l}, \frac{b}{2^l}\right) < +\infty \tag{12.97}$$

for all $a, b \in X$. Then there exists a unique additive mapping $A : X \to Y$ such that

$$\|f_n([x_{ij}]) - A_n([x_{ij}])\|_n \leq \sum_{i,j=1}^{n} \Phi(x_{ij}, x_{ij})$$

for all $x = [x_{ij}] \in M_n(X)$.

Proof. The proof is similar to that of Theorem 12.16, and hence it is omitted. $\square$

Corollary 12.9 ([176]). *Let r, θ be positive real numbers with $r > 1$. Let $f : X \to Y$ be a mapping satisfying (12.96). Then there exists a unique additive mapping $A : X \to Y$ such that*

$$\|f_n([x_{ij}]) - A_n([x_{ij}])\|_n \leq \sum_{i,j=1}^{n} \frac{2\theta}{2^r - 2} \|x_{ij}\|^r$$

for all $x = [x_{ij}] \in M_n(X)$.

Proof. Letting $\phi(a, b) = \theta(\|a\|^r + \|b\|^r)$ in Theorem 12.17, we obtain the result. This completes the proof. $\square$

We need the following lemma to prove our main results.

Lemma 12.2 ([176]). *If E is a L^∞-matrix normed space, then $\|[x_{ij}]\|_n \leq \|[\|x_{ij}\|]\|_n$ for all $[x_{ij}] \in M_n(E)$.*

Theorem 12.18 ([176]). *Let Y be a L^∞-normed Banach space. Let $f : X \to Y$ be a mapping and let $\phi : X \times X \to [0, \infty)$ be a function satisfying (12.93), and*

$$\|Df_n([x_{ij}], [y_{ij}])\|_n \leq \|[\phi(x_{ij}, y_{ij})]\|_n \tag{12.98}$$

for all $x - [x_{ij}], y - [y_{ij}] \in M_n(X)$. Then there exists a unique additive mapping $A : X \to Y$ such that

$$\|[f(x_{ij}) - A(x_{ij})]\|_n \leq \|[\Phi(x_{ij}, x_{ij})]\|_n \tag{12.99}$$

for all $x = [x_{ij}] \in M_n(X)$. Here Φ is given in Theorem 12.16.

Proof. By the same reasoning as in the proof of Theorem 12.16, there exists a unique additive mapping $A : X \to Y$ such that

$$\|f(a) - A(a)\| \leq \Phi(a, a)$$

for all $a \in X$. The mapping $A : X \to Y$ is given by

$$A(a) = \lim_{l \to \infty} \frac{1}{2^l} f\left(2^l a\right)$$

for all $a \in X$.

It is easy to show that if $0 \leq a_{ij} \leq b_{ij}$ for all i, j, then

$$\|[a_{ij}]\|_n \leq \|[b_{ij}]\|_n. \tag{12.100}$$

By Lemma 12.2 and (12.100), we have

$$\|[f(x_{ij}) - A(x_{ij})]\|_n \leq \|[\|f(x_{ij}) - A(x_{ij})\|]\|_n \leq \|\Phi(x_{ij}, x_{ij})\|_n$$

for all $x = [x_{ij}] \in M_n(X)$. So, we obtain the inequality (12.99). This completes the proof. □

Corollary 12.10 ([176]). *Let Y be a L^∞-normed Banach space. Let r, θ be positive real numbers with $r < 1$. Let $f : X \to Y$ be a mapping such that*

$$\|Df_n([x_{ij}], [y_{ij}])\|_n \leq \|[\theta\left(\|x_{ij}\|^r + \|y_{ij}\|^r\right)]\|_n \tag{12.101}$$

for all $x = [x_{ij}]$, $y = [y_{ij}] \in M_n(X)$. Then there exists a unique additive mapping $A : X \to Y$ such that

$$\|f_n([x_{ij}]) - A_n([x_{ij}])\|_n \leq \left\|\left[\frac{2\theta}{2 - 2^r}\|x_{ij}\|^r\right]\right\|_n$$

for all $x = [x_{ij}] \in M_n(X)$.

Proof. Letting $\phi(a, b) = \theta\left(\|a\|^r + \|b\|^r\right)$ in Theorem 12.18, we obtain the desired result. This completes the proof. □

Theorem 12.19 ([176]). *Let Y be a L^∞-normed Banach space. Let $f : X \to Y$ be a mapping and let $\phi : X \times X \to [0, \infty)$ be a function satisfying (12.97) and (12.98). Then there exists a unique additive mapping $A : X \to Y$ such that*

$$\|[f(x_{ij}) - A(x_{ij})]\|_n \leq \|[\Phi(x_{ij}, x_{ij})]\|_n$$

for all $x = [x_{ij}] \in M_n(X)$. Here Φ is given in Theorem 12.17.

Proof. The proof is similar to that of Theorem 12.18, and hence the details are omitted. □

Corollary 12.11 ([176]). *Let Y be a L^∞-normed Banach space. Let r, θ be positive real numbers with $r > 1$. Let $f : X \to Y$ be a mapping satisfying (12.101). Then there exists a unique additive mapping $A : X \to Y$ such that*

$$\|f_n([x_{ij}] - A_n([x_{ij}]))\|_n \le \left\| \left[\frac{2\theta}{2^r - 2} \|x_{ij}\|^r \right] \right\|_n$$

for all $x = [x_{ij}] \in M_n(X)$.

Proof. Letting $\phi(a, b) = \theta \left(\|a\|^r + \|b\|^r \right)$ in Theorem 12.19, we obtain the desired result. This completes the proof. $\square$

12.8.2. Hyers–Ulam stability of the quadratic functional equation in matrix normed spaces

In the following theorems, we obtain the Hyers–Ulam stability of the quadratic functional equation in matrix normed spaces.

For a mapping $f : X \to Y$, define $Df : X \times X \to Y$ and $Df_n : M_n(X \times X) \to M_n(Y)$ by

$$Df(a, b) := f(a + b) + f(a - b) - 2f(a) - 2f(b),$$

and

$$Df_n([x_{ij}], [y_{ij}]) := f_n([x_{ij} + y_{ij}]) + f_n([x_{ij} = y_{ij}]) - 2f_n([x_{ij}]) - 2f_n([y_{ij}])$$

for all $a, b \in X$, and all $x = [x_{ij}]$, $y = [y_{ij}] \in M_n(X)$.

Theorem 12.20 ([176]). *Let $f : X \to Y$ be a mapping, and let $\phi : X \times X \to [0, \infty)$ be a function such that*

$$\Phi(a, b) := \frac{1}{4} \sum_{l=0}^{\infty} \frac{1}{4^l} \phi \left(2^l a, 2^l b \right) < +\infty, \tag{12.102}$$

and

$$\|Df_n([x_{ij}], [y_{ij}])\|_n \le \sum_{i,j=1}^{n} \phi(x_{ij}, y_{ij}) \tag{12.103}$$

for all $a, b \subset X$, and all $x = [x_{ij}]$, $y = [y_{ij}] \in M_n(X)$. Then there exists a unique quadratic mapping $Q : X \to Y$ such that

$$\|f_n([x_{ij}]) - Q_n([x_{ij}])\|_n \le \sum_{i,j=1}^{n} \Phi(x_{ij}, x_{ij}) \tag{12.104}$$

for all $x = [x_{ij}] \in M_n(X)$.

Proof. Let $n = 1$ in (12.103); then we have

$$\|f(a + b) - f(a - b) - 2f(a) - 2f(b)\| \leq \phi(a, b)$$

for all $a, b \in X$. Then there exists a unique quadratic mapping $Q : X \to Y$ such that

$$\|f(a) - Q(a)\| \leq \Phi(a, a)$$

for all $a \in X$. Define $Q : X \to Y$ by

$$Q(a) = \lim_{l \to \infty} \frac{1}{4^l} f\left(2^l a\right)$$

for all $a \in X$. Then by Lemma 12.1, we have

$$\|f_n([x_{ij}]) - Q_n([x_{ij}])\|_n \leq \sum_{i,j=1}^{n} \|f(x_{ij}) - Q(x_{ij})\|$$

$$\leq \sum_{i,j=1}^{n} \Phi(x_{ij}, x_{ij})$$

for all $x = [x_{ij}] \in M_n(X)$. Thus, $Q : X \to Y$ is a unique quadratic mapping satisfying (12.104), as desired. This completes the proof. $\square$

Corollary 12.12 ([176]). *Let r, θ be positive real numbers with $r < 2$. Let $f : X \to Y$ be a mapping such that*

$$\|Df_n([x_{ij}], [y_{ij}])\|_n \leq \sum_{i,j=1}^{n} \theta\left(\|x_{ij}\|^r + \|y_{ij}\|^r\right) \tag{12.105}$$

for all $x = [x_{ij}], y = [y_{ij}] \in M_n(X)$. Then there exists a unique quadratic mapping $Q : X \to Y$ such that

$$\|f_n([x_{ij}]) - Q_n([x_{ij}])\|_n \leq \sum_{i,j=1}^{n} \frac{2\theta}{4 - 2^r} \|x_{ij}\|^r$$

for all $x = [x_{ij}] \in M_n(X)$.

Proof. Letting $\phi(a, b) = \theta\left(\|a\|^r + \|b\|^r\right)$ in Theorem 12.20, we obtain the result. The proof is now completed. $\square$

Theorem 12.21 ([176]). *Let $f : X \to Y$ be a mapping and let $\phi : X \times X \to [0, \infty)$ be a function satisfying (12.103), and*

$$\Phi(a, b) = \frac{1}{4} \sum_{l=1}^{\infty} 4^l \phi\left(\frac{a}{2^l}, \frac{b}{2^l}\right) < +\infty \tag{12.106}$$

for all $a, b \in X$. *Then there exists a unique quadratic mapping* $Q : X \to Y$ *such that*

$$\|f_n([x_{ij}]) - Q_n[(x_{ij})]\|_n \leq \sum_{i,j=1}^{n} \Phi(x_{ij}, x_{ij})$$

for all $x = [x_{ij}] \in M_n(X)$.

Proof. The proof is similar to that of Theorem 12.20, and the details are left to the reader. $\qquad\square$

Corollary 12.13 ([176]). *Let* r, θ *be positive real numbers with* $r > 2$. *Let* $f : X \to Y$ *be a mapping satisfying* (12.105). *Then there exists a unique quadratic mapping* $Q : X \to Y$ *such that*

$$\|f_n([x_{ij}]) - Q_n([x_{ij}])\|_n \leq \sum_{i,j=1}^{n} \frac{2\theta}{2^r - 4} \|x_{ij}\|^r$$

for all $x = [x_{ij}] \in M_n(X)$.

Proof. Letting $\phi(a, b) = \theta(\|a\|^r + \|b\|^r)$ in Theorem 12.21; then we obtain the desired result. This completes the proof. $\qquad\square$

Theorem 12.22 ([176]). *Let* $f : X \to Y$ *be a mapping and let* $\phi : X \times X \to [0, \infty)$ *be a function satisfying* (12.102), *and*

$$\|Df_n([x_{ij}], [y_{ij}])\|_n \leq \|[\phi(x_{ij}, y_{ij})]\|_n \qquad (12.107)$$

for all $x = [x_{ij}], y = [y_{ij}] \in M_n(X)$. *Then there exists a unique quadratic mapping* $Q : X \to Y$ *such that*

$$\|[f(x_{ij}) - Q(x_{ij})]\|_n \leq \|\Phi(x_{ij}, x_{ij})\|_n \qquad (12.108)$$

for all $x = [x_{ij}] \in M_n(X)$. *Here* Φ *is as given in Theorem 12.20.*

Proof. By the same reasoning as in the proof of Theorem 12.20, there exists a unique quadratic mapping $Q : X \to Y$ such that

$$\|f(a) - Q(a)\| \leq \Phi(a, a)$$

for all $a \in X$. Define $Q : X \to Y$ by

$$Q(a) = \lim_{l \to \infty} \frac{1}{4^l} f(2^l a)$$

for all $a \in X$, then by Lemma 12.2, and (12.100), we obtain

$$\|[f(x_{ij}) - Q(x_{ij})]\|_n \leq \|[\|f(x_{ij}) - Q(x_{ij})\|]\|_n \leq \|[\Phi(x_{ij}, x_x ij)]\|_n$$

for all $x = [x_{ij}] \in M_n(X)$. So, we obtain the inequality (12.108). The proof is now completed. $\qquad\square$

Corollary 12.14 ([176]). *Let r, θ be positive real numbers with $r < 2$. Let $f : X \to Y$ be a mapping such that*

$$\|Df_n([x_{ij}], [y_{ij}])\|_n \leq \|[\theta\,(\|x_{ij}\|^r + \|y_{ij}\|^r)]\|_n \qquad (12.109)$$

for all $x = [x_{ij}]$, $y = [y_{ij}] \in M_n(X)$. Then there exists a unique quadratic mapping $Q : X \to Y$ such that

$$\|f_n([x_{ij}]) - Q_n([x_{ij}])\|_n \leq \left\| \left[\frac{2\theta}{4 - 2^r} \|x_{ij}\|^r \right] \right\|_n$$

for all $x = [x_{ij}] \in M_n(X)$.

Proof. Letting $\phi(a, b) = \theta\,(\|a\|^r + \|b\|^r)$ in Theorem 12.22, we obtain the desired result. The proof is now completed. □

Theorem 12.23 ([176]). *Let $f : X \to Y$ be a mapping and let $\phi : X \times X \to [0, \infty)$ be a function satisfying (12.106) and (12.107). Then there exists a unique quadratic mapping $Q : X \to Y$ such that*

$$\|[f(x_{ij}) - Q(x_{ij})]\|_n \leq \|[\Phi(x_{ij}, x_{ij})]\|_n$$

for all $x = [x_{ij}] \in M_n(X)$. Here Φ is as given in Theorem 12.21.

Proof. The proof is similar to that of Theorem 12.22, and the details are left to the reader. □

Corollary 12.15 ([176]). *Let r, θ be positive real numbers with $r > 2$. Let $f : X \to Y$ be a mapping satisfying (12.109). Then there exists a unique quadratic mapping $Q : X \to Y$ such that*

$$\|f_n([x_{ij}]) - Q_n([x_{ij}])\|_n \leq \left\| \left[\frac{2\theta}{2^r - 4} \|x_{ij}\|^r \right] \right\|_n$$

for all $x = [x_{ij}] \in M_n(X)$.

Proof. Letting $\phi(a, b) = \theta\,(\|a\|^r + \|b\|^r)$ in Theorem 12.23, we obtain the desired result. This completes the proof. □

12.9. A fixed point approach to the stability of quadratic functional equation

Cădariu and Radu [40] applied the fixed point method to the investigation of Cauchy functional equation.

This section deals with the Hyers–Ulam–Rassias stability of the quadratic functional equation for a large class of functions between a vector space and a complete β-normed space, by adopting the idea of Cădariu and Radu.

An advantage of this result is that the range of relevant functions is extended to any complete (real or complex) β-normed space, while the existing results concern only the real Banach space as we see in the proceeding theorems.

12.9.1. *Preliminaries*

Here, we present the definition of generalized metric on a set.

Definition 12.20. Let X be a set. A function $d : X \to X \to [0, \infty]$ is called a Generalized metric on X if and only if d satisfies:

(1) $d(x, y) = 0$ if and only if $x = y$;
(2) $d(x, y) = d(y, x)$ for all $x, y \in X$;
(3) $d(x, z) \leq d(x, y) + d(y, z)$ for all $x, y, z \in X$.

Note that the only substantial difference of the generalized metric from the metric is that the range of generalized metric includes the infinity.

Definition 12.21. Let X and Y be metric spaces with a metric d. Suppose a function $f : X \to Y$ satisfies

$$d(f(x), f(y)) \leq Ld(x, y), \quad \text{for all } x, y \in X$$

and for a real constant $L \geq 0$. Then L is called as a Lipschitz constant for the function f.

The following theorem is one of the fundamental results of fixed point theory, which is useful in obtaining the Hyers–Ulam stability of functional equations. For the proof, one can refer to [65].

Theorem 12.24 ([65]). *Let (X, d) be a generalized complete metric space. Assume that $\Lambda : X \to X$ is a strictly contractive operator with the Lipschitz constant $L < 1$. If there exists a non-negative integer k such that $d\left(\Lambda^{k+1}x, \Lambda^k x\right) < \infty$ for some $x \in X$, then the followings conditions are true:*

(a) *The sequence $\{\Lambda^n x\}$ converges to a fixed point x^* of Λ.*
(b) *x^* is the unique fixed point of Λ in $X^* = \{y \in X | d(\Lambda^k x, y) < \infty\}$.*
(c) *If $y \in X^*$, then $d(y, x^*) \leq \frac{1}{1-L}d(\Lambda y, y)$.*

Throughout this section, we fix a real number β with $0 < \beta \leq 1$, and let $\mathbb{K}$ denote either $\mathbb{R}$ or $\mathbb{C}$. Suppose E is a vector space over $\mathbb{K}$. A function $\| \cdot \|_\beta : E \to [0, \infty)$ is called a β-norm if and only if it satisfies

(i) $\|x\|_\beta = 0$ if and only if $x = 0$;
(ii) $\|\lambda x\|_\beta = |\lambda|^\beta \|x\|_\beta$ for all $\lambda \in \mathbb{K}$ and all $x \in E$;
(iii) $\|x + y\|_\beta \leq \|x\|_\beta + \|y\|_\beta$ for all $x, y \in E$.

12.9.2. *Hyers–Ulam–Rassias stability of the quadratic functional equation using fixed point approach*

Theorem 12.25 ([150]). *Let E_1 and E_2 be vector spaces over $\mathbb{K}$. In particular, let E_2 be a complete β-normed space, where $0 < \beta \leq 1$. Suppose $\varphi : E_1 \times E_1 \rightarrow [0, \infty)$ is a given function, and there exists a constant L, $0 < L < 1$, such that*

$$\varphi(2x, 2x) \leq 4^\beta L \varphi(x, x) \tag{12.110}$$

for all $x \in E_1$. Furthermore, let $f : E_1 \rightarrow E_2$ be a function with $f(0) = 0$ which satisfies

$$\|f(x + y) + f(x - y) - 2f(x) - 2f(y)\|_\beta \leq \varphi(x, y) \tag{12.111}$$

for all $x, y \in E_1$. If φ satisfies

$$\lim_{n \to \infty} \frac{\varphi(2^n x, 2^n y)}{4^{n\beta}} = 0 \tag{12.112}$$

for any $x, y \in E_1$, then there exists a unique quadratic function $q : E_1 \rightarrow E_2$ such that

$$\|f(x) - q(x)\|_\beta \leq \frac{1}{4^\beta} \frac{1}{1 - L} \varphi(x, x) \tag{12.113}$$

for all $x \in E_1$.

Proof. If we define

$$X = \{h : E_1 \rightarrow E_2 \mid h(0) = 0\},$$

and introduce a generalized metric on X as follows:

$$d(g, h) = \inf\{C \in [0, \infty] : \|g(x) - h(x)\|_\beta \leq C\varphi(x, x) \text{ for all } x \in E_1\},$$

then (X, d) is complete. (See the proof in [40].)

We define an operator $\Lambda : X \rightarrow X$ by

$$(\Lambda h)(x) = \frac{1}{4} h(2x)$$

for all $x \in E_1$.

First, we assert that Λ is strictly contractive on X. Given $g, h \in X$, let $C \in [0, \infty]$ be an arbitrary constant with $d(g, h) \leq C$, i.e.,

$$\|g(x) - h(x)\|_\beta \leq C\varphi(x, x)$$

for all $x \in E_1$. If we replace x in the last inequality by $2x$, and make use of (12.110), then we have

$$\|(\Lambda g)(x) - (\Lambda h)(x)\|_\beta \leq LC\varphi(x, x)$$

for every $x \in E_1$, i.e., $d(\Lambda g, \Lambda h) \leq LC$. Hence, we conclude that $d(\Lambda g, \Lambda h) \leq Ld(g, h)$ for any $g, h \in X$.

Next, we assert that $d(\Lambda f, f) < \infty$. If we substitute x for y in (12.111), and we divide both sides by 4^β, then (12.110) establishes

$$\|(\Lambda f)(x) - f(x)\|_\beta \leq \frac{1}{4^\beta} \varphi(x, x)$$

for any $x \in E_1$, i.e.,

$$d(\Lambda f, f) \leq \frac{1}{4^\beta} < \infty. \tag{12.114}$$

Then, it follows from Theorem 12.24(a) that there exists a function $q : E_1 \to E_2$ with $q(0) = 0$, which is a fixed point of Λ, such that $\Lambda^n f \to q$, i.e.,

$$\lim_{n \to \infty} \frac{1}{4^n} f(2^n x) = q(x) \tag{12.115}$$

for all $x \in E_1$.

Since the integer k of Theorem 12.24 is 0, and $f \in X^*$ (see Theorem 12.24 for the definition of X^*), by Theorem 12.24 (c) and (12.114), we obtain

$$d(f, q) \leq \frac{1}{1 - L} d(\Lambda f, f) \leq \frac{1}{4^\beta} \frac{1}{1 - L}, \tag{12.116}$$

i.e., the inequality (12.113) is true for all $x \in E_1$.

Now, substitute $2^n x$ and $2^n y$ for x and y in (12.111), respectively. If we divide both sides of the resulting inequality by $4^{n\beta}$, and letting $n \to \infty$, it follows from (12.112) and (12.115) that q is a quadratic function.

Assume that inequality (12.113) is also satisfies with another quadratic function $q_1 : E_1 \to E_2$ besides q. (As q_1 is a quadratic function, q_1 satisfies $q_1(x) = \frac{1}{4} q_1(2x) = (\Lambda q_1)(x)$ for all $x \in E_1$, that is, q_1 is a fixed point of Λ.) In view of (12.113), and the definition of d, we know that

$$d(f, q_1) \leq \frac{1}{4^\beta} \frac{1}{1 - L} < \infty,$$

i.e., $q_1 \in X^* = \{y \in X | d(\Lambda f, y) < \infty\}$. (In view of (12.114), the integer k of Theorem 12.24 is 0.) Thus, Theorem 12.24(b) implies that $q = q_1$. This completes the proof of the theorem. $\square$

By removing the hypothesis $f(0) = 0$, the above theorem is generalized to get the following theorem.

Theorem 12.26 ([150]). *Let E_1 be a vector space over K, and let E_2 be a complete β-normed space over $\mathbb{K}$, where $0 < \beta \leq 1$. Suppose a function $\varphi : E_1 \times E_1 \to [0, \infty)$ satisfies the condition (12.112) for all $x, y \in E_1$, and there exists a constant $L, \frac{1}{4^\beta} \leq L < 1$, for which the inequality (12.110) holds for any $x \in E_1$. If a function*

$f : E_1 \to E_2$ *satisfies the inequality* (12.111) *for all* $x, y \in E_1$, *then there exists a unique quadratic function* $q : E_1 \to E_2$ *such that*

$$\|f(x) - f(0) - q(x)\|_\beta \leq \frac{1}{4^\beta} \frac{1}{1 - L} [\inf\{\varphi(z, 0) | z \in E_1\} + \varphi(x, x)]$$

for all $x \in E_1$.

Proof. Putting $y = 0$ in (12.111) yields

$$\|2f(0)\|_\beta \leq \varphi(x, 0)$$

for any $x \in E_1$. We define a function $g : E_1 \to E_2$ by $g(x) = f(x) - f(0)$. If we set

$$\psi(x, y) = \varphi_0 + \varphi(x, y)$$

for each $x, y \in E_1$, where $\varphi_0 = \inf\{\varphi(x, 0) | x \in E_1\}$, it then follows from (12.111) that

$$\|g(x + y) + g(x - y) - 2g(x) - 2g(y)\|_\beta \leq \psi(x, y)$$

for all $x, y \in E_1$.

Considering (12.110) and $L \geq \frac{1}{4^\beta}$, we see that

$$\psi(2x, 2x) = \varphi_0 + \varphi(2x, 2x) \leq \varphi_0 + 4^\beta L \varphi(x, x) \leq 4^\beta L \psi(x, x)$$

for any $x \in E_1$.

Moreover, we make use of (12.112) to verify that

$$\lim_{n \to \infty} \frac{\psi(2^n x, 2^n y)}{4^{n\beta}} = \lim_{n \to \infty} \frac{\varphi_0 + \varphi(2^n x, 2^n y)}{4^{n\beta}} = 0$$

for every $x, y \in E_1$.

According to Theorem 12.25, there exists a unique quadratic function $q : E_1 \to E_2$ satisfying the inequality (12.113) with g instead of f. This completes the proof of the theorem. $\square$

Similar to the proof of Theorem 12.25, and using Theorem 12.24, we have the following theorem.

Theorem 12.27 ([150]). *Let* E_1 *and* E_2 *be a vector space over* $\mathbb{K}$ *and a complete β-normed space over* $\mathbb{K}$, *respectively. Assume that* $\varphi : E_1 \times E_1 \to [0, \infty)$ *is a given* *function, and there exists a constant* $L, 0 < L < 1$, *such that*

$$\varphi(x, x) \leq \frac{1}{4^\beta} L \varphi(2x, 2x) \tag{12.117}$$

for all $x \in E_1$. *Furthermore, assume that* $f : E_1 \to E_2$ *is a given function with* $f(0) = 0$, *and satisfies the inequality (12.111) for all* $x, y \in E_1$. *If* φ *satisfies*

$$\lim_{n \to \infty} 4^{n\beta} \varphi\left(\frac{x}{2^n}, \frac{y}{2^n}\right) = 0$$

for every $x, y \in E_1$, *then there exists a unique quadratic function* $q : E_1 \to E_2$ *such that*

$$\|f(x) - q(x)\|_\beta \leq \frac{1}{4^\beta} \frac{L}{1 - L} \varphi(x, x) \tag{12.118}$$

for any $x \in E_1$.

Proof. We use the definitions for X and d, the generalized metric on X, as in the proof of Theorem 12.25. Then (X, d) is complete.

We define an operator $\Lambda : X \to X$ by

$$(\Lambda h)(x) = 4h\left(\frac{x}{2}\right)$$

for all $x \in E_1$. We apply the same argument as in the proof of Theorem 12.25 and prove that Λ is a strictly contractive operator. Moreover, we prove

$$d(\Lambda f, f) \leq \frac{1}{4^\beta} L \tag{12.119}$$

instead of (12.114).

According to Theorem 12.24(a), there exists a function $q : E_1 \to E_2$ with $q(0) = 0$, which is a fixed point of Λ, such that

$$\lim_{n \to \infty} 4^n f\left(\frac{x}{2^n}\right) = q(x)$$

for each $x \in E_1$.

Since the integer k of Theorem 12.24 is 0, and $f \in X^*$ (see Theorem 12.24 for the definition of X^*), using Theorem 12.24(c) and (12.119), we obtain

$$d(f, q) \leq \frac{1}{1 - L} d(\Lambda f, f) \leq \frac{1}{4^\beta} \frac{L}{1 - L},$$

which implies the validity of inequality (12.118).

In the last part of proof of Theorem 12.25, if we replace $2^n x$, $2^n y$ and $4^{n\beta}$ by $\frac{x}{2^n}$, $\frac{y}{2^n}$ and $\frac{1}{4^{n\beta}}$, respectively, then we can prove that q is a unique quadratic function satisfying inequality (12.118) for all $x \in E_1$. This completes the proof. □

Remark 12.1. Theorem 12.27 cannot be generalized to the case without the condition $f(0) = 0$. For example, if φ is continuous at $(0,0)$, then the condition (12.117) implies

$$\varphi(x, x) \geq \left(\frac{4^\beta}{L}\right)^n \varphi\left(\frac{x}{2^n}, \frac{x}{2^n}\right)$$

for any $n \in \mathbb{N}$. By letting $n \to \infty$, we conclude that $\varphi(0,0) = 0$. And if we put $x = y = 0$ in (12.111), then we obtain $f(0) = 0$.

Chapter 13

Functional Inequalities

13.1. Introduction

So far, we have discussed about the recent stability results of functional equations in various abstract spaces, however in this chapter, we deal with Hyers–Ulam stability of an additive functional inequality, Cauchy type additive functional inequality, Cauchy–Jensen type functional inequality, and quadratic functional inequality using direct and fixed point methods.

Functional inequalities play a very important role in mathematics and physics, and occur in several other fields, such as information theory, inner product spaces, geometry, complex analysis, trigonometry, and Cauchy, beta and gamma equations. The subject of Ulam's type stability is one of very active areas of research, involving the functional inequalities.

13.2. Stability involving a product of different powers of norms

In this section, we present stability results for functional equation using weaker conditions controlled by a product of different powers of norms.

Theorem 13.1 ([255–258]). *Let X be a normed linear space with norm $\| \cdot \|_1$, and let Y be a Banach space with norm $\| \cdot \|_2$. Assume in addition that $f : X \to Y$ is a mapping such that $f(tx)$ is continuous in t for each fixed x. If there exist $a, b : 0 \leq a + b < 1$, and $c_2 \geq 0$ such that*

$$\|f(x + y) - f(x) - f(y)\|_2 \leq c_2 \|x\|_1^a \|y\|_1^b \tag{13.1}$$

for all $x, y \in X$, then there exists a unique linear mapping $L : X \to Y$ such that

$$\|f(x) - L(x)\|_2 \leq c \|x\|_1^{a+b} \tag{13.2}$$

for all $x \in X$, where $c = \frac{c_2}{2 - 2^{a+b}}$.

Remark 13.1. If we take $a = b = 0$ in the above theorem, and follow the proof, we obtain an additive functional L such that $\|f(x) - L(x)\|_2 \leq c_2$, for all $x \in X$. This is Hyers' result.

233

Proof of Existence. Inequality (13.1), and $y = x$ imply

$$\|f(2x) - 2f(x)\|_2 \leq c_2 \|x\|_1^{a+b}$$

or

$$\left\|\frac{f(2x)}{2} - f(x)\right\|_2 \leq \frac{c_2}{2} \|x\|_1^{a+b}. \tag{13.3}$$

More generally, the following lemma holds.

Lemma 13.1 ([255–258]). *In the space X,*

$$\left\|\frac{f(2^n x)}{2^n} - f(x)\right\|_2 \leq c_2 \sum_{i=0}^{n-1} 2^{i(a+b-1)-1} \|x\|_1^{a+b} \tag{13.4}$$

for some $c_2 \geq 0$, and for any integer n.

Proof. To prove Lemma 13.1, we proceed by induction on n.

For $n = 1$, the result is obvious from (13.3). We assume then that (13.4) holds for $n = k$, and prove that (13.4) is true for $n = k+1$. Indeed, from (13.4) and $n = k$ and $2x = z$ we find:

$$\left\|\frac{f(2^k z)}{2^k} - f(z)\right\|_2 \leq c_2 \sum_{i=0}^{k-1} 2^{i(a+b-1)-1} \|z\|_1^{a+b}$$

or

$$\left\|\frac{f(2^{k+1} x)}{2^{k+1}} - \frac{f(2x)}{2}\right\|_2 \leq c_2 \sum_{i=1}^{k} 2^{i(a+b-1)-1} \|x\|_1^{a+b}. \tag{13.5}$$

Therefore from (13.3) and the induction argument, we obtain

$$\left\|\frac{f(2^{k+1} x)}{2^{k+1}} - f(x)\right\|_2 \leq \left\|\frac{f(2^{k+1} x)}{2^{k+1}} - \frac{f(2x)}{2}\right\|_2 + \left\|\frac{f(2x)}{2} - f(x)\right\|_2$$

$$\leq c_2 \sum_{i=1}^{k} 2^{i(a+b-1)-1} \|x\|_1^{a+b} + c_2 \|x\|_1^{a+b} 2^{-1}$$

$$= c_2 \sum_{i=0}^{k} 2^{i(a+b-1)-1} \|x\|_1^{a+b}$$

or (13.4) holds for $n = k + 1$, and this completes the proof of Lemma 13.1. □

Now

$$\sum_{i=0}^{n-1} 2^{i(a+b-1)} < \sum_{i=0}^{\infty} 2^{i(a+b-1)} = \frac{1}{1 - 2^{a+b-1}} = c_0. \tag{13.6}$$

Set

$$c = c_0 \frac{c_2}{2}. \tag{13.7}$$

Then using Lemma 13.1, (13.6), and (13.7), we have

$$\left\| \frac{f(2^n x)}{2^n} - f(x) \right\|_2 \le c \|x\|_1^{a+b} \tag{13.8}$$

for any $x \in X$, any positive integer n, and some $c \ge 0$.

Lemma 13.2 ([255–258]). *The sequence $\{\frac{f(2^n x)}{2^n}\}$ converges.*

Proof. We first use (13.8), and the completeness of Y to prove that the sequence $\{\frac{f(2^n x)}{2^n}\}$ is a Cauchy sequence. In fact, if $i > j > 0$, then

$$\left\| \frac{f(2^i x)}{2^i} - \frac{f(2^j x)}{2^j} \right\|_2 = 2^{-j} \left\| \frac{f(2^i x)}{2^{i-j}} - f(2^j x) \right\|_2, \tag{13.9}$$

and if we set $2^j x = h$ in (13.9) and employ (13.8), we get

$$\left\| \frac{f(2^i x)}{2^i} - \frac{f(2^j x)}{2^j} \right\|_2 = 2^{-j} \left\| \frac{f(2^{i-j} h)}{2^{i-j}} - f(h) \right\|_2 < 2^{j(a+b-1)} c \|x\|_1^{a+b},$$

or

$$\lim_{j \to \infty} \left\| \frac{f(2^i x)}{2^i} - \frac{f(2^j x)}{2^j} \right\|_2 = 0 \tag{13.10}$$

because $a, b : 0 \le u + b < 1$.

It is obvious now from (13.10) and the completeness of Y that the sequence $\{\frac{f(2^n x)}{2^n}\}$ converges, and therefore the proof of Lemma 13.2 is complete.

We set

$$L(x) = \lim_{n \to \infty} \frac{f(2^n x)}{2^n}. \tag{13.11}$$

It is clear from (13.1) and (13.11) that

$$\|f(2^n x + 2^n y) - f(2^n x) - f(2^n y)\|_2 \le c_2 \|2^n x\|_1^a \|2^n y\|_1^b,$$

or

$$2^{-n} \|f(2^n x + 2^n y) - f(2^n x) - f(2^n y)\|_2 \le c_2 2^{(a+b-1)n} \|x\|_1^a \|y\|_1^b,$$

or

$$\left\| \lim_{n \to \infty} \frac{f(2^n(x+y))}{2^n} - \lim_{n \to \infty} \frac{f(2^n x)}{2^n} - \lim_{n \to \infty} \frac{f(2^n y)}{2^n} \right\|_2 = 0,$$

or

$$L(x+y) = L(x) + L(y) \quad \text{for any } x, y \in X. \tag{13.12}$$

From (13.12), we get

$$L(qx) = qL(x) \tag{13.13}$$

for any $q \in Q$, where Q is the set of rationals.

If we take limits on both sides of (13.8) as $n \to \infty$, we obtain (13.2). Thus, the existence of a linear mapping $L : X \to Y$ is proved, which also satisfies (13.2).
Uniqueness. It remains to show that the uniqueness part of our theorem.
Let M be a linear mapping $M : X \to Y$, such that

$$\|f(x) - M(x)\|_2 \le c' \|x\|_1^{a'+b'}, \qquad c' \ge 0 \tag{13.14}$$

for any $x \in X$ where $a', b' : 0 \le a' + b' < 1$, and c' is a constant. If there exists a linear mapping $L : X \to Y$ such that (13.2) holds, then

$$L(x) = M(x) \tag{13.15}$$

for any $x \in X$. □

To prove (13.15), the following lemma is needed.

Lemma 13.3 ([255–258]). *If* (13.2) *and* (13.14) *hold, then*

$$\|L(x) - M(x)\|_2 \le m^{a+b-1} c \|x\|_1^{a+b} + m^{a'+b'-1} c' \|x\|_1^{a'+b'} \tag{13.16}$$

for any $x \in X$.

Proof. The required result (13.16) follows immediately if we use inequalities (13.2) and (13.14), the linearity of L and M, as well as the triangle inequality. In fact,

$$L(x) = \frac{L(mx)}{m}, \quad M(x) = \frac{M(mx)}{m},$$

$\|L(mx) - M(mx)\|_2 \le \|L(mx) - f(mx)\|_2 + \|f(mx) - M(mx)\|_2$. Then if we apply (13.2) and (13.14), we obtain inequality (13.16), and the proof of Lemma 13.3 is complete. □

It is clear now that (13.16) implies $\lim_{n\to\infty}\|L(x) - M(x)\|_2 = 0$ for any $x \in X$, completing the proof of (13.15). Thus the uniqueness part of our theorem is complete, as well.

Theorem 13.2 ([255–258]). *Let X be a real normed linear space, and Y be a real Banach space. Assume that $f : X \to Y$ is a mapping for which there exist constants $\theta \geq 0$, and $p, q \in \mathbb{R}$ such that $r = p + q \neq 1$, and f satisfies the functional inequality*

$$\|f(x + y) - f(x) - f(y)\| \leq \theta \|x\|^p \|y\|^q$$

for all $x, y \in X$. Then there exists a unique additive mapping $L : X \to Y$ satisfying

$$\|f(x) - L(x)\| \leq \frac{\theta}{|2^r - 2|} \|x\|^r$$

for all $x \in X$. If, in addition, $f : X \to Y$ is a mapping such that the transformation $t \to f(tx)$ is continuous in $t \in \mathbb{R}$ for each fixed $x \in X$, then L is an $\mathbb{R}$-linear mapping.

Proof. The proof follows from Lemmas 13.1–13.3, and the details are left to the reader. $\square$

13.3. Stability of an additive functional inequality in real Banach spaces

This section deals with the generalized Hyers–Ulam stability of the functional inequality

$$\|f(x) + f(y) + f(z) + f(w)\|$$
$$- \|f(x) + f(y + z + w)\| \leq \theta \left(\|x\|^p + \|y\|^p + \|z\|^p + \|w\|^p\right) \qquad (13.17)$$

in real Banach spaces, where θ, p are positive real numbers with $p \neq 1$.

Let X be a real normed linear space, and Y be a real Banach space in the following theorem.

Theorem 13.3 ([241]). *Assume that $f : X \to Y$ is an odd mapping for which there exist constants $\theta \geq 0$ and $p \in \mathbb{R}$ such that $p \neq 1$, and $f : X \to Y$ satisfies the functional inequality*

$$\|f(x) + f(y) + f(z) + f(w)\|$$
$$- \|f(x) + f(y + z + w)\| \leq \theta \left(\|x\|^p + \|y\|^p + \|z\|^p + \|w\|^p\right) \qquad (13.18)$$

for all $x, y, z, w \in X$. Then there exists a unique Cauchy additive mapping $A : X \to Y$ satisfying

$$\|f(x) - A(x)\| \leq \frac{2^p + 2}{|2^p - 2|} \theta \|x\|^p \qquad (13.19)$$

for all $x \in X$. If, in addition, $f : X \to Y$ is a mapping such that the transformation $t \to f(tx)$ is continuous in $t \in \mathbb{R}$ for each fixed $x \in X$, then A is an $\mathbb{R}$-linear mapping.

Proof. Since f is odd, $f(0) = 0$ and $f(-x) = -f(x)$, for all $x \in X$.
    Letting $x = 0$, $z = y$ and $w = -2y$ in (13.18), we obtain

$$\|2f(y) - f(2y)\| \le (2 + 2^p)\,\theta\,\|y\|^p \tag{13.20}$$

for all $y \in X$. So

$$\left\| f(x) - 2f\left(\frac{x}{2}\right) \right\| \le \frac{2 + 2^p}{2^p}\theta\,\|x\|^p$$

for all $x \in X$. Hence

$$\left\| 2^l f\left(\frac{x}{2^l}\right) - 2^m f\left(\frac{x}{2^m}\right) \right\| \le \frac{2 + 2^p}{2^p} \sum_{j=l}^{m-1} \frac{2^j}{2^{pj}}\theta\,\|x\|^p \tag{13.21}$$

for all non-negative integers m and l with $m > l$, and all $x \in X$.
    Assume that $p > 1$. It follows from (13.21) that the sequence $\left\{ 2^k f\left(\frac{x}{2^k}\right) \right\}$ is Cauchy for all $x \in X$. Since Y is complete, the sequence $\left\{ 2^k f\left(\frac{x}{2^k}\right) \right\}$ converges. So one can define the mapping $A : X \to Y$ by

$$A(x) = \lim_{n \to \infty} 2^k f\left(\frac{x}{2^k}\right)$$

for all $x \in X$. Letting $l = 0$ and $m \to \infty$ in (13.21), we have

$$\|f(x) - A(x)\| \le \frac{2^p + 2}{2^p - 2}\theta\,\|x\|^p$$

for all $x \in X$.
    It follows from (13.18) that

$$\left\| 2^k f\left(\frac{x}{2^k}\right) + 2^k f\left(\frac{y}{2^k}\right) + 2^k f\left(\frac{z}{2^k}\right) + 2^k f\left(\frac{w}{2^k}\right) \right\|$$

$$\le \left\| 2^k f\left(\frac{x}{2^k}\right) + 2^k f\left(\frac{y + z + w}{2^k}\right) \right\| + \frac{2^k \theta}{2^{pk}}\left(\|x\|^p + \|y\|^p + \|z\|^p + \|w\|^p\right) \tag{13.22}$$

for all $x, y, z, w \in X$. Letting $k \to \infty$ in (13.22), we get

$$\|A(x) + A(y) + A(z) + A(w)\| \le \|A(x) + A(y + z + w)\| \tag{13.23}$$

for all $x, y, z, w \in X$. It is easy to show that $A : X \to Y$ is odd. Letting $w = -y - z$, and $x = 0$ in (13.22), we get $A(y + z) = A(y) + A(z)$ for all $y, z \in X$. So there exists a Cauchy additive mapping $A : X \to Y$ satisfying (13.19) for the case $p > 1$.

Now, let $T : X \to Y$ be another Cauchy additive mapping satisfying (13.19). Then we have

$$\|A(x) - T(x)\| = 2^q \left\| A\left(\frac{x}{2^q}\right) - T\left(\frac{x}{2^q}\right) \right\|$$

$$\leq 2^q \left(\left\| L\left(\frac{x}{2^q}\right) - f\left(\frac{x}{2^q}\right) \right\| + \left\| T\left(\frac{x}{2^q}\right) - f\left(\frac{x}{2^q}\right) \right\| \right)$$

$$\leq \frac{2^p + 2}{2^p - 2} \cdot \frac{2 \cdot 2^q}{2^{pq}} \theta \|x\|^p$$

which tends to zero as $q \to \infty$ for all $x \in X$. So we can conclude that $A(x) = T(x)$, for all $x \in X$. This proves the uniqueness of A.

Assume that $p < 1$. It follows from (13.20) that

$$\left\| f(x) - \frac{1}{2} f(2x) \right\| \leq \frac{2^p + 2}{2} \theta \|x\|^p$$

for all $x \in X$. Hence

$$\left\| \frac{1}{2^l} f\left(2^l x\right) - \frac{1}{2^m} f\left(2^m x\right) \right\| \leq \frac{2^p + 2}{2} \sum_{j=l}^{m-1} \frac{2^{pj}}{2^j} \theta \|x\|^p \qquad (13.24)$$

for all non-negative integers m and l with $m > l$, and all $x \in X$.

It follows from (13.24) that the sequence $\left\{ \frac{1}{2^k} f\left(2^k x\right) \right\}$ is Cauchy for all $x \in X$. Since Y is complete, the sequence $\left\{ \frac{1}{2^k} f\left(2^k x\right) \right\}$ converges. So one can define the mapping $A : X \to Y$ by

$$A(x) = \lim_{k \to \infty} \frac{1}{2^k} f\left(2^k x\right)$$

for all $x \in X$.

Letting $l = 0$ and $m \to \infty$ in (13.24), we get

$$\|f(x) - A(x)\| \leq \frac{2^p + 2}{2 - 2^p} \theta \|x\|^p$$

for all $x \in X$.

The rest of the proof is similar to the above proof, and the details are omitted. This completes the proof of the theorem. $\qquad \square$

So there exists a unique Cauchy additive mapping $A : X \to Y$ satisfying

$$\|f(x) - A(x)\| \leq \frac{2^p + 2}{|2^p - 2|} \theta \|x\|^p, \quad \text{for all } x \in X. \qquad (13.25)$$

Assume that $f : X \to Y$ is a mapping such that the transformation $t \to f(tx)$ is continuous in $t \in \mathbb{R}$ for each fixed $x \in X$. By the same reasoning as in the proof of Theorem 13.2, one can prove that A is an $\mathbb{R}$-linear mapping.

13.4. Stability of an additive functional inequality using fixed point method

In this section, using fixed point method, we present a theorem to obtain the generalized Hyers–Ulam stability of the functional inequality (13.17) in Banach spaces.

Theorem 13.4 ([241]). *Let $f : X \to Y$ be an odd mapping for which there exists a function $\varphi : X^4 \to [0, \infty)$ such that there exists a $L < 1$ such that $\varphi(x, y, z, w) \leq \frac{1}{2}L\varphi(2x, 2y, 2z, 2w)$ for all $x, y, z, w \in X$, and*

$$\|f(x) + f(y) + f(z) + f(w)\| \leq \|f(x) + f(y + z + w)\| + \varphi(x, y, z, w) \quad (13.26)$$

for all $x, y, z, w \in X$. Then there exists a unique Cauchy additive mapping $A : X \to Y$ satisfying

$$\|f(x) - A(x)\| \leq \frac{L}{2 - 2L}\varphi(0, x, x, -2x) \quad (13.27)$$

for all $x \in X$.

Proof. Consider the set

$$S : \{g : X \to Y\},$$

and introduce the generalized metric on S:

$$d(g, h) = \inf\{K \in \mathbb{R}_+ : \|g(x) - h(x)\| \leq K\varphi(0, x, x, -2x), \forall\, x \in X\}.$$

It is easy to show that (S, d) is complete.

Now, we consider the linear mapping $J : S \to S$ such that

$$Jg(x) = 2g\left(\frac{x}{2}\right)$$

for all $x \in X$. It is easy to show that

$$d(Jg, Jh) \leq Ld(g, h)$$

for all $g, h \in S$.

Since $f : X \to Y$ is odd, $f(0) = 0$, and $f(-x) = -f(x)$ for all $x \in X$. Letting $z = y = x$, and $w = -2x$ in (13.26), we get

$$\|2f(x) - f(2x)\| = \|2f(x) + f(-2x)\| \leq \varphi(0, x, x, -2x) \quad (13.28)$$

for all $x \in X$. It follows from (13.28) that

$$\left\|f(x) - 2f\left(\frac{x}{2}\right)\right\| \leq \varphi\left(0, \frac{x}{2}, \frac{x}{2}, -x\right) \leq \frac{L}{2}\varphi(0, x, x, -2x)$$

for all $x \in X$. Hence $d(f, Jf) \leq \frac{L}{2}$.

By fundamental theorem of fixed point theory, there exists a mapping $A : X \to Y$ satisfying the following:

(1) A is a fixed point of J, i.e.,

$$A\left(\frac{x}{2}\right) = \frac{1}{2}A(x) \tag{13.29}$$

for all $x \in X$. Then $A : X \to Y$ is an odd mapping. The mapping A is a unique fixed point of J in the set

$$M = \{g \in S : d(f, g) < \infty\}.$$

This implies that A is a unique mapping satisfying (13.29) such that there exists a $K \in (0, \infty)$ satisfying

$$\|f(x) - A(x)\| \leq K\varphi(0, x, x - 2x)$$

for all $x \in X$.

(2) $d(J^n f, A) \to 0$ as $n \to \infty$. This implies the equality

$$\lim_{n \to \infty} 2^n f\left(\frac{x}{2^n}\right) = A(x) \tag{13.30}$$

for all $x \in X$.

(3) $d(f, A) \leq \frac{1}{1-L} d(f, Jf)$, which implies the inequality

$$d(f, A) \leq \frac{L}{2 - 2L}.$$

This implies that the inequality (13.27) holds.

It follows from (13.26) and (13.30) that

$$\|A(x) + A(y) + A(z) + A(w)\| \leq \|A(x) + A(y + z + w)\|$$

for all $x, y, z, w \in X$. By Theorem 13.3, the mapping $A : X \to Y$ is a Cauchy additive mapping. Therefore, there exists a unique Cauchy additive mapping $A : X \to Y$ satisfying (13.27), as desired. This completes the proof. $\square$

Corollary 13.1 ([241]). *Let $r > 1$ and θ be non-negative real numbers, and let $f : X \to Y$ be an odd mapping such that*

$$\|f(x) + f(y) + f(z) + f(w)\|$$
$$\leq \|f(x) + f(y + z + w)\| + \theta\left(\|x\|^r + \|y\|^r + \|z\|^r + \|w\|^r\right) \tag{13.31}$$

for all $x, y, z, w \in X$. Then there exists a unique Cauchy additive mapping $A : X \to Y$ such that

$$\|f(x) - A(x)\| \leq \frac{2^r + 2}{2^r - 2}\theta\|x\|^r$$

for all $x \in X$.

Proof. Taking $\varphi(x, y, z, w) = \theta \left(\|x\|^r + \|y\|^r + \|z\|^r + \|w\|^r \right)$, for all $x, y, z, w \in X$, in Theorem 13.4, and then choosing $L = 2^{1-r}$, we get the desired result. □

Remark 13.2 ([241]). Let $f : X \to Y$ be an odd mapping for which there exists a function $\varphi : X^4 \to [0, \infty)$ satisfying (13.26). By a similar method to the proof of Theorem 13.4, one can show that if there exists a $L < 1$ such that $\varphi(x, y, z, w) \leq 2L\varphi \left(\frac{x}{2}, \frac{y}{2}, \frac{z}{2}, \frac{w}{2} \right)$ for all $x, y, z, w \in X$, then there exists a unique Cauchy additive mapping $A : X \to Y$ satisfying

$$\|f(x) - A(x)\| \leq \frac{1}{2 - 2L} \varphi(0, x, x, -2x)$$

for all $x \in X$.

For the case $0 < r < 1$, one can obtain a similar result to Corollary 13.1: Let $0 < r < 1$ and $\theta \geq 0$ be real numbers, and let $f : X \to Y$ be an odd mapping satisfying (13.31). Then there exists a unique Cauchy additive mapping $A : X \to Y$ satisfying

$$\|f(x) - A(x)\| \leq \frac{2 + 2^r}{2 - 2^r} \theta \|x\|^r$$

for all $x \in X$.

13.5. Hyers–Ulam stability of Cauchy type additive functional inequality

In this section, we present a theorem to investigate the Hyers–Ulam stability of a Cauchy type additive functional inequality

$$f(x + y) + f(x - y) + f(y - x) \leq f(x) + f(y)$$

in the set of real numbers $\mathbb{R}$.

Definition 13.1 ([280]). A mapping $f : \mathbb{R} \to \mathbb{R}$ is called approximately Cauchy type additive, if the approximately Cauchy additive functional inequality

$$|f(x + y) + f(x - y) + f(y - x) - f(x) - f(y)| \leq \epsilon \tag{13.32}$$

holds for every $x, y \in \mathbb{R}$ with $\epsilon \geq 0$.

Theorem 13.5 ([280]). *Assume that* $f : \mathbb{R} \to \mathbb{R}$ *is an approximately Cauchy type additive mapping satisfying the inequality* (13.32). *Define,* $f_n(x) = 2^{-n} f(2^n x)$. *Then, there exists a unique Cauchy type additive mapping* $A : \mathbb{R} \to \mathbb{R}$ *such that*

$$A(x) = \lim_{n \to \infty} 2^{-n} f\left(2^{-n} x\right) \tag{13.33}$$

for all $x \in \mathbb{R}$ *and* $n \in \mathbb{N}$, *and*

$$|f(x) - A(x)| \leq 3\epsilon \tag{13.34}$$

for some fixed $\epsilon > 0$, and all $x \in \mathbb{R}$. If, moreover, $f(tx)$ is continuous in t for each fixed $x \in \mathbb{R}$, then $A(tx) = tA(x)$ for all $t, x \in \mathbb{R}$.

$A : \mathbb{R} \to \mathbb{R}$ *is a unique linear Cauchy type additive mapping satisfying the equation*

$$A(x + y) + A(x - y) + A(y - x) = A(x) + A(y). \tag{13.35}$$

Proof. *Step* 1. By substituting $x = y = 0$ and $x = y$ in (13.32), respectively, we can observe that

$$|f(0)| \leq \epsilon, \tag{13.36}$$

and

$$\left|f(x) - 2^{-1}f(2x)\right| \leq \frac{3}{2}\epsilon. \tag{13.37}$$

Hence, for $n \in \mathbb{N} - \{0\}$

$$|f(x) - 2^{-n}f(2^n x)| \leq |f(x) - 2^{-1}f(2x)| + |2^{-1}f(2x) - 2^{-2}f(2^2 x)|$$

$$+ \cdots + |2^{-(n-1)}f(2^{n-1}x) - 2^{-n}f(2^n x)|$$

$$\leq \frac{3}{2}\left(1 + \frac{1}{2} + \cdots + \frac{1}{2^{n-1}}\right)\epsilon$$

$$= 3\left(1 - 2^{-n}\right)\epsilon. \tag{13.38}$$

Step 2. We need to show that if there is a sequence $\{f_n\} : f_n(x) = 2^{-n}f(2^n x)$, then $\{f_n\}$ converges.

For every $m > n > 0$, we can obtain

$$|f_m(x) - f_n(x)| = \left|2^{-m}f(2^m x) - 2^{-n}f(2^n x)\right|$$

$$= 2^{-n}|2^{-(m-n)}f(2^m x) - f(2^n x)|$$

$$\leq 2^{-n}3\left(1 - 2^{-(m-n)}\right)\epsilon$$

$$< \frac{3\epsilon}{2} \to 0,$$

for $n \to \infty$. Since $\mathbb{R}$ is complete, we can conclude that $\{f_n\}$ is convergent. Thus, there is a well-defined $A : \mathbb{R} \to \mathbb{R}$ such that $A(x) = \lim_{n \to \infty} 2^{-n}f(2^n x)$.

Step 3. Observe that

$$|f(x) - f_n(x)| = \left|f(x) - 2^{-n}f(2^n x)\right| \leq 3\left(1 - 2^{-n}\right)\epsilon,$$

from which by letting $n \to \infty$, we obtain

$$|f(x) - A(x)| \leq 3\epsilon.$$

Step 4. By letting $x \to 2^n x$ and $y \to 2^n y$, from (13.32), we have:

$$|f(2^n(x+y)) + f(2^n(x-y)) + f(2^n(y-x)) - f(2^n x) - f(2^n y)| \leq \epsilon.$$

Next, by multiplying with 2^{-n} and by letting $n \to \infty$, we can conclude that truly exists an $A : \mathbb{R} \to \mathbb{R}$ such that $A(x) = \lim_{n \to \infty} 2^{-n} f(2^n x)$ satisfies the Cauchy-type additivity property

$$A(x + y) + A(x - y) + A(y - x) = A(x) + A(y). \tag{13.39}$$

Step 5. We need to prove that A is unique.

Observe, from (13.39), that

$$A(0) = 0, \quad \text{and} \quad A(2x) = 2A(x).$$

Therefore, by induction hypothesis we can show that

$$A(2^n x) = 2A(2^{n-1} x) = 2^n A(x)$$

or equivalently

$$A(x) = 2^{-n} A(2^n x).$$

Assume, now, the existence of $A' : \mathbb{R} \to \mathbb{R}$, such that $A'(x) = 2^{-n} A'(2^n x)$. With the aid of the triangular inequality,

$$|A(x) - A'(x)| \leq |2^{-n} A(2^n x) - 2^{-n} f(2^n x)| + |2^{-n} f(2^n x) - 2^{-n} A'(2^n x)|$$
$$\leq 2^{-n} 3\epsilon + 2^{-n} 3\epsilon$$
$$\to 0, \quad \text{as } n \to \infty.$$

Thus, the uniqueness of A is proved, and the stability of Cauchy-type additive mapping $A : \mathbb{R} \to \mathbb{R}$ is established.

Step 6. To complete the proof of Theorem 13.5, we only need to examine whether $A : \mathbb{R} \to \mathbb{R}$ is a linear Cauchy-type mapping. To be more precise, we need to show that:

(1) $A(x + y) + A(x - y) + A(y - x) = A(x) + A(y)$, and
(2) $A(rx) = rA(x), \forall r \in \mathbb{R}$.

Recall that we have shown already that (1) holds.

Therefore, we only need to show that (2) is valid for all $r \in \mathbb{R}$. For that, we will study four cases.

Case 1. Let $r = k \in \mathbb{N}$.

For $k = 0$, from (2), we have $A(0) = 0$. This is verified if we substitute $x = y = 0$ in (13.39).

Assume that $A((k - 1)x) = (k - 1)A(x)$ is true for all k.

Then, we need to prove that $A(kx) = kA(x)$.

Note that for $x = x$ and $y = 0$ from (13.39), we can easily obtain $A(-x) = (-1)A(x)$.

Let $x = x$ and $y = (k-1)x$ in (13.39). Then,

$$A(kx) + A(-(k-2)x) + A((k-2)x) = A(x) + A((k-1)x),$$

or

$$A(kx) = kA(x), \quad \forall k \in \mathbb{N}.$$

Case 2. Let $r = k \in \mathbb{Z}$.

We only need to observe that A is odd. Since, we have already proved that (2) is valid for all $k \in \mathbb{N}$, we can then conclude that

$$A(kx) = kA(x), \quad \forall k \in \mathbb{Z}.$$

Case 3. Let $r = \frac{k}{l} \in \mathbb{Q}$, for $k \in \mathbb{Z}, l \in \mathbb{Z} - \{0\}$.

Then, $A(x) = A\left(l \cdot \frac{1}{l}x\right) = lA\left(\frac{1}{l}x\right)$, for $l \in \mathbb{Z} - \{0\}$. Hence $A\left(\frac{1}{l}x\right) = \frac{1}{l}A(x)$.

Besides, for $k \in \mathbb{Z}$, $A\left(\frac{k}{l}x\right) = A\left(k \cdot \frac{1}{l}x\right) = kA\left(\frac{1}{l}x\right)$, from Case 2.

Thus, $A\left(\frac{k}{l}x\right) = \frac{k}{l}A(x)$, or $A(rx) = rA(x)$ for $r \in \mathbb{Q}$.

Case 4. Let $r \in \mathbb{R}$, where $r = q_n$: rational numbers.

Since $\mathbb{R}$ is a complete space, every sequence $\{q_n\}$ converges in $\mathbb{R}$, i.e., $\lim_{n \to \infty} q_n = q \in \mathbb{R}$.

Recall that $A(x) = \lim_{n \to \infty} 2^{-n} f(2^n x)$, and $f(tx)$ is continuous in t for each fixed $x \in \mathbb{R}$. Therefore, $A(tx)$ is continuous in t for each fixed x in $\mathbb{R}$. Besides,

$$\lim_{n \to \infty} A(q_n x) = A\left(\lim_{n \to \infty} q_n x\right) = A(qx), \tag{13.40}$$

and

$$\lim_{n \to \infty} A(q_n x) = \lim_{n \to \infty} q_n A(x) = qA(x). \tag{13.41}$$

From (13.40) and (13.41) Case 4 is now proved, which completes *Step* 6. Thus the proof of Theorem 13.5 is complete. □

13.6. Hyers–Ulam stability of Cauchy–Jensen type functional inequality

In this section, the Hyers–Ulam stability of Cauchy–Jensen type functional inequality

$$J\left(\frac{x+y}{2}\right) \le \frac{1}{2}[J(x) + J(y)]$$

in Banach spaces is obtained.

Definition 13.2 ([269]). Let X be a linear space, and let Y be a real complete linear space. Then a mapping $J_2 : X \to Y$ is called Cauchy–Jensen, if functional equation

$$J_2 \left(\frac{x_1 + x_2}{2} \right) = \frac{1}{2} \left[J_2(x_1) + J_2(x_2) \right] \qquad (13.42)$$

holds for all vectors $(x_1, x_2) \in X^2$ with initial condition

$$J_2(0) = 0. \qquad (13.43)$$

Note that substituting $x_1 = 0$, $x_2 = 2x$ into equation (13.42), and considering condition (13.43) one concludes that

$$J_2(x) = 2^{-1} J_2(2x). \qquad (13.44)$$

Similarly substitution of x with $2x$ into (13.44) yields

$$J_2(2x) = 2^{-1} J_2(2^2 x). \qquad (13.45)$$

Combining (13.44) and (13.45), one gets that

$$J_2(x) = 2^{-2} J_2 \left(2^2 x \right). \qquad (13.46)$$

Then by induction on $n \in \mathbb{N}$ with $x \to 2^{n-1} x$ one proves that the general identity

$$J_2(x) = 2^{-n} J_2 \left(2^n x \right), \qquad (13.47)$$

holds for all $x \in X$, and all $n \in \mathbb{N}$.

Theorem 13.6 ([269]). *Let X be a normed linear space, and let Y be a real complete normed linear space. Assume in addition that $f : X \to Y$ is an approximately Cauchy–Jensen mapping: that is, a mapping for which there exist constants c, c_0 (independent of x_1, x_2) ≥ 0 such that the Cauchy–Jensen functional inequality is given by*

$$\left\| f \left(\frac{x_1 + x_2}{2} \right) - \frac{1}{2} [f(x_1) + f(x_2)] \right\| \leq c \qquad (13.48)$$

for all vectors $(x_1, x_2) \in X^2$ with initial condition

$$\| f(0) \| \leq c_0. \qquad (13.49)$$

Then the limit

$$J_2(x) = \lim_{n \to \infty} 2^{-n} f \left(2^n x \right) \qquad (13.50)$$

exists for all $x \in X$, and $J_2 : X \to Y$ is the unique Cauchy–Jensen mapping satisfying equation (13.42) and initial condition (13.43), such that J_2 is near f: that *is, inequality*

$$\|f(x) - J_2(x)\| \le c_1 \qquad (= 2c + c_0) \tag{13.51}$$

holds for all $x \in X$ with constant c_1 (independent of x) ≥ 0. Moreover, the identity

$$J_2(x) = 2^{-n} J_2(2^n x) \tag{13.52}$$

holds for all $x \in X$, and all $n \in \mathbb{N}$.

Note. From (13.49) and (13.50), one obtains

$$\|J_2(0)\| = \lim_{n \to \infty} 2^{-n} \|f(0)\| \le \left(\lim_{n \to \infty} 2^{-n} \right) c_0 = 0,$$

or

$$J_2(0) = 0.$$

Proof. Substitution $x_1 = 0, x_2 = 2x$ into (13.48) yields

$$\|f(x) - 2^{-1}[f(0) + f(2x)]\| \le c \tag{13.53}$$

for all $x \in X$. Inequality (13.53), triangle inequality, and (13.49) imply

$$\begin{aligned}
\|f(x) - 2^{-1} f(2x)\| &\le \|f(x) - 2^{-1}[f(0) + f(2x)]\| + 2^{-1} \|f(0)\| \\
&\le c + 2^{-1} c_0 \\
&= \frac{c_1}{2} \left(= c_1 \left(1 - 2^{-1} \right) \right)
\end{aligned} \tag{13.54}$$

for all $x \in X$, where $c_1 = 2c + c_0 (\ge 0)$.

Thus, substituting x with $2x$ in (13.54), one gets that

$$\left\| f(2x) - 2^{-1} f\left(2^2 x \right) \right\| \le \frac{c_1}{2},$$

or

$$\left\| 2^{-1} f(2x) - 2^{-2} f\left(2^2 x \right) \right\| \le \frac{c_1}{2^2} (= c_1 2^{-(2-1)}(1 - 2^{-1})) \tag{13.55}$$

holds for all $x \in X$.

Inequalities (13.54), (13.55), and triangle inequality yield

$$\left\| f(x) - 2^{-2} f\left(2^2 x \right) \right\| \le \left\| f(x) - 2^{-1} f(2x) \right\| + \left\| 2^{-1} f(2x) - 2^{-2} f\left(2^2 x \right) \right\|,$$

or

$$\left\| f(x) - 2^{-2} f\left(2^2 x \right) \right\| \le c_1 \left(\frac{1}{2} + \frac{1}{2^2} \right) \qquad (= c_1 \left(1 - 2^{-2} \right)) \tag{13.56}$$

for all $x \in X$.

Similarly by induction on $n \in \mathbb{N}$ with $x \to 2^{n-1}x$ in (13.54), one concludes that

$$\left\| f\left(2^{n-1}x\right) - 2^{-1}f\left(2^{n}x\right) \right\| \leq c_1 \left(1 - 2^{-1}\right),$$

or

$$\left\| 2^{-(n-1)}f(2^{n-1}x) - 2^{-n}f(2^{n}x) \right\| \leq c_1 2^{-(n-1)}(1 - 2^{-1}) \tag{13.57}$$

holds for all $x \in X$. By induction hypothesis on $n \in \mathbb{N}$, the following inequality

$$\left\| f(x) - 2^{-(n-1)}f(2^{n-1}x) \right\| \leq c_1(1 - 2^{-(n-1)}) \tag{13.58}$$

holds for all $x \in X$.

Then inequalities (13.57), (13.58), and triangle inequality yield that

$$\left\| f(x) - 2^{-n}f\left(2^{n}x\right) \right\|$$
$$\leq \left\| f(x) - 2^{-(n-1)}f(2^{n-1}x) \right\| + \left\| 2^{-(n-1)}f(2^{n-1}x) - 2^{-n}f(2^{n}x) \right\|$$
$$\leq c_1(1 - 2^{-n}) \tag{13.59}$$

holds for all $x \in X$ and all $n \in \mathbb{N}$, with $c_1 = 2c + c_2 (\geq 0)$.

Claim that the sequence $\{2^{-n}f\left(2^{n}x\right)\}$ converges.

Note that from the general inequality (13.59) and the completeness of Y, one proves that the aforementioned sequence is a Cauchy sequence.

In fact, if $i > j > 0$, then

$$\left\| 2^{-i}f\left(2^{i}x\right) - 2^{-j}f\left(2^{j}x\right) \right\| = 2^{-j}\left\| 2^{-(i-j)}f(2^{i}x) - f(2^{j}x) \right\| \tag{13.60}$$

holds for all $x \in X$, and all $i, j \in \mathbb{N}$.

Setting $h = 2^{j}x$ in (13.60) and employing inequality (13.59) one obtains

$$\left\| 2^{-i}f\left(2^{i}x\right) - 2^{-j}f\left(2^{j}x\right) \right\| = 2^{-j}\left\| 2^{-(i-j)}f(2^{i-j}h) - f(h) \right\|$$
$$\leq 2^{-j}c_1 \left(1 - 2^{-(i-j)}\right)$$
$$\leq c_1 \left(2^{-j} - 2^{-i}\right) < c_1 2^{-j},$$

or

$$\lim_{j \to \infty} \left\| 2^{-i}f\left(2^{i}x\right) - 2^{-j}f\left(2^{j}x\right) \right\| = 0 \tag{13.61}$$

completing the proof that the sequence $\{2^{-n}f\left(2^{n}x\right)\}$ converges.

Hence the mapping $J_2 = J_2(x)$ is well-defined via formula

$$J_2(x) = \lim_{n \to \infty} 2^{-n}f\left(2^{n}x\right) \tag{13.62}$$

for all $x \in X$, and all $n \in \mathbb{N}$. This means that limit (13.62) or (13.50) exists for all $x \in X$.

In addition claim that J_2 satisfies the functional equation (13.42). In fact, it is clear from (13.48), and (13.62) that

$$2^{-n} \left\| f\left(\frac{2^n x_1 + 2^n x_2}{2}\right) - \frac{1}{2}[f(2^n x_1) + f(2^n x_2)] \right\| \leq 2^{-n} c$$

for all $x_1, x_2 \in X$ and all $n \in \mathbb{N}$.
Therefore

$$\left\| \lim_{n\to\infty} 2^{-n} f\left(2^n \frac{x_1 + x_2}{2}\right) - \frac{1}{2}\left[\lim_{n\to\infty} 2^{-n} f(2^n x_1) + \lim_{n\to\infty} 2^{-n} f(2^n x_2)\right] \right\|$$

$$\leq \left(\lim_{n\to} 2^{-n}\right) c = 0,$$

or

$$\left\| J_2\left(\frac{x_1 + x_2}{2}\right) - \frac{1}{2}[J_2(x_1) + J_2(x_2)] \right\| = 0,$$

or J_2 satisfies the functional equation (13.42) for all $(x_1, x_2) \in X^2$. Thus, J_2 is a Cauchy–Jensen mapping.

It is clear now from general inequality (13.59), $n \to \infty$, and formula (13.62) that inequality (13.51) holds in X, completing the *existence proof* of Theorem 13.6.

Let $J_2' : X \to Y$ be another Cauchy–Jensen mapping satisfying functional equation (13.42), and initial condition (13.43), such that inequality

$$\|f(x) - J_2'(x)\| \leq c_1 \tag{13.63}$$

holds for all $x \in X$ with constant c_1 (independent of $x \in X$)≥ 0.

If there exists a Cauchy–Jensen mapping $J_2 : X \to Y$ satisfying equation (13.42), and initial condition (13.43), then

$$J_2(x) = J_2'(x) \tag{13.64}$$

holds for all $x \in X$.

To prove the aforementioned *uniqueness* employ (13.47) for J_2', as well, so that

$$J_2'(x) = 2^{-n} J_2'(2^n x) \tag{13.65}$$

holds for all $x \in X$, and all $n \in \mathbb{N}$.

Moreover, triangle inequality, and (13.51) imply that

$$\|J_2(2^n x) - J_2'(2^n x)\| \leq \|J_2(2^n x) - f(2^n x)\| + \|f(2^n x) - J_2'(2^n x)\|,$$

or

$$\|J_2(2^n x) - J_2'(2^n x)\| \leq c_1 + c_1 = 2c_1 \tag{13.66}$$

for all $x \in X$, and all $n \in \mathbb{N}$.

Then from (13.47), (13.65), and (13.66) one proves that

$$\|J_2(x) - J_2'(x)\| = \left\|2^{-n}J_2\left(2^n x\right) - 2^{-n}J_2'\left(2^n x\right)\right\|,$$

or

$$\|J_2(x) - J_2'(x)\| \leq 2^{1-n}c_1 \tag{13.67}$$

holds for all $x \in X$, and all $n \in \mathbb{N}$.

Therefore from inequality (13.66), and $n \to \infty$, one obtains that

$$\lim_{n \to \infty} \|J_2(x) - J_2'(x)\| \leq \left(\lim_{n \to \infty} 2^{1-n}\right) c_1 = 0,$$

or

$$\|J_2(x) - J_2'(x)\| = 0,$$

or

$$J_2(x) = J_2'(x)$$

for all $x \in X$, completing the proof of *uniqueness* and thus the proof of the theorem is complete. $\qquad\qquad\square$

13.7. Hyers–Ulam stability of a quadratic functional inequality

In this section, the Hyers–Ulam stability of a quadratic functional inequality is solved.

Definition 13.3 ([265, 273]). Let X and Y be real linear spaces. Then a mapping $Q : X \to Y$ is called **quadratic**, if the functional equation

$$Q(a_1 x_1 + a_2 x_2) + Q(-a_1 x_1 + a_2 x_2) = 2\left[a_1^2 Q(x_1) + a_2^2 Q(x_2)\right] \tag{13.68}$$

holds for every vector $(x_1, x_2) \in X^2$, and any fixed pair (a_1, a_2) of reals $a_i (i = 1, 2)$: $0 < a_2 < l = \sqrt{a_1^2 + a_2^2} < 1$ (respectively, $l > a_2 > 1$ or $l = \sqrt{2} > a_2 = 1 = a_1$).

Theorem 13.7 ([265, 273]). *Let X and Y be linear spaces. Assume that Y is complete. Assume in addition that $f : X \to Y$ is a mapping for which there exists a constant $c \geq 0$ (independent of x_1, x_2), such that the quadratic functional inequality*

$$\left\|f(a_1 x_1 + a_2 x_2) + f(-a_1 x_1 + a_2 x_2) - 2\left[a_1^2 f(x_1) + a_2^2 f(x_2)\right]\right\| \leq c \tag{13.69}$$

holds for every vector $(x_1, x_2) \in X^2$, and any fixed pair (a_1, a_2) of reals $a_i (i = 1, 2)$: $0 < a_2 < l = \sqrt{a_1^2 + a_2^2} < 1$ (respectively $l > a_2 > 1$, or $l = \sqrt{2} > a_2 = 1 = a_1$).

Define

$$f_n(x) = \begin{cases} a_2^{2n} f\left(a_2^{-n}x\right) & \text{if } 0 < a_2 < l < 1, \\ 2^{-2n} f\left(2^n x\right) & \text{if } l = \sqrt{2} > a_2 = 1 = a_1, \\ a_2^{-2n} f\left(2_2^n x\right) & \text{if } l > a_2 > 1. \end{cases}$$

Then the limit

$$Q(x) = \lim_{n \to} f_n(x) \tag{13.70}$$

exists for every $x \in X$, and $Q : X \to Y$ is the unique quadratic mapping such that

$$\|f(x) - Q(x)\| \le \begin{cases} \dfrac{c}{2\left(1 - l^2\right)} & \text{if } 0 < a_2 < l < 1, \\ \dfrac{c}{2} & \text{if } l = \sqrt{2} > a_2 = 1 = a_1, \\ \dfrac{\left[2l^2 - \left(a_2^2 + 1\right)\right] c}{2\left(l^2 - 1\right)\left(a_2^2 - 1\right)} & \text{if } l > a_2 > 1, \end{cases} \tag{13.71}$$

holds for every $x \in X$.

Proof. It is useful for the following, to observe that, from (13.69) with $x_1 = x_2 = 0$, we obtain

$$\|f(0)\| \le \begin{cases} \dfrac{c}{2(1 - l^2)} & \text{if } 0 < l < 1, \\ \dfrac{c}{2(l^2 - 1)} & \text{if } l > 1. \end{cases} \tag{13.72}$$

Now, claim that for $n \in \mathbb{N}$

$$\|f(x) - f_n(x)\| \le \begin{cases} \dfrac{c}{2(1 - l^2)}\left(1 - a_2^{2n}\right) & \text{if } 0 < a_2 < l < 1, \\ \dfrac{c}{2}\left(1 - 2^{-2n}\right) & \text{if } l = \sqrt{2} > a_2 = 1 a_1, \\ \dfrac{\left[2l^2 - \left(a_2^2 + 1\right)\right] c}{2(l^2 - 1)(a_2^2 - 1)}\left(1 - a_2^{-2n}\right) & \text{if } l > a_2 > 1. \end{cases} \tag{13.73}$$

For $n = 0$, it is trivial.

From (13.69), with $x_1 = 0$, $x_2 = a_2^{-1}x$, we obtain

$$\left\|f(x) - \left[a_1^2 f(0) + a_2^2 f\left(a_2^{-1}\right)\right]\right\| \le \frac{c}{2}. \tag{13.74}$$

From (13.72), and (13.73), as well as the triangle inequality, we have

$$\left\|f(x) - a_2^2 f\left(a_2^{-1}x\right)\right\| \le \frac{c}{2(1 - l^2)}\left(1 - a_2^2\right) \tag{13.75}$$

which is (13.73) for $n = 1$ and $0 < a_2 < l < 1$ with $f_1(x) = a_2^2 f\left(a_2^{-1}x\right)$.

Similarly, with $x_1 = 0$, $x_2 = x$ and (13.72), we obtain

$$\left\| f(a_2 x) - [a_1^2 f(0) + a_2^2 f(x)] \right\| \le \frac{c}{2},$$

or

$$\left\| f(x) - a_2^{-2} f(a_2 x) \right\| \le \frac{\left[2l^2 - (a_2^2 + 1) \right] c}{2(l^2 - 1)(a_2^2 - 1)} \left(1 - a_2^{-2} \right) \qquad (13.76)$$

which is (13.71) for $n = 1$, and $l > a_2 > 1$ with $f_1(x) = a_2^{-2} f(a_2 x)$.

Also, with $x_1 = x_2 = x$ and $a_1 = a_2 = 1$, we find

$$\left\| f(2x) + f(0) - 4f(x) \right\| \le c,$$

or

$$\left\| f(x) - 2^{-2} f(2x) \right\| \le \frac{3}{8} c = \frac{c}{2} \left(1 - 2^{-2} \right) \qquad (13.77)$$

which is (13.71) for $n = 1$, and $l = \sqrt{2} > a_2 = 1 = a_1$ with $f_1(x) = 2^{-2} f(2x)$.

Assume (13.71) is true. From (13.75), with $a_2^{-n} x$ in place of x, and the triangle inequality, we have

$$\left\| f(x) - a_2^{2(n+1)} f(a_2^{-n(n+1)} x) \right\|$$

$$\le \left\| f(x) - a_2^{2n} f\left(a_2^{-n} x\right) \right\| + \left\| a_2^{2n} f(a_2^{-n} x) - a_2^{2(n+1)} f(a_2^{-(n+1)} x) \right\|$$

$$\le \frac{c}{2(1 - l^2)} \left(1 - a_2^{2(n+1)} \right). \qquad (13.78)$$

Similarly, assume (13.73) is true. From (13.76), with $a_2^n x$ in place of x, we have

$$\left\| f(x) - a_2^{-2(n+1)} f(a_2^{n+1} x) \right\|$$

$$\le \left\| f(x) - a_2^{-2n} f\left(a_2^n x\right) \right\| + \left\| a_2^{-2n} f(a_2^n x) - a_2^{-2(n+1)} f(a_2^{n+1} x) \right\|$$

$$\le \frac{\left[2l^2 - (a_2^2 + 1) \right] c}{2(l^2 - 1)(a_2^2 - 1)} \left(1 - a_2^{-2(n+1)} \right). \qquad (13.79)$$

Also, assume (13.71) is true. From (13.77), with $2^n x$ in place of x, we have

$$\left\| f(x) - 2^{-2(n+1)} f(2^{n+1} x) \right\|$$

$$\le \left\| f(x) - 2^{-2n} f\left(2^n x\right) \right\| + \left\| 2^{-2n} f(2^n x) - 2^{-2(n+1)} f(2^{n+1} x) \right\|$$

$$\le \frac{c}{2} \left(1 - 2^{-2(n+1)} \right). \qquad (13.80)$$

These inequalities (13.78), (13.79), and (13.80) by induction, prove inequality (13.73).

Now, we claim that the sequence $\{ f_n(x) \}$ converges.

To do this, it is sufficient to prove that it is a Cauchy sequence. Inequality (13.73) is involved. In fact, if $i > j > 0$, and $h_1 = a_2^{-j}x$ with $0 < a_2 < l < 1$, we have

$$\|f_i(x) - f_j(x)\| = \|a_2^{2i} f(a_2^{-i}x) - a_2^{2j} f(a_2^{-j}x)\|$$

$$= a_2^{2j} \|a_2^{2(i-j)} f(a_2^{-(i-j)}h_1) - f(h_1)\|$$

$$\leq \frac{c}{(1-l^2)} a_2^{2j}. \tag{13.81}$$

Similarly, if $h_2 = a_2^j x$ with $l > a_2 > 1$, we have

$$\|f_i(x) - f_j(x)\| = \|a_2^{-2j} f(a_2^i x) - a_2^{-2j} f(a_2^j x)\|$$

$$= a_2^{-2j} \|a_2^{-2(i-j)} f(a_2^{i-j}h_2) - f(h_2)\|$$

$$\leq \frac{[2l^2 - (a_2^2 + 1)] c}{2(l^2 - 1)(a_2^2 - 1)} a_2^{-2j}. \tag{13.82}$$

Also, if $h_3 = 2^j x$ with $l = \sqrt{2} > a_2 = 1 = a_1$, we have

$$\|f_i(x) - f_j(x)\| = \|2^{-2i} f(2^i x) - 2^{-2j} f(2^j x)\|$$

$$= 2^{-2j} \|2^{-(i-j)} f(2^{-j}h_3) - f(h_3)\|$$

$$\leq \frac{c}{2} 2^{-2j}. \tag{13.83}$$

Then inequalities (13.81), (13.82), and (13.83) define a mapping $Q : X \to Y$. Claim that from (13.69), and (13.70), we can obtain (13.68).

In fact, it is clear from the quadratic functional inequality (13.69), and the limit (13.70), that the following functional inequality

$$a_2^{2n} \left\| f\left(a_1 a_2^{-n} x_1 + a_2 a_2^{-n} x_2\right) + f\left(-a_1 a_2^{-n} x_1 + a_2 a_2^{-n} x_2\right) \right.$$

$$\left. - 2\left[a_1^2 f\left(a_2^{-n} x_1\right) + a_2^2 f\left(a_2^{-n} x_2\right)\right] \right\| \leq a_2^{2n} c$$

holds for all vectors $(x_1, x_2) \in X^2$, all $n \in \mathbb{N}$, and $0 < a_2 < l = \sqrt{a_1^2 + a_2^2} < 1$ with $f_n(x) = a_2^{2n} f(a_2^{-n}x)$. Therefore

$$\left\| \lim_{n \to \infty} a_2^{2n} f\left(a_2^{-n}(a_1 x_1 + a_2 x_2)\right) + \lim_{n \to \infty} a_2^{2n} f\left(a_2^{-n}(-a_1 x_1 + a_2 x_2)\right) \right.$$

$$\left. - 2\left[a_1^2 \lim_{n \to \infty} a_2^{2n} f\left(a_2^{-n} x_1\right) + a_2^2 \lim_{n \to \infty} a_2^{2n} f\left(a_2^{-n} x_2\right)\right] \right\| \leq \left(\lim_{n \to \infty} a_2^{2n}\right) c = 0,$$

or

$$\|Q(a_1 x_1 + a_2 x_2) + Q(-a_1 x_1 + a_2 x_2) - 2\left[a_1^2 Q(x_1) + a_2^2 Q(x_2)\right]\| = 0, \tag{13.84}$$

or mapping Q satisfies the quadratic equation (13.68) for $0 < a_2 < l < 1$.

Similarly, from (13.69), and (13.70), we obtain that

$$a_2^{-2n} \| f(a_1 a_2^n x_1 + a_2 a_2^n x_2) + f(-a_1 a_2^n x_1 + a_2 a_2^n x_2)$$
$$- 2 \left[a_1^2 f(a_2^n x_1) + a_2^2 f(a_2^n x_2) \right] \| \leq a_2^{-2n} c$$

holds for all vectors $(x_1, x_2) \in X^2$, all $n \in \mathbb{N}$, and $l > a_2 > 1$ with $f_n(x) = a_2^{-2n} f(a_2^n x)$. Therefore

$$\left\| \lim_{n \to \infty} a_2^{-2n} f(a_2^n(a_1 x_1 + a_2 x_2)) + \lim_{n \to \infty} a_2^{-2n} f(a_2^n(-a_1 x_1 + a_2 x_2)) \right.$$
$$\left. - 2 \left[a_1^2 \lim_{n \to \infty} a_2^{-2n} f(a_2^n x_1) + a_2^2 \lim_{n \to \infty} a_2^{-2n} f(a_2^n x_2) \right] \right\|$$
$$\leq \left(\lim_{n \to \infty} a_2^{-2n} \right) c = 0,$$

or (13.84) holds for $l > a_2 > 1$ or mapping Q satisfies (13.68) for $l > a_2 > 1$.

Also, from (13.69), and (13.70) we have

$$2^{-2n} \| f(a_1 2^n x_1 + a_2 2^n x_2) + f(-a_1 2^n x_1 + a_2 2^n x_2)$$
$$- 2 \left[a_1^2 f(2^n x_1) + a_2^2 f(2^n x_2) \right] \| \leq 2^{-2n} c$$

for all vectors $(x_1, x_2) \in X^2$, all $n \in \mathbb{N}$ and $l = \sqrt{2} > a_2 = 1 = a_1$ with $f_n(x) = 2^{-2n} f(2^n x)$. Therefore

$$\left\| \lim_{n \to \infty} 2^{-2n} f(2^n(a_1 x_1 + a_2 x_2)) + \lim_{n \to \infty} 2^{-2n} f(2^n(-a_1 x_1 + a_2 x_2)) \right.$$
$$\left. - 2 \left[a_1^2 \lim_{n \to \infty} 2^{-2n} f(2^n x_1) + a_2^2 \lim_{n \to \infty} 2^{-2n} f(2^n x_2) \right] \right\|$$
$$\leq \left(\lim_{n \to \infty} 2^{-2n} \right) c = 0,$$

or (13.84) holds for $l = \sqrt{2} > a_2 = 1 = a_1$ or mapping Q satisfies (13.68) for $l = \sqrt{2} > a_2 = 1 = a_1$.

Thus (13.84) holds for $0 < a_2 < l < 1$ (respectively, $l > a_2 > 1$, and $l = \sqrt{2} > a_2 = 1 = a_1$) or mapping Q satisfies (13.68) for $0 < a_2 < l < 1$ (respectively, $l > a_2 > 1$, and $l = \sqrt{2} > a_2 = 1 = a_1$), completing the proof that Q is a well-defined quadratic mapping in X.

It is now clear from (13.73) with $n \to \infty$, and formula (13.70) that inequality (13.71) holds in X. This completes the existence proof of Theorem 13.7.

Let $Q' : X \to Y$ be a quadratic mapping satisfying (13.71). Then both Q, and Q' satisfy for $n \in \mathbb{N}$

$$Q(x) = \begin{cases} a_2^{2n} Q(a_2^{-n} x) & \text{if } 0 < a_2 < l < 1, \\ 2^{-2n} Q(2^n x) & \text{if } l = \sqrt{2} > a_2 = 1 a_1, \\ a_2^{-2n} Q(a_2^n x) & \text{if } l > a_2 > 1. \end{cases}$$

Then for $0 < a_2 < l < 1$ and every $x \in X$, and $n \in \mathbb{N}$

$$
\begin{aligned}
\|Q(x) - Q'(x)\| &= \left\|a_2^{2n} Q\left(a_2^{-n}x\right) - a_2^{2n} Q'\left(a_2^{-n}x\right)\right\| \\
&\leq a_2^{2n} \left(\left\|Q\left(a_2^{-n}x\right) - f\left(a_2^{-n}x\right)\right\| + \left\|Q'\left(a_2^{-n}x\right) - f\left(a_2^{-n}x\right)\right\|\right) \\
&\leq a_2^{2n} \frac{c}{1 - l^2}.
\end{aligned}
\tag{13.85}
$$

Similarly, for $l > a_2 > 1$, we establish

$$
\begin{aligned}
\|Q(x) - Q'(x)\| &= \left\|a_2^{-2n} Q\left(a_2^{n}x\right) - a_2^{-2n} Q'\left(a_2^{n}x\right)\right\| \\
&\leq a_2^{-2n} \left(\left\|Q\left(a_2^{n}x\right) - f\left(a_2^{n}x\right)\right\| + \left\|Q'\left(a_2^{n}x\right) - f\left(a_2^{n}x\right)\right\|\right) \\
&\leq a_2^{-2n} \frac{\left[2l^2 - \left(a_2^2 + 1\right)\right] c}{\left(l^2 - 1\right)\left(a_2^2 - 1\right)}.
\end{aligned}
\tag{13.86}
$$

Also, for $l = \sqrt{2} > a_2 = 1 = a_1$, we get

$$
\begin{aligned}
\|Q(x) - Q'(x)\| &= \left\|2^{-2n} Q\left(2^{n}x\right) - 2^{-2n} Q'\left(2^{n}x\right)\right\| \\
&\leq 2^{-2n} \left(\left\|Q\left(2^{n}x\right) - f\left(2^{n}x\right)\right\| + \left\|Q'\left(2^{n}x\right) - f\left(2^{n}x\right)\right\|\right) \\
&\leq 2^{-2n} c.
\end{aligned}
\tag{13.87}
$$

Thus from inequalities (13.85), (13.86), and (13.87), and $n \to \infty$ we find $Q(x) = Q'(x)$, for every $x \in X$. This completes the proof of uniqueness and the stability of equation (13.68). This completes the proof of the theorem. $\qquad\square$

Note. If we take $a_1 = a_2 = \frac{1}{2}$, then Theorem 13.7 takes the following form.

Theorem 13.8 ([265, 273]). *Let X and Y be normed linear spaces. Assume that Y is complete. Assume in addition that $f : X \to Y$ is a mapping for which there exists a constant $c \geq 0$ (independent of x_1, x_2), such that the quadratic functional inequality*

$$
\left\| f\left(\frac{x_1 + x_2}{2}\right) + f\left(\frac{-x_1 + x_2}{2}\right) - \frac{f(x_1) + f(x_2)}{2} \right\| \leq c
$$

holds for every vector $(x_1, x_2) \in X^2$. Then the limit

$$
Q(x) = \lim_{n \to \infty} 2^{-2n} f\left(2^{n}x\right)
$$

exists for every $x \in X$ and $Q : X \to Y$ is the unique quadratic mapping, such that

$$
\|f(x) - Q(x)\| \leq c
$$

holds for every $x \in X$.

Example 13.1 ([265, 273]). Let $f : \mathbb{R} \to \mathbb{R}$ be a real function, such that $f(x) = x^2 + k$ satisfies

$$\left\| [(a_1x_1 + a_2x_2)^2 + k] + [(-a_1x_1 + a_2x_2)^2 + k] - 2\left[a_1^2\left(x_1^2 + k\right) + a_2^2\left(x_2^2 + k\right)\right] \right\| \leq c$$

with

$$|k| \leq \begin{cases} \dfrac{c}{2(1 - l^2)} & \text{if } 0 < a_2 < l < 1, \\[2mm] \dfrac{c}{2} & \text{if } l > a_2 = 1 = a_1, \\[2mm] \dfrac{c}{2(l^2 - 1)} & \text{if } l > a_2 > 1. \end{cases}$$

Then $Q : \mathbb{R} \to \mathbb{R}$ is such that $Q(x) = x^2$, for all $x \in \mathbb{R}$.

Chapter 14

Ulam–Hyers Stabilities of Functional Equations in Felbin's Normed Spaces

14.1. Introduction

In the previous chapter, we have discussed the stability results of additive functional inequality, Cauchy–Jensen type functional inequality, and quadratic functional inequality. This chapter contains some preliminaries related to the theory of fuzzy reals numbers, the generalized Hyers–Ulam stability of a general cubic functional equation in Felbin's type fuzzy normed linear spaces using direct method, and the generalized Hyers–Ulam stability of Cauchy functional equation in Felbin's type fuzzy normed linear spaces using fixed point method.

14.2. Preliminaries

A fuzzy number is a fuzzy set on the real axis, i.e., a mapping $x : \mathbb{R} \to [0,1]$ associating with each real number t its grade of membership $\eta(t)$. A fuzzy number x is convex if $x(t) \geq \min(x(s), x(r))$ where $s \leq t \leq r$. As obtained by Zadeh [345], x is convex if and only if each of its α-level sets $[x]_\alpha$,

$$[x]_\alpha = \{t | x(t) \geq \alpha\}, \quad 0 < \alpha \leq 1, \tag{14.1}$$

is a convex set in $\mathbb{R}$. If there exists a $t_0 \in \mathbb{R}$ such that $x(t_0) = 1$, then x is called normal.

Let η be a fuzzy subset on $\mathbb{R}$, i.e., a mapping $\eta : \mathbb{R} \to [0,1]$ associating with each real number t its grade of membership $\eta(t)$.

Definition 14.1 ([74]). A fuzzy subset η on $\mathbb{R}$ is called a fuzzy real number, whose α-level set is denoted by $[\eta]_\alpha$, i.e., $[\eta]_\alpha = \{t : \eta(t) \geq \alpha\}$, if it satisfies two axioms:

(N1) There exists $t_0 \in \mathbb{R}$ such that $\eta(t_0) = 1$.
(N2) For each $\alpha \in (0,1]$, $[\eta]_\alpha = [\eta_\alpha^-, \eta_\alpha^+]$ where $-\infty < \eta_\alpha^- \leq \eta_\alpha^+ < +\infty$.

The set of all fuzzy real numbers is denoted by $f(\mathbb{R})$. If $\eta \in F(\mathbb{R})$, and $\eta(t) = 0$ whenever $t < 0$, then η is called a non-negative fuzzy real number, and $F^*(\mathbb{R})$ denotes the set of all non-negative fuzzy real numbers.

The number $\bar{0}$ stands for the fuzzy real number as follows:

$$\bar{0}(t) = \begin{cases} 1, & t = 0, \\ 0, & t \neq 0. \end{cases}$$

Clearly, $\bar{0} \in F^*(\mathbb{R})$. Also the set of all real numbers can be embedded in $f(\mathbb{R})$ because if $r \in (-\infty, \infty)$, then $\bar{r} \in F(\mathbb{R})$ satisfies $\bar{r}(t) = \bar{0}(t - r)$.

Definition 14.2 ([74]). Fuzzy arithmetic operations $\oplus, \ominus, \otimes, \oslash$ on $F(\mathbb{R}) \times F(\mathbb{R})$ can be defined as follows:

(1) $(\eta \oplus \delta)(t) = \sup_{s \in \mathbb{R}} \{\eta(s) \wedge \delta(t - s)\}, t \in \mathbb{R}$,
(2) $(\eta \ominus \delta)(t) = \sup_{s \in \mathbb{R}} \{\eta(s) \wedge \delta(t - s)\}, t \in \mathbb{R}$,
(3) $(\eta \otimes \delta)(t) = \sup_{s \in \mathbb{R}, s \neq 0} \{\eta(s) \wedge \delta(t/s)\}, t \in \mathbb{R}$,
(4) $(\eta \oslash \delta)(t) = \sup_{s \in \mathbb{R}} \{\eta(st) \wedge \delta(s)\}, t \in \mathbb{R}$.

The additive and multiplicative identities in $F(\mathbb{R})$ are $\bar{0}$ and $\bar{1}$, respectively. Let $\ominus \eta$ be defined as $\bar{0} \ominus \eta$. It is clear that $\eta \ominus \delta = \eta \oplus (\ominus \delta)$.

Definition 14.3 ([74]). For $k \in \mathbb{R} \backslash \{0\}$, fuzzy scalar multiplication $k \odot \eta$ is defined as $(k \odot \eta)(t) = \eta(t/k)$, and $0 \odot \eta$ is defined to be $\bar{0}$.

Lemma 14.1 ([315]). *Let η, δ be fuzzy real numbers. Then*

$$\forall t \in \mathbb{R}, \eta(t) = \delta(t) \Leftrightarrow \forall \alpha \in (0, 1], [\eta]_\alpha = [\delta]_\alpha.$$

Proof. The proof of $(\Rightarrow)$ is clear. For $(\Leftarrow)$, suppose there exists $t_0 \in \mathbb{R}$ such that $\eta(t_0) \neq \delta(t_0)$ and for all $\alpha \in (0, 1]$, $[\eta]_\alpha = [\delta]_\alpha$. Since $\eta(t_0) \neq \delta(t_0)$, suppose $\eta(t_0) < \delta(t_0)$, then there exists $\alpha \in (0, 1)$ such that $\eta(t_0) < \alpha < \delta(t_0)$. From $\alpha < \delta(t_0)$, we have $\delta_\alpha^- \leq t_0 \leq \delta_\alpha^+$ and since $\eta_\alpha^\pm = \delta_\alpha^\pm$. So $\eta(t_0) \geq \alpha$, which contradicts $\eta(t_0) < \alpha$. $\square$

Lemma 14.2 ([72]). *Let $\eta, \delta \in F(\mathbb{R})$, and $[\eta]_\alpha = [\eta_\alpha^-, \eta_\alpha^+]$, $[\delta]_\alpha = [\delta_\alpha^-, \delta_\alpha^+]$. Then*

(1) $[\eta \oplus \delta]_\alpha = [\eta_\alpha^- + \delta_\alpha^-, \eta_\alpha^+ + \delta_\alpha^+]$,
(2) $[\eta \ominus \delta]_\alpha = [\eta_\alpha^- - \delta_\alpha^+, \eta_\alpha^+ - \delta_\alpha^-]$,
(3) $[\eta \otimes \delta]_\alpha = [\eta_\alpha^- \delta_\alpha^-, \eta_\alpha^+ \delta_\alpha^+], \eta, \delta \in F(\mathbb{R})$,
(4) $[\bar{1} \oslash \delta]_\alpha = [1/\delta_\alpha^+, 1/\delta_\alpha^-], \delta_\alpha^- > 0$.

Proof. The proof is obtained directly by the definition of Lemma 14.1. $\square$

Definition 14.4 ([315]). Let η be a non-negative fuzzy real number, and $p \neq 0$ be a real number. Define η^p as follows:

$$\eta^p(t) = \begin{cases} \eta(t^{\frac{1}{p}}), & t \geq 0, \\ 0, & t < 0. \end{cases}$$

Set $\eta^p = \bar{1}$, in case $p = 0$.

We show that η^p is a non-negative fuzzy real number, i.e., $\eta^p \in F^*(\mathbb{R})$, $\forall p \in \mathbb{R}$. We need to investigate conditions (N1), and (N2) in the definition of fuzzy real numbers.

For condition (N1), since η is a fuzzy real number, there exists $t_0 \in [0, +\infty)$ such that $\eta(t_0) = 1$. Set $t' = t_0^p$. Then $\eta^p(t') = \eta((t')^{\frac{1}{p}}) = \eta((t_0^p)^{\frac{1}{p}}) = \eta(t_0) = 1$.

For condition (N2), since for all $\alpha \in (0, 1]$, $[\eta]_\alpha = [\eta_\alpha^-, \eta_\alpha^+]$, we have

$$\begin{aligned} [\eta^p]_\alpha &= \{t : \eta^p(t) \geq \alpha\} \\ &= \{t : \eta(t^{\frac{1}{p}})\} = \{s^p : \eta(s) \geq \alpha\} \\ &= (\{s : \eta(s) \geq \alpha\})^p \\ &= ([\eta_\alpha^-, \eta_\alpha^+])^p. \end{aligned}$$

Therefore, η^p is a fuzzy real number. Also it is clear that if $p > 0$ then $[\eta^p]_\alpha = [(\eta_\alpha^-)^p, (\eta_\alpha^+)^p]$, and if $p < 0$ then $[\eta^p]_\alpha = [(\eta_\alpha^+)^p, (\eta_\alpha^-)^p]$.

Theorem 14.1 ([315]). *Let η be a non-negative fuzzy real number, and p, q be non-zero integers. Then*

(1) $p > 0 \Rightarrow \eta^p = \bigotimes_{i=1}^p \eta$,

(2) $p < 0 \Rightarrow \eta^p = \bar{1} \oslash (\bigotimes_{i=1}^{-p} \eta)$,

(3) $\eta^p \otimes \eta^q = \eta^{p+q}; pq > 0$,

(4) $(\eta^p)^q = \eta^{pq}$.

Proof. We prove this theorem by using Lemmas 14.1 and 14.2.

(1) Let $p > 0$ and $\alpha \in (0, 1]$. Then

$$[\eta^p]_\alpha = ([\eta_\alpha^-, \eta_\alpha^+])^p = [(\eta_\alpha^-)^p, (\eta_\alpha^+)^p] = \left[\prod_{i=1}^p \eta_\alpha^-, \prod_{i=1}^p \eta_\alpha^+\right] = \left[\bigotimes_{i=1}^p \eta\right]_\alpha.$$

Hence $\eta^p = \bigotimes_{i=1}^p \eta$.

(2) For all $\alpha \in (0, 1]$ and $p < 0$, we have

$$[\eta^p]_\alpha = ([\eta_\alpha^-, \eta_\alpha^+])^p = [(\eta_\alpha^+)^p, (\eta_\alpha^-)^p]$$

$$= \left[\frac{1}{(\eta_\alpha^+)^{-p}}, \frac{1}{(\eta_\alpha^-)^{-p}} \right] = \left[\frac{1}{\prod_{i=1}^{-p} \eta_\alpha^+}, \frac{1}{\prod_{i=1}^{-p} \eta_\alpha^-} \right]$$

$$= \left[\bar{1} \oslash \left(\bigotimes_{i=1}^{-p} \eta \right) \right]_\alpha .$$

Thus $\eta^p = \bar{1} \oslash (\bigotimes_{i=1}^{-p} \eta)$.

(3) For all $\alpha \in (0,1]$ and $p, q \geq 0$,

$$[\eta^p \otimes \eta^q]_\alpha = \left[(\eta^p)_\alpha^- (\eta^q)_\alpha^-, (\eta^p)_\alpha^+ (\eta^q)_\alpha^+ \right]$$

$$= \left[(\eta_\alpha^-)^p (\eta_\alpha^-)^q, (\eta_\alpha^+)^p (\eta_\alpha^+)^q \right]$$

$$= \left[(\eta_\alpha^-)^{p+q}, (\eta_\alpha^+)^{p+q} \right] = [\eta^{p+q}]_\alpha.$$

So $\eta^p \otimes \eta^q = \eta^{p+q}$. In case that $p, q < 0$, the proof is similar.

(4) For all $\alpha \in (0,1]$ and $p, q \geq 0$,

$$[(\eta^p)^q]_\alpha = \left[(\eta^p)_\alpha^-, (\eta^p)_\alpha^+ \right]^q = \left(\left[\eta_\alpha^-, \eta_\alpha^+ \right]^p \right)^q = \left[\eta_\alpha^-, \eta_\alpha^+ \right]^{pq} = [\eta^{pq}]_\alpha.$$

Hence $(\eta^p)^q = \eta^{pq}$. It is easy to check the other cases. This completes the proof.

$\square$

Definition 14.5 ([74]). Define a partial ordering $\preceq$ in $F(\mathbb{R})$ by $\eta \preceq \delta$ if and only if $\eta_\alpha^- \leq \delta_\alpha^-$ and $\eta_\alpha^+ \leq \delta_\alpha^+$ for all $\alpha \in (0,1]$. The strict inequality in $F(\mathbb{R})$ is defined by $\eta \prec \delta$ if and only if $\eta_\alpha^- < \delta_\alpha^-$ and $\eta_\alpha^+ < \delta_\alpha^+$ for all $\alpha \in (0,1]$.

Definition 14.6 ([341]). Let X be a real linear space; L and R (respectively, left norm and right norm) be symmetric and non-decreasing mappings from $[0,1] \times [0,1]$ into $[0,1]$ satisfying $L(0,0) = 0$, $R(1,1) = 1$. Then $\|\cdot\|$ is called a fuzzy norm, and $(X, \|\cdot\|, L, R)$ is a fuzzy normed linear space (abbreviated to FNLS) if the mapping $\|\cdot\|$ from X into $F^*(\mathbb{R})$ satisfies the following axioms, where $[\|x\|]_\alpha = [\|x\|_\alpha^-, \|x\|_\alpha^+]$ for $x \in X$ and $\alpha \in (0,1]$:

(A1) $\|x\| = \bar{0}$ if and only if $x = 0$.

(A2) $\|rx\| = |r| \odot \|x\|$ for all $x \in X$ and $r \in (-\infty, \infty)$.

(A3) For all $x, y \in X$:

(A3L) if $s \leq \|x\|_1^-, t \leq \|y\|_1^-$ and $s + t \leq \|x+y\|_1^-$, then $\|x+y\|(s+t) \geq L(\|x\|(s), \|y\|(t))$;

(A3R) if $s \geq \|x\|_1^-, t \geq \|y\|_1^-$ and $s + t \geq \|x+y\|_1^-$, then $\|x+y\|(s+t) \leq R(\|x\|(s), \|y\|(t))$.

Lemma 14.3 ([342]). *Let $(X, \|\cdot\|, L, R)$ be an FNLS, and suppose that*

$(R-1)$ $R(a,b) \leq \max(a,b)$,

$(R-2)$ $\forall \alpha \in (0,1], \exists \beta \in (0,\alpha]$ such that $R(\beta, y) \leq \alpha$ for all $y \in (0,\alpha)$,

$(R-3)$ $\lim_{a \to 0+} R(a,a) = 0$.

Then $(R-1) \Rightarrow (R-2) \Rightarrow (R-3)$ but not conversely.

Proof. Suppose that $R(a, b) \leq \max(a, b)$. Then for each $\alpha \in (0, 1]$ there exists $\beta = \frac{\alpha}{2} \in (0, \alpha]$ such that $R(\beta, \gamma) \leq \max\{\alpha/2, \gamma\} < \alpha$ for all $\gamma \in (0, \alpha)$, showing $(R - 1) \Rightarrow (R - 2)$.

Suppose that $(R - 2)$ holds. Then for each $\epsilon \in (0, 1)$ there exist $\beta \in (0, \epsilon]$ and $\gamma = \beta/2$ such that $R(\beta, \beta/2) < \epsilon$. Thus, $0 \leq R(a, a) \leq R(\beta, \beta/2) < \epsilon$ for all $\alpha \in (0, \beta/2]$. Hence, $\lim_{a \to 0^-} R(a, a) = 0$, showing $(R - 2) \Rightarrow (R - 3)$. $\square$

Lemma 14.4 ([341]). *Let $\eta \in f(\mathbb{R})$. Then $\eta(t)$ is non-decreasing right continuous on $(-\infty, \eta_1^+)$, and non-increasing left continuous on $(\eta_1^-, +\infty)$.*

Proof. Suppose that $t_1, t_2 \in (-\infty, \eta_1^+)$, $t_1 < t_2$, but $\eta(t_1) > \eta(t_2)$. Then there exists $\alpha \in (0, 1)$ such that $\eta(t_1) > \alpha > \eta(t_2)$; i.e., $t_1 \in [\eta]_\alpha$ and $t_2 \notin [\eta]_\alpha$. Thus $t_2 < \eta_\alpha^- \leq t_1$, which is a contradiction. Suppose that $t \in (-\infty, \eta_1^+)$, $\lim_{\epsilon \to 0^+} \eta(t + \epsilon) > \eta(t)$. Then there exists $\alpha \in (0, 1)$ such that $\lim_{\epsilon \to 0^+} \eta(t + \epsilon) > \alpha > \eta(t)$; i.e., there is $\epsilon_0 > 0$ such that $t \notin [\eta]_\alpha$ and $t + \epsilon \in [\eta]_\alpha$ for each $\epsilon \in (0, \epsilon_0)$. Thus $t < \eta_\alpha^- \leq t + \epsilon$. Letting $\epsilon \to 0^+$ we have $t < \eta_\alpha^- \leq t$, which is a contradiction. Likewise, the case of $(\eta_1^-, +\infty)$ follows as above. This completes the proof. $\square$

Lemma 14.5 ([342]). *Let $(X, \|\cdot\|, L, R)$ be an FNLS. Then we have the following conditions:*

(1) *If $R(a, b) \leq \max(a, b)$, then $\forall \alpha \in (0, 1]$, $\|x + y\|_\alpha^+ \leq \|x\|_\alpha^+ + \|y\|_\alpha^+$ for all $x, y \in X$.*

(2) *If $(R - 2)$ then for each $\alpha \in (0, 1]$ there is $\beta \in (0, \alpha]$ such that $\|x + y\|_\alpha^+ \leq \|x\|_\beta^+ + \|y\|_\alpha^+$ for all $x, y \in X$.*

(3) *If $\lim_{a \to 0^+} R(a, a) = 0$, then for each $\alpha \in (0, 1]$ there is $\beta \in (0, \alpha]$ such that $\|x + y\|_\alpha^+ \leq \|x\|_\beta^+ + \|y\|_\beta^+$ for all $x, y \in X$.*

Proof. Only (2) requires proof. Suppose that $(R - 2)$ holds, that is, for each $\alpha \in (0, 1]$, there exists $\beta = \beta(x) \in (0, \alpha]$ such that $R(\beta, \gamma) < \alpha$ for all $\gamma \in (0, \alpha)$. By Lemma 14.4, it will suffice to prove that $\|x + y\|_\alpha^+ \leq \|x\|_\beta^+ + \|y\|_\gamma^+$. Assume that $\|x + y\|_\alpha^+ > \|x\|_\beta^+ + \|y\|_\gamma^+$. Then there exist s and t: $s + t = \|x + y\|_\alpha^+ \geq \|x + y\|_1^-$, $s > \|x\|_\beta^+ \geq \|x\|_1^-$, $t \geq \|y\|_\gamma^+ \geq \|y\|_1^-$. By (A3R), we have

$$\alpha \leq \|x + y\|(s + t) \leq R(\|x\|(s), \|y\|(t)) \leq R(\beta, \gamma) < \alpha.$$

This contradiction completes the proof. $\square$

Lemma 14.6 ([315]). *Let $(X, \|\cdot\|, L, R)$ be an FNLS, and suppose that*
$(L - 1)$ $L(a, b) \geq \min(a, b)$,
$(L - 2)$ $\forall \alpha \in (0, 1]$, $\exists \beta \in (\alpha, 1]$ such that $L(\beta, \gamma) \geq \alpha$ for all $\gamma \in [\alpha, 1]$,
$(L - 3)$ $\lim_{a \to 1^-} L(a, a) = 1$.
Then $(L - 1) \Rightarrow (L - 2) \Rightarrow (L - 3)$.

Proof. Suppose that $L(a, b) \geq \min(a, b)$. Then for each $\alpha \in (0, 1]$ there exists $\beta = \frac{1+\alpha}{2} \in [\alpha, 1]$ such that

$$L(\beta, \gamma) = L\left(\frac{1+\alpha}{2}, \gamma\right) \geq \min\left(\frac{1+\alpha}{2}, \gamma\right) \geq \alpha; \quad \forall \gamma \in [\alpha, 1].$$

It means $(L-1) \Rightarrow (L-2)$. Suppose $(L-2)$ holds; then

$$\forall \epsilon \in (0, 1), \exists \beta \in [\epsilon, 1]; \quad L(\beta, \gamma) \geq \epsilon, \quad \gamma \in [\epsilon, 1].$$

Set $\gamma = \frac{1+\beta}{2}$. For all $\alpha \in [\frac{1+\beta}{2}, 1]$, we have

$$\epsilon \leq L(\beta, \gamma) = L\left(\beta, \frac{1+\beta}{2}\right) < L(\alpha, \alpha).$$

Hence $\lim_{a \to 1^-} L(a, a) = 1$. The proof is now completed. $\quad\square$

Lemma 14.7 ([315]). *Let $(X, \|\cdot\|, L, R)$ be an FNLS. Then we have the following conditions:*

(1) *If $L(a, b) \geq \min(a, b)$, then $\forall \alpha \in (0, 1]$, $\|x + y\|_\alpha^- \leq \|x\|_\alpha^- + \|y\|_\alpha^-$ for all $x, y \in X$.*

(2) *If $(L-2)$ then for each $\alpha \in (0, 1]$, there is $\beta \in [\alpha, 1]$ such that $\|x + y\|_\alpha^- \leq \|x\|_\beta^- + \|y\|_\alpha^-$ for all $x, y \in X$.*

(3) *If $\lim_{a \to 1^-} L(a, a) = 1$, then for each $\alpha \in (0, 1]$, there is $\beta \in [\alpha, 1]$ such that $\|x + y\|_\alpha^- \leq \|x\|_\beta^- + \|y\|_\beta^-$ for all $x, y \in X$.*

Proof. (1) It has been proved in [341]. (2) By $(L-2)$, for each $\alpha \in (0, 1]$ there exist $\beta \in [\alpha, 1]$ such that $L(\beta, \gamma) \geq \alpha$ for all $\gamma \in [\alpha, 1]$. Suppose that $\|x + y\|_\alpha^- > \|x\|_\beta^- + \|y\|_\alpha^-$. Then there exists s, t such that $s = \|x\|_\beta^- \leq \|x\|_1^-$, $t = \|y\|_\alpha^- \leq \|y\|_1^-$ and $s + t = \|x\|_\beta^- + \|y\|_\alpha^- < \|x + y\|_\alpha^- \leq \|x + y\|_1^-$. Thus by (A3L),

$$\alpha > \|x + y\|(s + t) \geq L(\|x\|(s), \|y\|(t)) \geq L(\beta, \alpha) \geq \alpha,$$

which is a contradiction.

(3) By $\lim_{a \to 1^-} L(a, a) = 1$, for each $\alpha \in (0, 1]$ there exists $\beta \in [\alpha, 1]$ such that $L(\beta, \beta) \geq \alpha$. Suppose $\|x + y\|_\alpha^- > \|x\|_\beta^- + \|y\|_\beta^-$. Then there exist s and t such that $s = \|x\|_\beta^- \leq \|x\|_1^-$, $t = \|y\|_\beta^- \leq \|y\|_1^-$ and $s + t = \|x\|_\beta^- + \|y\|_\beta^- < \|x + y\|_\alpha^- \leq \|x + y\|_1^-$. Thus by (A3L),

$$\alpha > \|x + y\|(s + t) \geq L(\|x\|(s), \|y\|(t)) \geq L(\beta, \beta) \geq \alpha,$$

which is impossible. This completes the proof. $\quad\square$

Lemma 14.8 ([341]). *Let $(X, \|\cdot\|, L, R)$ be an FNLS. Then we have the following conditions:*

(1) *If $R(a, b) \geq \max(a, b)$, and $\forall \alpha \in (0, 1]$, $\|x + y\|_\alpha^+ \leq \|x\|_\alpha^+ + \|y\|_\alpha^+$ for all $x, y \in X$, then (A3R).*

(2) *If $L(a,b) \leq \min(a,b)$, and $\forall \alpha(0,1]$, $\|x+y\|_\alpha^- \leq \|x\|_\alpha^- + \|y\|_\alpha^-$ for all $x,y \in X$, then* (A3L).

Proof. (1) Let $R \geq \max(a,b)$, $s \geq \|x\|_1^-$, $t \geq \|y\|_1^-$, $s+t \geq \|x+y\|_1^-$ and $\alpha = \|x+y\|(s+t)$. Then $s+t \leq \|x+y\|_\alpha^+ \leq \|x\|_\alpha^+ + \|y\|_\alpha^+$. Hence, $s \leq \|x\|_\alpha^+$ or $t \leq \|y\|_\alpha^+$; i.e., $\|x\|(s) \geq \alpha$ or $\|y\|(t) \geq \alpha$. Therefore, $R(\|x\|(s), \|y\|(t)) \geq \max(\|x\|(s), \|y\|(t)) \geq \alpha = \|x+y\|(s+t)$.

(2) Let $L(a,b) \leq \min(a,b)$, and $s \leq \|x\|_1^-$, $t \leq \|y\|_1^-$, $s+t \leq \|x+y\|_1^-$, and $\alpha = \min(\|x\|(s), \|y\|(t))$. Then $\|x\|(s) \geq \alpha$, $\|y\|(t) \geq \alpha$; i.e., $s \geq \|x\|_\alpha^-$, $t \geq \|y\|_\alpha^-$, $s+t \geq \|x\|_\alpha^- + \|y\|_\alpha^- \geq \|x+y\|_\alpha^-$. Therefore, $\|x+y\|(s+t) \geq \alpha = \min(\|x\|(s), \|y\|(t)) \geq L(\|x\|(s), \|y\|(t))$. This completes the proof. $\square$

Theorem 14.2 ([313]). *Let $(X, \|\cdot\|, L, R)$ be an FNLS, and $\lim_{a \to 0^+} R(a,a) = 0$. Then $(X, \|\cdot\|, L, R)$ is a Hausdorff topological vector space, whose neighborhood base of origin θ is $\{N(\epsilon, \alpha) : \epsilon > 0, \alpha \in (0,1]\}$, where $N(\epsilon, \alpha) = \{x : \|x\|_\alpha^+ \leq \epsilon\}$.*

Proof. For $W_1 = N(\epsilon_1, \alpha_1)$ and $W_2 = N(\epsilon_2, \alpha_2)$ there exists $W_0 = N(\epsilon_0, \alpha_0)$ such that $W_0 \subseteq W_1 \cap W_2$, where $\alpha_0 = \min\{\alpha_1, \alpha_2\}$, $\epsilon_0 = \min\{\epsilon_1, \epsilon_2\}$. So the set $\{N(\epsilon, \alpha) : \epsilon > 0, \alpha \in (0,1]\}$ is the base of the topology on X. Now, we show that, by the above topology, X is topological vector space. To do that we should verify the following conditions:

(a) The topological space X is a Hausdorff space.

(b) The vector space operations are continuous with respect to the topology.

For (a), if $x \in X$, $x \neq 0$, then $\|x\| = \bar{0}$, so there is $\alpha_0 \in (0,1]$ with $\|x\|_{\alpha_0}^+ \neq 0$. Hence, there exists $\epsilon_0 > 0$ such that $\|x\|_\alpha^+ > \epsilon_0$, i.e., $x \notin N(\epsilon_0, \alpha_0)$. So X is Hausdorff space. For (b), we prove that the vector space operations (the addition and scalar multiplication) are continuous with respect to the defined topology on X. Let $W = N(\epsilon, \alpha)$. By $\lim_{a \to 0^+} R(a,a) = 0$ and Lemma 14.4, there exists $W_0 = N\left(\frac{\epsilon}{2}, \alpha_0\right)$ such that $\|x+y\|_{\alpha_0}^+ \leq \|x\|_{\alpha_0}^+ + \|y\|_{\alpha_0}^+ < \epsilon$, for $x,y \in W_0$, which shows that $W_0 + W_0 \subseteq W$. To show that scalar multiplication is continuous, we should find $N(\epsilon_0, \alpha_0)$ and $N_\lambda(r)$ and $x \in N(\epsilon_0, \alpha_0)$, then

$$\|bx\|_\alpha^+ = |b| \|x\|_\alpha^+ = |b - r + r| \|x\|_\alpha^+ < (|b-r| + |r|) \|x\|_\alpha^+ < (\lambda + |r|) \|x\|_\alpha^+ < \epsilon$$

when $\alpha_0 = \alpha$ and $\epsilon_0 = \frac{\epsilon}{2}|r|$ and $\lambda = |r|$. Therefore, $(X, \|\cdot\|, L, R)$ is a topological vector space. This completes the proof of the theorem. $\square$

Definition 14.7 ([315]). Let $(X, \|\cdot\|, L, R)$ be an FNLS. A sequence $\{x_n\}_{n=1}^\infty \subseteq X$ converges to $x \in X$, denoted by $\lim_{n \to \infty} x_n = x$, if $\lim_{n \to \infty} \|x_n - x\|_\alpha^+ = 0$ for every $\alpha \in (0,1]$, and is called a Cauchy sequence if $\lim_{m,n \to \infty} \|x_m - x_n\|_\alpha^+ = 0$ for every $\alpha \in (0,1]$. A subset $A \subseteq X$ is said to be complete if every Cauchy sequence in A converges in A. The fuzzy normed space $(X, \|\cdot\|, L, R)$ is said to be a fuzzy Banach space if it is complete.

14.3. Stability of a general cubic functional equation in Felbin's type fuzzy normed linear spaces

In the following theorem, the generalized Hyers–Ulam stability of a general cubic functional equation

$$f(x + ky) - kf(x + y) + kf(x - y) - f(x - ky) = 2k(k^2 - 1)f(y) \qquad (14.2)$$

with $k \in \mathbb{N}$, $k \neq 1$, in Felbin's type fuzzy normed linear spaces is obtained. For the sake of convenience, we use the following abbreviation for a given mapping $f : X \to Y$:

$$D_k f(x, y) = f(x + ky) - kf(x + y) + kf(x - y) - f(x - ky) - 2k(k^2 - 1)f(y)$$

for $x, y \in X$.

Theorem 14.3 ([72]). *Let X be a linear space and $(Y, \|\cdot\|, L, R)$ be a fuzzy Banach space satisfying $(R-1)$. Let $f : X \to Y$ be a mapping for which there exists a function $\varphi : X \times X \to F^*(\mathbb{R})$ such that*

$$\lim_{n \to \infty} \frac{1}{k^{3n}} \varphi\left(k^n x, k^n y\right)_\alpha^+ = 0, \qquad (14.3)$$

and

$$\widetilde{\varphi_\alpha}(x) = \sum_{i=0}^\infty \frac{1}{k^{3i}} \varphi\left(0, k^i x\right)_\alpha^+ < \infty, \qquad (14.4)$$

$$\|D_k f(x, y)\| \preceq \varphi(x, y) \qquad (14.5)$$

for $x, y \in X$, and $\alpha \in (0, 1]$. Then there exists a unique cubic mapping $C_k : X \to Y$ such that

$$\forall \alpha \in (0, 1], \exists \beta \in (0, \alpha]; \|f(x) - f(-x) - C_k(x)\|_\alpha^+ \leq \frac{1}{2k^3} \left[\widetilde{\varphi_\beta}(x) + \widetilde{\varphi_\beta}(-x)\right] \qquad (14.6)$$

for $x \in X$.

Proof. Replacing $x = 0$ in (14.5), we obtain

$$\left\|g(ky) - kg(y) - 2k(k^2 - 1)f(y)\right\| \preceq \varphi(0, y) \qquad (14.7)$$

for all $y \in X$, where $g(y) = f(y) - f(-y)$. Replacing y in (14.7) by $-y$, we obtain

$$\left\|g(ky) - kg(y) + 2k(k^2 - 1)f(-y)\right\| \preceq \varphi(0, -y) \qquad (14.8)$$

for all $y \in X$. It follows from (14.7) and (14.8) that

$$\left\|g(ky) - kg(y) - 2k(k^2 - 1)f(y)\right\|_\alpha^+ \leq \varphi(0, y)_\alpha^+, \qquad (14.9)$$

or

$$\|g(ky) - kg(y) + 2k(k^2 - 1)f(-y)\|_\alpha^+ \leq \varphi(0, -y)_\alpha^+ \tag{14.10}$$

for all $\alpha \in (0, 1]$. It follows from (14.9), (14.10), and Lemma 14.4 that

$$\|g(ky) - k^3 g(y)\|_\alpha^+ \leq \frac{1}{2}\widetilde{\varphi}_\alpha(y) \tag{14.11}$$

for all $y \in X$ and $\alpha \in (0, 1]$, where $\widetilde{\varphi}_\alpha(y) = \varphi(0, y)_\alpha^+ + \varphi(0, -y)_\alpha^+$. Replacing y by $k^n x$, and dividing both sides of (14.11) by $k^{3(n+1)}$, we have

$$\left\|\frac{1}{k^{3(n+1)}}g\left(k^{n+1}x\right) - \frac{1}{k^{3n}}g\left(k^n x\right)\right\|_\alpha^+ \leq \frac{1}{2k^{3(n+1)}}\widetilde{\varphi}_\alpha\left(k^n x\right) \tag{14.12}$$

for all $x \in X$, $\alpha \in (0, 1]$, and all non-negative integers n. By Lemma 14.4, and inequality (14.12), we conclude that for all $\alpha \in (0, 1]$ there exists $\beta \in (0, \alpha]$ such that

$$\left\|\frac{1}{k^{3(n+1)}}g\left(k^{n+1}x\right) - \frac{1}{k^{3m}}g\left(k^m x\right)\right\|_\alpha^+ \leq \sum_{i=m}^{n}\left\|\frac{1}{k^{3(i+1)}}g\left(k^{i+1}x\right) - \frac{1}{k^{3i}}g\left(k^i x\right)\right\|_\beta^+$$

$$\leq \frac{1}{2k^3}\sum_{i=m}^{n}\frac{1}{k^{3i}}\widetilde{\varphi}_\beta\left(k^i x\right) \tag{14.13}$$

for all $x \in X$, and all non-negative integers m and n with $n \geq m$. Passing the limit $n, m \to \infty$ in (14.13), we obtain

$$\lim_{n,m\to\infty}\left\|\frac{1}{k^{3n}}g\left(k^n x\right) - \frac{1}{k^{3m}}g\left(k^m x\right)\right\|_\alpha^+ = 0.$$

Therefore, the sequence $\left\{\frac{1}{k^{3n}}g\left(k^n x\right)\right\}$ is a Cauchy sequence in Y for all $x \in X$. Since Y is complete, the sequence $\left\{\frac{1}{k^{3n}}g\left(k^n x\right)\right\}$ converges for all $x \in X$. So one can define the mapping $C_k : X \to Y$ by

$$C_k(x) = \lim_{n\to\infty}\frac{1}{k^{3n}}g\left(k^n x\right) \tag{14.14}$$

for all $x \in X$. Letting $m = 0$, and passing the limit $n \to \infty$ in (14.13), by continuity of $\|\cdot\|^+$, we obtain (14.6).

Now, we show that C_k is cubic. It follows from (14.5) and (14.14) that

$$\|D_k C_k(x, y)\|_\alpha^+ = \lim_{n\to\infty}\left\|\frac{1}{k^{3n}}D_k g\left(k^n x, k^n y\right)\right\|_\alpha^+$$

$$= \lim_{n\to\infty}\frac{1}{k^{3n}}\|D_k g\left(k^n x, k^n y\right)\|_\alpha^+$$

$$\leq \lim_{n\to\infty}\frac{1}{k^{3n}}[\varphi\left(k^n x, k^n y\right)_\alpha^+ + \varphi\left(-k^n x, -k^n y\right)_\alpha^+] = 0 \tag{14.15}$$

for all $x, y \in X$. Therefore, we get that the mapping $C_k : X \to Y$ is cubic.

To prove the uniqueness of C_k, let $C' : X \to Y$ be another cubic mapping which satisfies inequality (14.6), we have

$$\|C_k(x) - C'(x)\|_\alpha^+ = \lim_{n \to \infty} \frac{1}{k^{3n}} \|g(k^n x) - C'(k^n x)\|_\alpha^+$$

$$\leq \lim_{n \to \infty} \frac{1}{2k^3} [\widetilde{\varphi_\beta}(k^n x) + \widetilde{\varphi_\beta}(-k^n x)] = 0. \qquad (14.16)$$

This completes the proof. $\qquad\qquad\qquad\qquad\qquad\qquad\qquad\qquad\qquad\qquad\qquad\square$

Corollary 14.1 ([72]). *Let μ be non-negative fuzzy real number, and p, q be non-negative real numbers such that $p, q < 3$. Let X be a linear space and $(Y, \|\cdot\|, L, R)$ be a fuzzy Banach space satisfying $(R - 1)$. Suppose that $f : X \to$ be a mapping satisfies the inequality*

$$\|D_k f(x, y)\| \preceq \mu \otimes (\|x\|_X^p \oplus \|y\|_X^q)$$

for all $x, y \in X$. Then there exists a unique cubic mapping $C_k : X \to Y$ such that

$$\forall \alpha \in (0, 1], \exists \beta \in (0, \alpha]; \|f(x) - f(-x) - C_k(x)\|_\alpha^+ \leq \frac{\mu_\beta^+}{k^3 - k^q}(\|x\|_\beta^+)^q$$

for all $x \in X$.

Proof. The proof follows from Theorem 14.3 by taking

$$\varphi(x, y) = \mu \otimes (\|x\|_X^p \oplus \|y\|_X^q).$$

$$\qquad\qquad\qquad\qquad\qquad\qquad\qquad\qquad\qquad\qquad\qquad\qquad\qquad\qquad\qquad\square$

Theorem 14.4 ([72]). *Let X be a linear space, and $(Y, \|\cdot\|, L, R)$ be a fuzzy Banach space satisfying $(R - 1)$. Let $f : X \to Y$ be a mapping for which there exists a function $\varphi : X \times X \to F^*(\mathbb{R})$ such that*

$$\lim_{n \to \infty} k^{3n} \varphi \left(\frac{x}{k^{n+1}}, \frac{y}{k^{n+1}} \right)_\alpha^+, \qquad (14.17)$$

and

$$\widetilde{\varphi_\alpha}(x) = \sum_{i=0}^{\infty} k^{3i} \varphi \left(0, \frac{x}{k^{i+1}} \right)_\alpha^+, \qquad (14.18)$$

$$\|D_k f(x, y)\| \preceq \varphi(x, y) \qquad (14.19)$$

for $x, y \in X$ and $\alpha \in (0, 1]$. Then there exists a unique cubic mapping $C_k : X \to Y$ such that

$$\forall \alpha \in (0, 1], \exists \beta \in (0, \alpha]; \|f(x) - f(-x) - C_k(x)\|_\alpha^+ \leq \frac{1}{2} [\widetilde{\varphi_\beta}(x) + \widetilde{\varphi_\beta}(-x)] \quad (14.20)$$

for $x \in X$.

Proof. Similar to the proof of Theorem 14.3, we have

$$\left\| g(ky) - k^3 g(y) \right\|_\alpha^+ \leq \frac{1}{2} \widetilde{\varphi_\alpha}(y) \tag{14.21}$$

for all $y \in X$ and $\alpha \in (0,1]$, where $g(y) = f(y) - f(-y)$ and $\widetilde{\varphi_\alpha}(y) = \varphi(0,y)_\alpha^+ + \varphi(0,-y)_\alpha^+$. Replacing y by $\frac{x}{k^{n+1}}$, and then multiplying both sides of (14.21) by k^{3n}, we obtain

$$\left\| k^{3(n+1)} g\left(\frac{x}{k^{n+1}}\right) - k^{3n} g\left(\frac{x}{k^n}\right) \right\|_\alpha^+ \leq \frac{k^{3n}}{2} \widetilde{\varphi_\alpha}\left(\frac{x}{k^{n+1}}\right) \tag{14.22}$$

for all $x \in X$, $\alpha \in (0,1]$ and all non-negative integers n. By Lemma 14.4, and inequality (14.22), we conclude that for all $\alpha \in (0,1]$, there exists $\beta \in (0,\alpha]$ such that

$$\left\| k^{3(n+1)} g\left(\frac{x}{k^{n+1}}\right) - k^{3m} g\left(\frac{x}{k^m}\right) \right\|_\alpha^+ \leq \sum_{i=m}^{n} \left\| k^{3(i+1)} g\left(\frac{x}{k^{i+1}}\right) - k^{3i} g\left(\frac{x}{k^i}\right) \right\|_\beta^+$$

$$\leq \frac{1}{2} \sum_{i=m}^{n} k^{3i} \widetilde{\varphi_\beta}\left(\frac{x}{k^{i+1}}\right) \tag{14.23}$$

for all $x \in X$, and all non-negative integers m and n with $n \geq m$. Passing the limit $n, m \to \infty$ in (14.23), we obtain

$$\lim_{n,m\to\infty} \left\| k^{3(n+1)} g\left(\frac{x}{k^{n+1}}\right) - k^{3m} g\left(\frac{x}{k^m}\right) \right\|_\alpha^+ = 0.$$

Therefore, the sequence $\left\{ k^{3n} g\left(\frac{x}{k^n}\right) \right\}$ is a Cauchy sequence in Y for all $x \in X$. Since Y is complete, the sequence $\left\{ k^{3n} g\left(\frac{x}{k^n}\right) \right\}$ converges for all $x \in X$. So one can define the mapping $C_k : X \to Y$ by

$$C_k(x) = \lim_{n\to\infty} k^{3n} g\left(\frac{x}{k^n}\right) \tag{14.24}$$

for all $x \in X$. Letting $m = 0$, and passing the limit $n \to \infty$ in (14.23) by continuity of $\| \cdot \|^+$, we obtain (14.20). The rest of the proof is similar to that of Theorem 14.3. The proof is now complete. $\qquad\square$

Corollary 14.2 ([72]). *Let μ be non-negative fuzzy real number, and p, q be non-negative real numbers such that $p, q > 3$. Let X be a linear space and $(Y, \| \cdot \|, L, R)$ be a fuzzy Banach space satisfying $(R-1)$. Suppose that $f : X \to Y$ be a mapping satisfies the inequality*

$$\| D_k f(x,y) \| \preceq \mu \otimes \left(\|x\|_X^p \oplus \|y\|_X^q \right)$$

for all $x, y \in X$. Then there exists a unique cubic mapping $C_k : X \to Y$ such that

$$\forall \alpha \in (0,1], \exists \beta \in (0,\alpha]; \|f(x) - f(-x) - C_k(x)\|_\alpha^+ \leq \frac{\mu_\beta^+}{k^q - k^3} (\|x\|_\beta^+)^q$$

for all $x \in X$.

Proof. The proof follows from Theorem 14.4 by taking

$$\varphi(x, y) = \mu \otimes \left(\|x\|_X^p \oplus \|y\|_X^q \right).$$ □

14.4. Stability of Cauchy functional equation in Felbin's type fuzzy normed linear spaces using fixed point method

In this section, the generalized Hyers–Ulam stability of Cauchy functional equation in Felbin's type fuzzy normed linear spaces is obtained using fixed point method.

Theorem 14.5 ([315]). *Let X be a linear space, and $(Y, \|\cdot\|)$ be a fuzzy Banach space satisfying $(R-1)$. Let $f : X \to Y$ be a mapping for which there exists a function $\varphi : X \times X \to F^*(\mathbb{R})$ such that*

$$\lim_{i \to \infty} \frac{1}{2^i} \varphi \left(2^i x, 2^i y \right)_\alpha^+ = 0 \tag{14.25}$$

$$\|f(x+y) - f(x) - f(y)\| \preceq \varphi(x, y) \tag{14.26}$$

for all $x, y \in X$, and all $\alpha \in (0, 1]$. If there exists an $L < 1$ such that $\varphi(x, x) \preceq 2L \odot \varphi \left(\frac{x}{2}, \frac{x}{2} \right)$ for all $x \in X$, then there exists a unique additive mapping $T : X \to Y$ such that

$$\|f(x) - T(x)\|_\alpha^+ \leq \frac{1}{2 - 2L} \varphi(x, x)_\alpha^+$$

for all $x \in X$ and all $\alpha \in (0, 1]$.

Proof. Consider the set

$$K = \{g : X \to Y\},$$

and introduce the generalized metric on K:

$$d(g, h) = \inf\{C \in \mathbb{R}_+ : \|g(x) - h(x)\|_\alpha^+ \leq C\varphi(x, x)_\alpha^+, \forall x \in X, \forall \alpha \in (0, 1]\}.$$

It is easy to show that (K, d) is complete. Now we consider the linear mapping $J : K \to K$ such that

$$Jg(x) = \frac{1}{2} g(2x)$$

for all $x \in X$. For any $g, h \in K$, we have

$$d(g, h) < C \Rightarrow \|g(x) - h(x)\|_\alpha^+ \leq C\varphi(x, x)_\alpha^+, \quad \forall x \in X, \forall \alpha \in (0, 1]$$

$$\Rightarrow \left\| \frac{1}{2} g(2x) - \frac{1}{2} h(2x) \right\|_\alpha^+ \leq \frac{1}{2} C\varphi(2x, 2y)_\alpha^+ \leq LC\varphi(x, x)_\alpha^+$$

$$\Rightarrow d(Jg, Jh) \leq LC.$$

Therefore, we see that

$$d(Jg, Jh) \leq Ld(g, h), \quad \forall g, h \in K.$$

This means J is a strictly contractive self-mapping of K, with the Lipschitz constant L. Letting $y = x$ in (14.26), we get

$$\|f(2x) - 2f(x)\| \preceq \varphi(x, x) \tag{14.27}$$

for all $x \in X$. So

$$\left\| f(x) - \frac{1}{2}f(2x) \right\| \preceq \odot \varphi(x, x).$$

Therefore, we can conclude that

$$\left\| f(x) - \frac{1}{2}f(2x) \right\|_\alpha^+ \leq \frac{1}{2}\varphi(x, x)_\alpha^+ \tag{14.28}$$

for all $x \in X$ and all $\alpha \in (0, 1]$. Hence $d(f, Jf) \leq \frac{1}{2}$.

By fundamental theorem of fixed point theory, there exists a mapping $T : X \to Y$ such that

(1) T is a fixed point of J, i.e.,

$$T(2x) = 2T(x) \tag{14.29}$$

for all $x \in X$. The mapping T is a unique fixed point of J in the set

$$\Omega = \{g \in X : d(f, g) < \infty\}.$$

This implies that T is a unique mapping such that there exists $C \in (0, \infty)$ satisfying

$$\|T(x) - f(x)\|_\alpha^+ \leq C\varphi(x, x)_\alpha^+$$

for all $x \in X$.

(2) $d(J^n f, T) \to 0$ as $n \to \infty$. This implies the equality

$$\lim_{n \to \infty} \frac{f(2^n x)}{2^n} = T(x) \tag{14.30}$$

for all $x \in X$.

(3) $d(f, T) \leq \frac{1}{1-L}d(f, Jf)$, which implies the inequality

$$d(f, T) \leq \frac{1}{2 - 2L}.$$

It follows from (14.25) and (14.30), and continuity of $\|\cdot\|_\alpha^+$ that

$$\|T(x + y) - T(x) - T(y)\|_\alpha^+ = \lim_{n \to \infty} \frac{1}{2^n} \|f(2^n(x + y)) - f(2^n x) - f(2^n x)\|_\alpha^+$$

$$\leq \lim_{n \to \infty} \frac{1}{2^n} \varphi(2^n x, 2^n y)_\alpha^+ = 0$$

for all $x, y \in X$. Therefore, the mapping T is additive. This completes the proof. $\square$

Corollary 14.3 ([315]). *Let μ be a non-negative fuzzy real number, and let X be a linear space, and $(Y, \|\cdot\|, L, R)$ be a fuzzy Banach space satisfying $(R-1)$. Suppose that the mapping $f : X \to Y$ satisfies the inequality*

$$\|f(x+y) - f(x) - f(y)\| \preceq \mu,$$

for all $x, y \in X$. Then there exists a unique additive mapping $T : X \to Y$ such that

$$\|f(x) - T(x)\| \preceq \mu$$

for all $x \in X$.

Proof. Let $\varphi(x, y) = \mu$ for all $x, y \in X$ and use Theorem 14.5. Then take $L = \frac{1}{2}$ to get the desired result. □

Corollary 14.4 ([315]). *Let μ be a non-negative fuzzy real number, and let p, q be non-negative real numbers such that $0 < p, q < 1$. Let X be a fuzzy normed linear space and $(Y, \|\cdot\|, L, R)$ be a fuzzy Banach space satisfying $(R-1)$. Suppose that a mapping $f : X \to Y$ satisfies the inequality*

$$\|f(x+y) - f(x) - f(y)\|_Y \preceq \mu \otimes (\|x\|_X^p \oplus \|y\|_X^q) \tag{14.31}$$

for all $x, y \in X$. Then there exists a unique additive mapping $T : X \to Y$ such that

$$\|f(x) - T(x)\|_\alpha^+ \leq \frac{1}{2 - 2^{\max(p,q)}} \mu_\alpha^+ (\|x\|_X^p \oplus \|x\|_X^q)_\alpha^+$$

for all $x \in X$, and all $\alpha \in (0, 1]$.

Proof. The proof follows from Theorem 14.5 by taking $\varphi(x, y) = \mu \otimes (\|x\|_X^p \oplus \|y\|_X^q)$, for all $x, y \in X$. We can choose $L = 2^{\max(p,q)-1}$ to get the desired result. □

Corollary 14.5 ([315]). *Let μ be a non-negative fuzzy real number, and let p, q be non-negative real numbers such that $\lambda = p + q \in (0, 1)$. Let X be a fuzzy normed linear space, and $(Y, \|\cdot\|, L, R)$ be a fuzzy Banach space satisfying $(R-1)$. Suppose that a mapping $f : X \to Y$ satisfies the inequality*

$$\|f(x+y) - f(x) - f(y)\|_Y \preceq \mu \otimes \|x\|_X^p \otimes \|y\|_X^q \tag{14.32}$$

for all $x, y \in X$. Then there exists a unique additive mapping $T : X \to Y$ such that

$$\|f(x) - T(x)\|_\alpha^+ \leq \frac{1}{2 - 2^{p+q}} (\mu \otimes \|x\|_X^p \otimes \|x\|_X^q)_\alpha^+$$

for all $x \in X$, and all $\alpha \in (0, 1]$.

Proof. The proof follows from Theorem 14.5 by taking $\varphi(x, y) = \mu \otimes \|x\|_X^p \otimes \|y\|_X^q$, for all $x, y \in X$. We can choose $L = 2^{p+q-1}$ to get the desired result. □

Theorem 14.6 ([315]). *Let X be a linear space, and $(Y, \| \cdot \|, L, R)$ be a fuzzy Banach space satisfying $(R-1)$. Let $f : X \to Y$ be a mapping for which there exists a function $\varphi : X \times X \to F^*(\mathbb{R})$ such that*

$$\lim_{i \to \infty} 2^i \varphi \left(\frac{x}{2^i}, \frac{y}{2^i} \right) = 0 \tag{14.33}$$

$$\| f(x+y) - f(x) - f(y) \| \preceq \varphi(x,y) \tag{14.34}$$

for all $x, y \in X$, and all $\alpha \in (0,1]$. If there exists an $L < 1$ such that $\varphi(x,x) \preceq \frac{L}{2} \odot \varphi(2x, 2x)$ for all $x \in X$, then there exists a unique additive mapping $T : X \to Y$ such that

$$\| f(x) - T(X) \|_\alpha^+ \leq \frac{L}{2 - 2L} \varphi(x,x)_\alpha^+$$

for all $x \in X$, and all $\alpha \in (0,1]$.

Proof. Similar to proof of Theorem 14.5, we consider the linear mapping $J : K \to K$ such that

$$Jg(x) = 2g \left(\frac{x}{2} \right)$$

for all $x \in X$. It is easy to see that, J is a strictly contractive self-mapping of K, with the Lipschitz constant L. Letting $y = x = \frac{x}{2}$ in (14.34), we obtain

$$\left\| f(x) - 2f \left(\frac{x}{2} \right) \right\| \preceq \varphi \left(\frac{x}{2}, \frac{x}{2} \right) \preceq \frac{L}{2} \odot \varphi(x,x) \tag{14.35}$$

for all $x \in X$. Then we have

$$\left\| f(x) - 2f \left(\frac{x}{2} \right) \right\|_\alpha^+ \leq \varphi \left(\frac{x}{2}, \frac{x}{2} \right)_\alpha^+ \leq \frac{L}{2} \varphi(x,x)_\alpha^+ \tag{14.36}$$

for all $x \in X$, and all $\alpha \in (0,1]$. Hence $d(f, Jf) \leq \frac{L}{2}$.

By fundamental theorem of fixed point theory, there exists a mapping $T : X \to$ such that

(1) T is a fixed point of J, i.e.,

$$T \left(\frac{x}{2} \right) = \frac{1}{2} T(x) \tag{14.37}$$

for all $x \in X$. The mapping T is a unique fixed point of J in the set

$$\Omega = \{ g \in X : d(f, g) < \infty \}.$$

This implies that T is a unique mapping such that there exists $C \in (0, \infty)$ satisfying

$$\| T(x) - f(x) \|_\alpha^+ \leq C \varphi(x,x)_\alpha^+$$

for all $x \in X$.

(2) $d(J^n f, T) \to 0$ as $n \to \infty$. This implies the equality

$$\lim_{n \to \infty} 2^n f\left(\frac{x}{2^n}\right) = T(x) \qquad (14.38)$$

for all $x \in X$.

(3) $d(f, T) \leq \frac{1}{1-L} d(f, Jf)$, which implies the inequality

$$d(f, T) \leq \frac{L}{2 - 2L}.$$

The rest of the proof is similar to the proof of Theorem 14.5. This completes the proof. $\qquad \square$

Corollary 14.6 ([315]). *Let μ be a non-negative fuzzy real number, and let p, q be non-negative real numbers such that $p, q > 1$. Let X be a fuzzy normed linear space, and $(Y, \| \cdot \|, L, R)$ be a fuzzy Banach space satisfying $(R-1)$. Suppose that a mapping $f : X \to Y$ satisfies the inequality*

$$\|f(x + y) - f(x) - f(y)\| \preceq \mu \otimes (\|x\|_X^p \oplus \|y\|_X^q) \qquad (14.39)$$

for all $x, y \in X$. Then there exists a unique additive mapping $T : X \to Y$ such that

$$\|f(x) - T(x)\|_\alpha^+ \leq \frac{1}{2^{\min(p,q)} - 2} \mu_\alpha^+ (\|x\|_X^p \oplus \|y\|_X^q)_\alpha^+$$

for all $x \in X$, and all $\alpha \in (0, 1]$.

Proof. The proof follows from Theorem 14.6 by taking

$$\varphi(x, y) = \mu \otimes (\|x\|_X^p \oplus \|y\|_X^q)$$

for all $x, y \in X$. We can choose $L = 2^{1-\min(p,q)}$ to get the desired result. $\qquad \square$

Corollary 14.7 ([315]). *Let μ be a non-negative fuzzy real number, and let p, q be non-negative real numbers such that $\lambda = p + q \in (1, \infty)$. Let X be a fuzzy normed linear space, and $(Y, \| \cdot \|, L, R)$ be a fuzzy Banach space satisfying $(R-1)$. Suppose that a mapping $f : X \to Y$ satisfies the inequality*

$$\|f(x + y) - f(x) - f(y)\|_y \preceq \mu \otimes \|x\|_X^p \otimes \|y\|_X^q \qquad (14.40)$$

for all $x, y \in X$. Then there exists a unique additive mapping $T : X \to Y$ such that

$$\|f(x) - T(x)\|_\alpha^+ \leq \frac{1}{2^{p+q} - 2} (\mu \otimes \|x\|_X^p \otimes \|y\|_X^q)$$

for all $x \in X$, and all $\alpha \in (0, 1]$.

Proof. The proof follows from Theorem 14.6 by taking $\varphi(x, y) = \mu \otimes \|x\|_X^p \otimes \|y\|_X^q$ for all $x, y \in X$. We can choose $L = 2^{1-(p+q)}$ to get the desired result. $\qquad \square$

Chapter 15

Stabilities of Functional Equations on C^*-algebras and Lie C^*-algebras

15.1. Introduction

In the previous chapter, we have discussed about the generalized Hyers–Ulam stability of a general cubic functional equation, and Cauchy functional equation in Felbin's type fuzzy normed linear spaces using direct method and fixed point method, respectively. This chapter is assigned for the investigation of Cauchy–Ulam stability of Jensen equation in C^*-algebras, and Hyers–Ulam–Rassias stability of isomorphisms in C^*-algebras using direct method. Using fixed point approach, the generalized Hyers–Ulam stability of generalized additive functional equation in C^*-algebras, and Lie C^*-algebras are also dealt in this chapter.

15.2. Stability of the Jensen equation in C^*-algebras

First we introduce some basic terminology from *functional analysis*. An *algebra* $\mathbb{R}$ is a linear space over $\mathbb{C}$ together with a multiplication such that for $x, y, z \in \mathbb{R}$, and $\lambda \in \mathbb{C}$,

(1) $x(yz) = (xy)z$,
(2) $x(y + z) = xy + xz$,
(3) $(x + y)z = xz + yz$,
(4) $\lambda(xy) = (\lambda x)y = x(\lambda y)$.

A Banach space $\mathbb{R}$ is called a *Banach algebra* (or *normed ring*) if $\|xy\| \leq \|x\| \|y\|$ satisfies for all $x, y \in \mathbb{R}$.

When a Banach algebra contains a unity element e with respect to the multiplication, we call it *unital* and we can suppose $\|e\| = 1$.

An *involution* in a Banach algebra $\mathbb{R}$ is an operation $x \to x^*$ from $\mathbb{R}$ into itself that satisfies the following properties, for $x, y \in \mathbb{R}$, and $\lambda \in \mathbb{C}$,

(1) $(x + y)^* = x^* + y^*$;
(2) $(\lambda x)^* = \bar{\lambda} x^*$;

(3) $(xy)^* = y^*x^*$;

(4) $(x^*)^* = x$.

A Banach algebra with an involution such that $\|x^*\| = \|x\|$ is called a *Banach ∗-algebra*.

A mapping $h : A \to B$ between two Banach ∗-algebras A and B is a *homomorphism* if $h(xy) = h(x)h(y)$ for all $x, y \in A$.

A ∗-homomorphism $h : A \to B$ between two Banach ∗-algebras A and B is a homomorphism which preservers involutions, i.e., $h(x^*) = h(x)^*$.

A Banach ∗-algebra $\mathbb{R}$ satisfying $\|x^*x\| = \|x\|^2$ for all $x \in \mathbb{R}$ is called a C^*-*algebra*.

A $\mathbb{C}$-linear mapping $\delta : A \to A$ in a C^*-algebra A is called a *derivation* in A if its domain $D(\delta)$ is a dense subalgebra of A, and $\delta(xy) = x\delta(y) + \delta(x)y$ for all $x, y \in D(\delta)$. If, moreover, $x \in D(\delta)$ implies $x^* \in D(\delta)$ and $\delta(x^*) = \delta(x)^*$, then δ is a ∗-*derivation*.

Finally, a mapping $h : A \to B$ is *an almost unital mapping* if an element $\lim_{n \to \infty} 2^n h\left(2^{-n}e\right)$ in B is invertible.

Throughout this section, we assume that A and B are unital C^*-algebras with unit e. Besides we denote with $U(A)$ the set of all unitary elements. We note that $U(A) = \{u \in A | u^*u = uu^* = 1\}$ is the unitary group in A.

Theorem 15.1 ([274]). *Let X be a real normed linear space, and let Y be a real complete normed linear space. Assume in addition that $f : X \to Y$ is a mapping for which there exist constants $\theta \geq 0$, and $p \in \mathbb{R} - \{1\}$ such that*

$$\|f(x + y) - [f(x) + f(y)]\| \leq \theta \|x\|^{\frac{p}{2}} \|y\|^{\frac{p}{2}}$$

for all $x, y \in X$. Then there exists a unique additive mapping $L : X \to Y$ satisfying

$$\|f(x) - L(x)\| \leq \frac{\theta}{|2^p - 2|} \|x\|^p$$

for all $x \in X$. If in addition $f : X \to Y$ is a mapping such that the transformation $t \to f(tx)$ is continuous in $t \in \mathbb{R}$ for each fixed $x \in X$, then L is $\mathbb{R}$-linear mapping.

Let us consider two real linear spaces X and Y, and the Jensen functional equation

$$2f\left(\frac{x + y}{2}\right) = f(x) + f(y) \tag{15.1}$$

for $f : X \to Y$ and all $x, y \in X$, and the additive Cauchy functional equation

$$f(x + y) = f(x) + f(y) \tag{15.2}$$

for all $x, y \in X$.

Theorem 15.2 ([274]). *Let X be a real normed linear space, and let Y be a real complete normed linear space. Assume in addition that $f : X \to Y$ with $f(0) = 0$ is a mapping for which there exists a constant $\theta \geq 0$ such that*

$$\left\| 2f\left(\frac{x+y}{2}\right) - [f(x) + f(y)] \right\| \leq \theta \tag{15.3}$$

for all $x, y \in X$. Then there exists a unique Jensen mapping $J : X \to Y$ satisfying the functional equation (15.1), and the functional inequality

$$\|f(x) - J(x)\| \leq \theta \tag{15.4}$$

for all $x \in X$.

Proof. Substituting $x = 0$ and $y = 2x$ in the above inequality (15.3), and employing condition $f(0) = 0$, we obtain

$$\|2f(x) - [f(0) + f(2x)]\| \leq \theta,$$

or

$$\|f(x) - 2^{-1}f(2x)\| \leq \frac{\theta}{2} = \theta\left(1 - 2^{-1}\right)$$

for all $x \in X$. Thus, in general, we obtain

$$\|f(x) - 2^{-n}f(2^n x)\| \leq \theta\left(1 - 2^{-n}\right) \to \theta \qquad \text{as } n \to \infty$$

for all $x \in X$. Thus taking the limiting form $J(x) = \lim_{n \to \infty} 2^{-n}f(2^n x)$, and employing classical techniques on these concepts, we prove the inequality (15.4) for a unique mapping $J : X \to Y$ of the above limiting form. The rest of the proof is similar to that of theorems proved in previous sections. This completes the proof.
$\square$

Note. If we replace $x = 0$, and $y = x$ in (15.3), then

$$\|f(x) - 2f\left(2^{-1}x\right)\| \leq \theta$$

and, in general,

$$\|f(x) - 2^n f\left(2^{-n}x\right)\| \leq \theta\left(1 + 2 + \cdots + 2^{n-1}\right) = \theta\left(\frac{1 - 2^n}{1 - 2}\right) \to \infty, \text{ as } n \to \infty.$$

Thus the inequality (15.4) does not hold.

Proposition 15.1 ([274]). *A function $f : X \to Y$ between two real linear spaces X and Y satisfies the Jensen equation (15.1) for all $x, y \in X$ if and only if there exists an additive Cauchy mapping $C : X \to Y$ satisfying (15.2), and such that*

$$f(x) = C(x) + f(0) \tag{15.5}$$

for all $x \in X$.

Proof. *Necessary Part.* Let us assume that a mapping $f : X \to Y$ satisfies the Jensen equation (15.1). We consider, $C : X \to Y$ and $g : X \to Y$, two functions given by the formulas

$$C(x) = \frac{1}{2}[f(x) - f(-x)], \quad \text{and} \quad g(x) = \frac{1}{2}[f(x) + f(-x)] - f(0), \quad (15.6)$$

respectively, for all $x \in X$. Therefore

$$f(x) = C(x) + g(x) + f(0) \quad (15.7)$$

for all $x \in X$. Claim that C satisfies the additive Cauchy equation (15.2), and that $g(x) = 0$ in X. In fact,

$$2C\left(\frac{x+y}{2}\right) = C(x) + C(y), \quad (15.8)$$

and

$$2g\left(\frac{x+y}{2}\right) = g(x) + g(y) \quad (15.9)$$

for all $x, y \in X$, because f satisfies (15.1) for all $x, y \in X$, and equations (15.6)–(15.7) hold. By virtue of (15.6), we find the relations

$$C(-x) = -C(x) \quad \text{and} \quad g(-x) = g(x) \quad (15.10)$$

for all $x \in X$. Setting $x = 0$ in the first equation of (15.6), and $y = 0$ in (15.8), we obtain $C(0) = 0$, and

$$2C\left(2^{-1}x\right) = C(x) \quad (15.11)$$

for all $x \in X$. From (15.8) and (15.11), we obtain

$$\begin{aligned}
C(x+y) - C(x) - C(y) &= 2C\left(2^{-1}(x+y)\right) - C(x) - C(y) \\
&= [C(x) + C(y)] - C(x) - C(y) \\
&= 0,
\end{aligned}$$

or

$$C(x+y) = C(x) + C(y) \quad (15.12)$$

for all $x, y \in X$. Therefore, C is an additive Cauchy mapping.

On the other hand, substituting $x = 0$ in the second equation of (15.6), and $y = 0$ in (15.9), we find $g(0) = 0$, and

$$2g\left(2^{-1}x\right) = g(x) \quad (15.13)$$

for all $x \in X$. Taking into account (15.9) and (15.13), we obtain

$$g(x+y) = g(x) + g(y) \quad (15.14)$$

for all $x, y \in X$. Putting $y = -x$ in (15.14) yields $g(0) = g(x) + g(-x)$. From the evenness of g, by the second equation of (15.10), and $g(0) = 0$, we obtain $g(x) = 0$ for all $x \in X$.

Sufficient Part. The converse is trivial, which completes the proof of the Proposition 15.1. □

15.2.1. *The Jensen equation and almost unital mappings*

Let us denote $A_\mu(x, y) = 2h\left(\frac{\mu}{2}(x + y) - \mu[h(x) + h(y)]\right)$ for given mapping $h : A \to B$, any $\mu \in L' = \{\lambda \in \mathbb{C} : |\lambda| = 1\}$, and for all $x, y \in A$.

Theorem 15.3 ([274]). *Let $h : A \to B$ be an almost unital mapping such that $h(0) = 0$, and $h(2^n xu) = h(x)h(2^n u)$ for all $x \in A$, all $u \in U(A)$, and all sufficiently large integers n. If the condition*

$$\|A_\mu h(x, y)\| \leq \theta, \qquad \theta \geq 0, \tag{15.15}$$

holds for all L', and all $x, y \in A$ then h is a homomorphism. If, in addition, the condition

$$\left\|h\left(2^n u^*\right) - h\left(2^n u\right)^*\right\| \leq \theta \tag{15.16}$$

holds for all $u \in U(A)$, and all sufficiently large integers n, then h is a $*$-homomorphism.

Proof. Setting $\mu = 1$ in (15.15), and employing Theorem 15.1 and Proposition 15.1 with $f(0) = 0$, one proves that there exists a unique additive mapping $L : A \to B$ defined by

$$L(x) = \lim_{n \to \infty} 2^{-n} h\left(2^n x\right)$$

for all $x \in X$. Placing $y = x$ in (15.15), we find the inequality

$$\|A_\mu(x, x)\| = 2\|h(\mu x) - \mu h(x)\| \leq \theta$$

for all $\mu \in L'$, and all $x \in A$. Substituting x in this inequality with $2^n x$, we obtain

$$2^{-n}\left\|h\left(\mu 2^n x\right) - \mu h\left(2^n x\right)\right\| \leq 2^{-(n+1)}\theta,$$

or

$$\lim_{n \to \infty} 2^{-n}\left\|h\left(\mu 2^n x\right) - \mu h\left(2^n x\right)\right\| = 0,$$

or

$$\lim_{n \to \infty} h\left(\mu 2^n x\right) = \mu \lim_{n \to \infty} 2^{-n} h\left(2^n x\right)$$

for all $\mu \in L'$, and all $x \in A$.

Therefore $L(\mu x) = \lim_{n \to \infty} 2^{-n} h\left(2^n \mu x\right) = \mu \lim_{n \to \infty} 2^{-n} h\left(2^n x\right) = \mu L(x)$, for all $\mu \in L'$, and all $x \in A$. But it is well known that if an additive mapping $L : A \to B$

satisfies $L(\mu x) = \mu L(x)$ for all $\mu \in L'$, then L is a $\mathbb{C}$-linear mapping. Thus L is $\mathbb{C}$-linear. We now claim that L is a homomorphism. In fact, from the hypothesis that $h(2^n xu) = h(x)h(2^n u)$, we obtain

$$L(xu) = \lim_{n\to\infty} 2^{-n}h(2^n xu) = h(x) \lim_{n\to\infty} 2^{-n}h(2^n u) = h(x)L(u).$$

Thus from $\mathbb{C}$-linearity of L, we find $L(xu) = 2^{-n}L(2^n xu) = 2^{-n}h(2^n u)L(u)$. Therefore

$$\lim_{n\to\infty} L(xu) = \left[\lim_{n\to\infty} 2^{-n}h(2^n x)\right] L(u),$$

or

$$L(xu) = L(x)L(u)$$

for all $x \in A$ and all $u \in U(A)$.

But any element in a C^*-algebra is a finite linear combination of unitary elements in A, and so any $y \in A$ is of the form $y = \sum_{j=1}^{m} \eta_j(xv_j) = L(x)L(\sum_{j=1}^{m} \eta_j v_j) = L(x)L(y)$ for all $x, y \in A$, yielding that L is a homomorphism. Besides $h(x)L(e) = L(xe) = L(x)L(e)$ for all $x \in A$, because e is a unitary element of A.

Therefore the identity

$$L(x) = h(x) \tag{15.17}$$

holds for all $x \in A$, because $L(e) = \lim_{n\to\infty} 2^{-n}h(2^n e)$ is invertible as h is almost unital. Hence h is a homomorphism. From (15.16), we find

$$\lim_{n\to\infty} 2^{-n} \left\| h(2^n u^*) - h(2^n u)^* \right\| \leq \theta \left(\lim_{n\to\infty} 2^{-n} \right) = 0,$$

or

$$\lim_{n\to\infty} 2^{-n} h(2^n u^*) = \lim_{n\to\infty} 2^{-n} h(2^n u)^*,$$

or

$$L(u^*) = L(u)^*$$

for all $u \in U(A)$. But, in addition, the $\mathbb{C}$-linear map L is a homomorphism. We now claim that L is a $*$-homomorphism. In fact, any $x \in A$ is of the form $x = \sum_{i=1}^{k} \xi_i u_i$ for $\xi_i \in \mathbb{C}$ and $u_i \in U(A)$. Thus

$$L(x^*) = L\left(\left(\sum_{i=1}^{k} \xi_i u_i\right)^*\right) = \sum_{i=1}^{k} \bar{\xi}_i L(u_i)^*$$

$$= \left(\sum_{i=1}^{k} \bar{\xi}_i L(u_i)\right)^* = L(x)^* \tag{15.18}$$

for all $x \in A$, $\xi_i \in \mathbb{C}$ and $u_i \in U(A)$, yielding L is a $*$-homomorphism. Therefore from (15.17) and (15.18), we obtain $h(x^*) = h(x)^*$ for all $x \in A$, and thus h is also a $*$-homomorphism, completing the proof of the theorem. $\square$

15.2.2. *Stability and $*$-homomorphisms*

Let us denote

$$H_\mu h(x, y, z, w) = 2h\left(\frac{\mu}{2}(x+y) + \frac{1}{2}zw\right) - \mu[h(x) + h(y)] - h(z)h(w)$$

for any $\mu \in L'$, and for all $x, y, z, w \in A$, and for a given mapping $h : A \to B$ from a unital C^*-algebra A to a unital C^*-algebra B.

Theorem 15.4 ([274]). *Let $h : A \to B$ be a mapping such that $h(0) = 0$. If two conditions*

$$\|H_\mu h(x, y, z, w)\| \leq \theta, \qquad \theta \geq 0, \tag{15.19}$$

and

$$\left\|h\left(2^n u^*\right) - h\left(2^n u\right)^*\right\| \leq \theta \tag{15.20}$$

hold for all $\mu \in L'$, all $u \in U(A)$, and all $x, y, z, w \in A$, all sufficiently large integers n, then there exists a unique $*$-homomorphism $L : A \to B$ satisfying

$$\|h(x) - L(x)\| \leq \theta \tag{15.21}$$

for all $x \in A$.

Proof. Setting $\mu = 1$ and $z = w = 0$ in (15.19), and employing the ideas from the proof of Theorems 15.1–15.3, we get that there exists a unique $\mathbb{C}$-linear mapping $L : A \to B$ satisfying (15.21), and is given by $L(x) = \lim_{n \to \infty} 2^{-n} h(2^n x)$ for all $x \in A$. Therefore, from (15.20), we find $L(x^*) = L(x)^*$ for all $x \in A$. We claim that $L : A \to B$ is a homomorphism. In fact, putting $x = y = 0$, and $h(0) = 0$ in (15.19), we obtain

$$\|H_\mu h(0, 0, z, w)\| \leq \theta,$$

or

$$\left\|2h\left(\frac{zw}{2}\right) - h(z)h(w)\right\| \leq \theta$$

for all $z, w \in A$. Now, replacing z and w by $2^n z$ and $2^n w$, respectively, and multiplying by 2^{-2n}, we obtain

$$\left\|2^{-2n} \cdot 2h\left(2^{2n}\frac{zw}{2}\right) - 2^{-n}h\left(2^n z\right) 2^{-n}h\left(2^n w\right)\right\| \leq 2^{-2n}\theta, \qquad \text{for all } z, w \in A.$$

Therefore

$$\lim_{n\to\infty} 2^{-2n} \cdot 2h\left(2^{2n}\frac{zw}{2}\right) = \lim_{n\to\infty} 2^{-n}h\left(2^n z\right) \lim_{n\to\infty} 2^{-n}h\left(2^n w\right), \qquad \text{for all } z, w \in A.$$

But

$$L(x) = \lim_{n\to\infty} 2^{-n}h\left(2^n x\right) = \lim_{n\to\infty} 2^{-2n}h\left(2^{2n}x\right), \qquad \text{for all } x \in A.$$

Thus

$$L(zw) = 2L\left(\frac{zw}{2}\right) = 2\lim_{n\to\infty} 2^{-2n}h\left(2^{2n}\frac{zw}{2}\right) = \lim_{n\to\infty} 2^{-2n} \cdot 2h\left(2^{2n}\frac{zw}{2}\right)$$
$$= \left[\lim_{n\to\infty} 2^{-n}h\left(2^n z\right)\right]\left[\lim_{n\to\infty} 2^{-n}h\left(2^n w\right)\right]$$
$$= L(z)L(w), \qquad \text{for all } z, w, \in A.$$

Hence, L is a unique $*$-homomorphism satisfying the identity (15.17), completing the proof of the theorem. $\qquad\square$

15.2.3. *Stability and $*$-derivations*

Let us denote

$$D_\mu h(x, y, z, w) = 2h\left(\frac{\mu}{2}(x + y) + \frac{zw}{2}\right) - \mu[h(x) + h(y)] - zh(w) - h(z)w$$

for any $\mu \in L'$, and for all $x, y, z, w \in A$, and for a given mapping $h : A \to A$ from a unitial C^*-algebra A to itself.

Theorem 15.5 ([274]). *Let $h : A \to A$ be a mapping such that $h(0) = 0$. If conditions*

$$\|D_\mu h(x, y, z, w)\| \le \theta, \qquad \theta \ge 0, \tag{15.22}$$

and (15.20) hold for all $\mu \in L'$, and all $x, y, z, w \in A$, then there exists a unique $*$-derivation $\delta : A \to A$ satisfying

$$\|h(x) - \delta(x)\| \le \theta \tag{15.23}$$

for all $x \in A$.

Proof. Setting $\mu = 1$ and $z = w = 0$ in (15.23), and employing ideas from the proof of Theorems 15.1–15.3, we get that there exists a unique $\mathbb{C}$-linear mapping $\delta : A \to A$ satisfying (15.23), and is given by $\delta(x) = \lim_{n\to\infty} 2^{-n}h\left(2^n x\right)$, for all $x \in A$. Therefore $\delta(x^*) = \delta(x)^*$ for all $x \in A$. We claim that $\delta : A \to A$ is a derivation. In fact, putting $x = y = 0$, and $h(0) = 0$ in (15.22), we obtain

$$\|D_\mu h(0, 0, z, w)\| \leq \theta,$$

or

$$\|h(zw) - [zh(w) + h(z)w]\| \leq \theta$$

for all $z, w \in A$. Thus placing $2^n z$ and $2^n w$ on z and w, respectively, and multiplying by 2^{-2n}, we obtain

$$\left\| 2^{-2n} h\left(2^{2n} zw\right) - \left[2^{-n} zh\left(2^n w\right) + h\left(2^n z\right) 2^{-n} w\right] \right\|$$

$$= \left\| 2^{-2n} \cdot 2h\left(2^{2n} \frac{zw}{2}\right) - \left[z\left(2^{-n} h\left(2^n w\right)\right) + \left(2^{-n} h\left(2^n z\right)\right) w\right] \right\|$$

$$\leq 2^{-2n}\theta$$

for all $z, w \in A$. Therefore

$$\lim_{n\to\infty} 2^{-2n} \cdot 2h\left(2^{2n} \frac{zw}{2}\right) = \lim_{n\to\infty} \left[z\left(2^{-n} h\left(2^n w\right)\right) + \left(2^{-n} h\left(2^n z\right) w\right)\right]$$

$$= z\left[\lim_{n\to\infty} 2^{-n} h\left(2^n w\right)\right] + \left[\lim_{n\to\infty} 2^{-n} h\left(2^n z\right)\right] w$$

for all $z, w \in A$. But

$$\delta(x) = \lim_{n\to\infty} 2^{-n} h\left(2^n x\right) = \lim_{n\to\infty} 2^{-2n} h\left(2^{2n} x\right), \qquad \text{for all } x \in A.$$

Thus

$$\delta(zw) = 2\delta\left(\frac{zw}{2}\right) = 2 \lim_{n\to\infty} 2^{-2n} \cdot 2h\left(2^{2n} \frac{zw}{2}\right) = \lim_{n\to\infty} 2^{-2n} \cdot 2h\left(2^{2n} \frac{zw}{2}\right)$$

$$= z\left[\lim_{n\to\infty} 2^{-n} h\left(2^n w\right)\right] + \left[\lim_{n\to\infty} 2^{-n} h\left(2^n z\right)\right] w$$

$$= z\delta(w) + \delta(z)w$$

for all $z, w \in A$. Hence, δ is a unique $*$-derivation satisfying the identity (15.23) completing the proof of Theorem 15.5. $\qquad\square$

15.3. Hyers–Ulam–Rassias stability of isomorphisms in C^*-algebras

15.3.1. *Approximate isomorphisms in C^*-algebras*

In this section, we investigate the Hyers–Ulam–Rassias stability of the Jensen functional equation in unital C^*-algebras. Throughout this section, let us assume that A is a unital C^*-algebra with unit e and norm $\|\cdot\|_A$, and that B is a unital C^*-algebra with unit e', and norm $\|\cdot\|_B$. Let $U(A)$ be the group of unitary elements in A and $\mathbb{Z}_+$ the set of non-negative integers.

For a given mapping $f : A \to B$, define

$$C_\lambda f(x, y) = 2f\left(\frac{\lambda x + \lambda y}{2}\right) - \lambda f(x) - \lambda f(y)$$

for all $\lambda \in \mathbb{T}' = \{\mu \in \mathbb{C} | \, |\mu| = 1\}$ and all $x, y \in A$.

Theorem 15.6 ([225]). *Let p and θ be positive real numbers with $p < 1$, and let $f : A \to B$ be a multiplicative bijective mapping with $f(0) = 0$ such that*

$$\|C_\lambda f(x, y)\|_B \le \theta \left(\|x\|_A^p + \|y\|_A^p\right), \tag{15.24}$$

$$\left\|f\left(3^n u^*\right) - f\left(3^n u\right)^*\right\|_B \le 2 \cdot 3^{pn}\theta, \tag{15.25}$$

and

$$\lim_{n \to \infty} \frac{1}{3^n} f\left(3^n e\right) = e' \tag{15.26}$$

for all $\lambda \in \mathbb{T}'$, all $x, y \in A$, all $u \in U(A)$, and all $n \in \mathbb{Z}_+$. Then the multiplicative bijective mapping $f : A \to B$ is a C^*-algebra isomorphism.

Proof. Let $\lambda = 1$ in (15.24). Using Theorem 1 of [133], it is easy to show that there is a unique additive mapping $L : A \to B$ such that

$$\|f(x) - L(x)\|_B \le \frac{3 + 3^p}{3 - 3^p} \theta \|x\|_A^p \tag{15.27}$$

for all $x \in A$. The additive mapping $L : A \to B$ is given by

$$L(x) = \lim_{n \to \infty} \frac{1}{3^n} f\left(3^n x\right) \tag{15.28}$$

for all $x \in A$.

By the same method as in the proof of [220, Theorem 2.1], we can show that the mapping $L : A \to B$ is a C^*-algebra homomorphism, and that $L = f$. Thus the multiplicative bijective mapping $f : A \to B$ is a C^*-algebra isomorphism.   □

Theorem 15.7 ([225]). *Let p and θ be positive real numbers with $p > 1$, and let $f : A \to B$ be a multiplicative bijective mapping satisfying $f(0) = 0$ and (15.24) such that*

$$\left\|f\left(\frac{u^*}{3^n}\right) - f\left(\frac{u}{3^n}\right)^*\right\|_B \le \frac{2\theta}{3^{pn}}, \tag{15.29}$$

and

$$\lim_{n \to \infty} 3^n f\left(\frac{e}{3^n}\right) = e' \tag{15.30}$$

for all $u \in U(A)$, and all $n \in \mathbb{Z}_+$. Then the multiplicative bijective mapping $f : A \to B$ is a C^*-algebra isomorphism.

Proof. Let $\lambda = 1$ in (15.24). By Theorem 6 of [133], there is a unique additive mapping $L : A \to B$ such that

$$\|f(x) - L(x)\|_B \leq \frac{3^p + 3}{3^p - 3} \theta \|x\|_A^p \tag{15.31}$$

for all $x \in A$. The additive mapping $L : A \to B$ is given by

$$L(x) = \lim_{n \to \infty} 3^n f\left(\frac{x}{3^n}\right) \tag{15.32}$$

for all $x \in A$.

By the same method as in the proof of [220, Theorem 2.1], we can show that the mapping $L : A \to B$ is a C^*-algebra homomorphism, and that $L = f$. Thus the multiplicative bijective mapping $f : A \to B$ is a C^*-algebra isomorphism. This completes the proof. □

Theorem 15.8 ([225]). *Let p and θ be positive real numbers with $p < \frac{1}{2}$, and let $f : A \to B$ be a multiplicative bijective mapping satisfying $f(0) = 0$, and (15.26) such that*

$$\|C_\lambda f(x,y)\|_B \leq \theta \|x\|_A^p \|y\|_B^p, \tag{15.33}$$

and

$$\|f(3^n u^*) - f(3^n u)^*\|_B \leq 3^{2pn} \theta \tag{15.34}$$

for all $\lambda \in \mathbb{T}'$, all $x, y \in A$, all $u \in U(A)$, and all $n \in \mathbb{Z}_+$. Then the multiplicative bijective mapping $f : A \to B$ is a C^*-algebra isomorphism.

Proof. Let $\lambda = 1$ in (15.33). By Theorem 1 of [133], there is a unique additive mapping $L : A \to B$ such that

$$\|f(x) - L(x)\|_B \leq \frac{1 + 3^p}{3 - 3^{2p}} \theta \|x\|_A^{2p} \tag{15.35}$$

for all $x \in A$. The additive mapping $L : A \to B$ is given by

$$L(x) = \lim_{n \to \infty} \frac{1}{3^n} f(3^n x) \tag{15.36}$$

for all $x \in A$.

By the same method as in the proof of [220, Theorem 2.1], we can show that the mapping $L : A \to B$ is a C^*-algebra homomorphism, and that $L = f$. Thus the multiplicative bijective mapping $f : A \to B$ is a C^*-algebra isomorphism. □

Theorem 15.9 ([225]). *Let p and θ be positive real numbers with $p > \frac{1}{2}$, and let $f : A \to B$ be a multiplicative bijective mapping satisfying $f(0) = 0$, (15.30), and*

(15.33) *such that*

$$\left\| f\left(\frac{1}{3^n}u^*\right) - f\left(\frac{1}{3^n}u\right)^* \right\|_B \le \frac{\theta}{3^{2pn}} \tag{15.37}$$

for all $u \in U(A)$, and all $n \in \mathbb{Z}_+$. Then the bijective mapping $f : A \to B$ is a C^*-algebra isomorphism.

Proof. Let $\lambda = 1$ in (15.33). By Theorem 6 of [133], there is a unique additive mapping $L : A \to B$ such that

$$\|f(x) - L(x)\|_B \le \frac{3^p + 1}{3^{2p} - 3}\theta \|x\|_A^{2p} \tag{15.38}$$

for all $x \in A$. The additive mapping $L : A \to B$ is given by

$$L(x) = \lim_{n \to \infty} 3^n f\left(\frac{x}{3^n}\right) \tag{15.39}$$

for all $x \in A$.

By the same method as in the proof of [220, Theorem 2.1], we can show that the mapping $L : A \to B$ is a C^*-algebra homomorphism, and that $L = f$. Thus the multiplicative bijective mapping $f : A \to B$ is a C^*-algebra isomorphism. $\qquad\square$

15.3.2. *Approximate isomorphisms in Lie C^*-algebras*

In 1932, Jordan observed that $\mathcal{L}(\mathcal{H})$ is a (non-associative) algebra via the *anticommutator product* $x \circ y = (xy + yx)/2$. A commutative algebra X with product $x \circ y$ is called a *Jordan algebra*. A Jordan C^*-subalgebra of a C^*-algebra, endowed with the anticommutator product, is called a *JC^*-algebra*. A C^*-algebra $\mathcal{C}$, endowed with the Lie product $[x, y] = (xy - yx)/2$ on $\mathcal{C}$, is called a *Lie C^*-algebra*.

Throughout this section, assume that A is a Lie C^*-algebra with unit e, and norm $\|\cdot\|_A$, and that B is a Lie C^*-algebra with unit e', and norm $\|\cdot\|_B$.

Definition 15.1 ([225]). A $\mathbb{C}$-linear bijective mapping $L : A \to B$ is called a *Lie C^*-algebra homomorphism* if $L : A \to B$ satisfies

$$L([x, y]) = [L(x), L(y)],$$

and

$$L(x^*) = L(x)^*$$

for all $x, y \in A$.

Theorem 15.10 ([225]). *Let p and θ be positive real numbers with $p < 1$, and let $f : A \to B$ be a bijective mapping satisfying $f(0) = 0$, (15.24), (15.25), (15.26), and*

$f(3^n ey) = f(3^n e)f(y)$ *for all* $y \in A$ *and all* $n \in \mathbb{Z}_+$, *such that*

$$\|f([x, y]) - [f(x), f(y)]\|_B \leq \theta \left(\|x\|_A^p + \|y\|_A^p\right) \tag{15.40}$$

for all $x, y \in A$. *Then the bijective mapping* $f : A \to B$ *is a Lie* C^*-*algebra isomorphism.*

Proof. Let $\lambda = 1$ in (15.24). By Theorem 1 of [133], there is a unique additive mapping $L : A \to B$ satisfying (15.27).

It is easy to show that the mapping $L : A \to B$ is a Lie C^*-algebra homomorphism, and that $L = f$. Thus the bijective mapping $f : A \to B$ is a Lie C^*-algebra isomorphism. $\square$

Theorem 15.11 ([225]). *Let p and θ be positive real numbers with $p > 2$, and let $f : A \to B$ be a bijective mapping satisfying $f(0) = 0$, (15.24), (15.29), (15.30), (15.40), and $f\left(\frac{1}{3^n}ey\right) = f\left(\frac{1}{3^n e}\right)f(y)$ for all $y \in A$ and all $n \in \mathbb{Z}_+$. Then the bijective mapping $f : A \to B$ is a Lie C^*-algebra isomorphism.*

Proof. Let $\lambda = 1$ in (15.24). By Theorem 6 of [133], there is a unique additive mapping $L : A \to B$ satisfying (15.31).

It is easy to show that the mapping $L : A \to B$ is a C^*-algebra homomorphism and that $L = f$. Thus the bijective mapping $f : A \to B$ is a Lie C^*-algebra isomorphism. $\square$

Theorem 15.12 ([225]). *Let p and θ be positive real numbers with $p < \frac{1}{2}$, and let $f : A \to B$ be a bijective mapping satisfying $f(0) = 0$, (15.27), (15.34), (15.35), and $f(3^n ey) = f(3^n e)f(y)$ for all $y \in A$ and all $n \in \mathbb{Z}_+$. Then the bijective mapping $f : A \to B$ is a Lie C^*-algebra isomorphism.*

Proof. The proof is similar to that of Theorems 15.9 and 15.11, and the details are left to the reader. $\square$

Theorem 15.13 ([225]). *Let p and θ be positive real numbers with $p > 1$, and let $f : A \to B$ be a bijective mapping satisfying $f(0) = 0$, (15.30), (15.33), (15.38), (15.40), and $f\left(\frac{1}{3^n}ey\right) = f\left(\frac{1}{3^n}e\right)f(y)$ for all $y \in A$ and all $n \in \mathbb{Z}_+$. Then the bijective mapping $f : A \to B$ is a Lie C^*-algebra isomorphism.*

Proof. The proof is similar to that of Theorems 15.9 and 15.12, and the details are left to the reader. $\square$

15.3.3. *Stability of derivations on a C^*-algebra*

Throughout this section, we assume that A is a C^*-algebra with norm $\|\cdot\|_A$.

Definition 15.2 ([225]). A $\mathbb{C}$-linear involutive mapping $D : A \to A$ is called a *derivation* if $D : A \to A$ satisfies

$$D(xy) = D(x)y + xD(y),$$

and

$$D(x^*) = D(x)^*$$

for all $x, y \in A$.

Theorem 15.14 ([225]). *Let p and θ be positive real numbers with $p < 1$, and let $f : A \to A$ be a mapping with $f(0) = 0$ such that*

$$\|C_\lambda f(x,y)\|_A \leq \theta \left(\|x\|_A^p + \|y\|_A^p \right), \tag{15.41}$$

$$\|f(3^n u^*) - f(3^n u)^*\|_A \leq 2 \cdot 3^{pn} \theta, \tag{15.42}$$

and

$$\|f(xy) - f(x)y - xf(y)\|_A \leq \theta \left(\|x\|_A^p + \|y\|_A^p \right) \tag{15.43}$$

for all $\lambda \in \mathbb{T}'$, all $x, y \in A$, all $u \in U(A)$, and all $n \in \mathbb{Z}_+$. Then there exists a unique derivation $D : A \to A$ such that

$$\|f(x) - D(x)\|_A \leq \frac{3 + 3^p}{3 - 3^p} \theta \|x\|_A^p \tag{15.44}$$

for all $x \in A$.

Proof. By the same reasoning as in the proof of Theorem 15.7, there is a unique $\mathbb{C}$-linear involutive mapping $D : A \to A$ satisfying (15.44). The mapping $D : A \to A$ is given by

$$D(x) = \lim_{n \to \infty} \frac{1}{3^n} f(3^n x) \tag{15.45}$$

for all $x \in A$. It is easy to prove that the mapping $D : A \to A$ is a derivation satisfying (15.44). This completes the proof. $\square$

Theorem 15.15 ([225]). *Let p and θ be positive real numbers with $p > 2$, and let $f : A \to A$ be a mapping satisfying $f(0) = 0$, (15.41), and (15.43) such that*

$$\left\| f\left(\frac{1}{3^n} u^* \right) - f\left(\frac{1}{3^n} u \right)^* \right\|_A \leq \frac{2\theta}{3^{pn}} \tag{15.46}$$

for all $u \in U(A)$ and all $n \in \mathbb{Z}_+$. Then there exists a unique derivation $D : A \to A$ such that

$$\|f(x) - D(x)\|_A \leq \frac{3^p + 3}{3^p - 3} \theta \|x\|_A^p \tag{15.47}$$

for all $x \in A$.

Proof. By the same reasoning as in the proof of Theorem 15.8, there is a unique $\mathbb{C}$-linear involutive mapping $D : A \to A$ satisfying (15.47). The mapping $D : A \to A$ is given by

$$D(x) = \lim_{n\to\infty} 3^n f\left(\frac{x}{3^n}\right) \tag{15.48}$$

for all $x \in A$. It is easy to prove that the mapping $D : A \to A$ is a derivation satisfying (15.47). This completes the proof. $\quad\square$

Theorem 15.16 ([225]). *Let p and θ be positive real numbers with $p < \frac{1}{2}$, and let $f : A \to A$ be a mapping with $f(0) = 0$ such that*

$$\|C_\lambda f(x,y)\|_A \le \theta \|x\|_A^p \|y\|_A^p, \tag{15.49}$$

$$\|f(3^n u^*) - f(3^n u)^*\|_A \le 3^{2pn}\theta, \tag{15.50}$$

and

$$\|f(xy) - f(x)y - xf(y)\|_A \le \theta \|x\|_A^p \|y\|_A^p \tag{15.51}$$

for all $\lambda \in \mathbb{T}'$, all $x, y \in A$, all $u \in U(A)$, and all $n \in \mathbb{Z}_+$. Then there exists a unique derivation $D : A \to A$ such that

$$\|f(x) - D(x)\|_A \le \frac{1 + 3^p}{3 - 3^{2p}}\theta \|x\|_A^{2p} \tag{15.52}$$

for all $x \in A$.

Proof. By the same reasoning as in the proof of Theorem 15.9, there is a unique $\mathbb{C}$-linear involutive mapping $D : A \to A$ satisfying (15.52). The mapping $D; A \to A$ is given by

$$D(x) = \lim_{n\to\infty} \frac{1}{3^n} f(3^n x) \tag{15.53}$$

for all $x \in A$. It is easy to prove that the mapping $D : A \to A$ is a derivation satisfying (15.52). The proof is now complete. $\quad\square$

Theorem 15.17 ([225]). *Let p and θ be positive real numbers with $p > 1$, and let $f : A \to A$ be a mapping satisfying $f(0) = 0$, (15.49), and (15.51) such that*

$$\left\| f\left(\frac{1}{3^n} u^*\right) - f\left(\frac{1}{3^n} u\right)^* \right\|_B \le \frac{\theta}{3^{2pn}} \tag{15.54}$$

for all $u \in U(A)$ and all $n \in \mathbb{Z}_+$. Then there exists a unique derivation $D : A \to A$ such that

$$\|f(x) - D(x)\|_A \le \frac{3^p + 1}{3^{2p} - 3}\theta \|x\|_A^{2p} \tag{15.55}$$

for all $x \in A$.

Proof. By the same reasoning as in the proof of Theorem 15.9, there is a unique $\mathbb{C}$-linear involutive mapping $D : A \to A$ satisfying (15.55). The mapping $D : A \to A$ is given by

$$D(x) = \lim_{n \to \infty} 3^n f\left(\frac{x}{3^n}\right) \tag{15.56}$$

for all $x \in A$. It is easy to prove that the mapping $D : A \to A$ is a derivation satisfying (15.55). The proof is now complete.   □

15.3.4. *Stability of derivations on a Lie C^*-algebra*

Throughout this section, we assume that A is a Lie C^*-algebra with norm $\|\cdot\|_A$.

Definition 15.3 ([225]). A $\mathbb{C}$-linear mapping $D : A \to A$ is called a *Lie derivation* if $D : A \to A$ satisfies

$$D([x, y]) = [D(x), y] + [x, D(y)]$$

for all $x, y \in A$.

Theorem 15.18 ([225]). *Let p and θ be positive real numbers with $p < 1$, and let $f : A \to A$ be a mapping satisfying $f(0) = 0$, and (15.41) such that*

$$\|f([x, y]) - [f(x), y] - [x, f(y)]\|_A \leq \theta \left(\|x\|_A^p + \|y\|_A^p\right) \tag{15.57}$$

for all $x, y \in A$. Then there exists a unique Lie derivation $D : A \to A$ such that

$$\|f(x) - D(x)\|_A \leq \frac{3 + 3^p}{3 - 3^p} \theta \|x\|_A^p \tag{15.58}$$

for all $x \in A$.

Proof. By the same reasoning as in the proof of Theorem 15.11, there is a unique $\mathbb{C}$-linear mapping $D : A \to A$ satisfying (15.58). The mapping $D : A \to A$ is given by

$$D(x) = \lim_{n \to \infty} \frac{1}{3^n} f(3^n x) \tag{15.59}$$

for all $x \in A$.

By the same reasoning as in the proof of [220, Theorem 4.1], the mapping $D : A \to A$ is a Lie derivation satisfying (15.58). This completes the proof.   □

Theorem 15.19 ([225]). *Let p and θ be positive real numbers with $p > 2$, and let $f : A \to A$ be a mapping satisfying $f(0) = 0$, (15.41), and (15.57). Then there exists a unique Lie derivation $D : A \to A$ such that*

$$\|f(x) - D(x)\|_A \leq \frac{3^p + 3}{3^p - 3} \theta \|x\|_A^p \tag{15.60}$$

for all $x \in A$.

Proof. By the same meaning as in the proof of Theorem 15.12, there is a unique $\mathbb{C}$-linear mapping $D : A \to A$ satisfying (15.60). The mapping $D : A \to A$ is given by

$$D(x) = \lim_{n \to \infty} 3^n f\left(\frac{x}{3^n}\right) \tag{15.61}$$

for all $x \in A$.

By the same reasoning as in the proof of [220, Theorem 4.1], the mapping $D : A \to A$ is a Lie derivation satisfying (15.60). The proof is now complete. $\quad\square$

Theorem 15.20 ([225]). *Let p and θ be positive real numbers with $p < \frac{1}{2}$, and let $f : A \to A$ be a mapping satisfying $f(0) = 0$, and (15.41) such that*

$$\|f([x,y]) - [f(x), y] - [x, f(y)]\|_A \le \theta \, \|x\|_A^p \, \|y\|_A^p \tag{15.62}$$

for all $x, y \in A$. Then there exists a unique Lie derivation $D : A \to A$ such that

$$\|f(x) - D(x)\|_A \le \frac{1 + 3^p}{3 - 3^{2p}} \theta \, \|x\|_A^{2p} \tag{15.63}$$

for all $x \in A$.

Proof. By the same reasoning as in the proof of Theorem 15.13, there is a unique $\mathbb{C}$-linear mapping $D : A \to A$ satisfying (15.63). The mapping $D : A \to A$ is given by

$$D(x) = \lim_{n \to \infty} \frac{1}{3^n} f(3^n x) \tag{15.64}$$

for all $x \in A$.

By the same reasoning as in the proof of [220, Theorem 4.1], the mapping $D : A \to A$ is a Lie derivation satisfying (15.63). The proof is now complete. $\quad\square$

Theorem 15.21 ([225]). *Let p and θ be positive real numbers with $p > 1$, and let $f : A \to A$ be a mapping satisfying $f(0) = 0$, (15.49), and (15.62). Then there exists a unique Lie derivation $D : A \to A$ such that*

$$\|f(x) - D(x)\|_A \le \frac{3^p + 1}{3^{2p} - 3} \theta \, \|x\|_A^{2p} \tag{15.65}$$

for all $x \in A$.

Proof. By the same reasoning as in the proof of Theorem 15.13, there is a unique $\mathbb{C}$-linear mapping $D : A \to A$ satisfying (15.65). The mapping $D : A \to A$ is given by

$$D(x) = \lim_{n \to \infty} 3^n f\left(\frac{x}{3^n}\right) \tag{15.66}$$

for all $x \in A$.

By the same reasoning as in the proof of [220, Theorem 4.1], the mapping $D : A \to A$ is a Lie derivation satisfying (15.65). This completes the proof. $\quad\square$

15.4. Stability of generalized additive functional equation in C^*-algebras: A fixed point approach

This section deals with the generalized Hyers–Ulam stability of homomorphisms in C^*-algebras and Lie C^*-algebras, and also the derivations on C^*-algebras and Lie C^*-algebras for the following functional equation

$$\sum_{1 \leq i < j \leq n} f\left(\frac{x_i + x_j}{2} + \sum_{l=1, k_l \neq i, j}^{n-2} x_{k_l}\right) = \frac{(n-1)^2}{2} \sum_{i=1}^{n} f(x_i), \qquad (15.67)$$

where $n \in \mathbb{N}$ is a fixed integer with $n \geq 3$. The functional equation (15.67) was introduced by Rassias [276] in the year 2009.

15.4.1. *Stability of homomorphisms in C^*-algebras*

Throughout this section, A is a C^*-algebra with norm $\|\cdot\|_A$, and B is a C^*-algebra with norm $\|\cdot\|_B$.

Definition 15.4 ([291]). A $\mathbb{C}$-linear mapping $H : A \to B$ is called a homomorphism in C^*-algebra if H satisfies $H(xy) = H(x)H(y)$, and $H(x^*) = H(x)^*$ for all $x, y \in A$.

For a given mapping $f : A \to B$, we define

$$D_\mu f(x_1, x_2, \ldots, x_n) = \sum_{1 \leq i < j \leq n} f\left(\frac{\mu x_i + \mu x_j}{2} + \sum_{l=1, k_l \neq i, j}^{n-2} \mu x_{k_l}\right)$$

$$- \frac{(n-1)^2}{2} \sum_{i=1}^{n} \mu f(x_i)$$

for all $\mu \in \mathbb{T}' = \{\nu \in C : |\nu| = 1\}$ and all $x_1, x_2, \ldots, x_n \in A$, where $n \in \mathbb{N}$ is a fixed integer with $n \geq 3$.

In the following theorem, we obtain the Hyers–Ulam–Rassias stability of homomorphisms in C^*-algebras for functional equation $D_\mu f(x_1, x_2, \ldots, x_n) = 0$, and we set $\lambda = n - 1$.

Theorem 15.22 ([291]). *Suppose that $s \in \{-1, 1\}$, and let $f : A \to B$ be a mapping satisfying $f(0) = 0$ for which there exists a function $\phi : A^n \to [0, \infty)$ such that*

$$\lim_{j \to \infty} \lambda^{2js} \phi\left(\frac{x_1}{\lambda^{js}}, \frac{x_2}{\lambda^{js}}, \ldots, \frac{x_n}{\lambda^{js}}\right) = 0, \qquad (15.68)$$

$$\|D_\mu f(x_1, x_2, \ldots, x_n)\|_B \leq \phi(x_1, x_2, \ldots, x_n), \qquad (15.69)$$

$$\|f(xy) - f(x)f(y)\|_B \leq \phi(x, y, \underbrace{0, \ldots, 0}_{(n-2) \; times}), \qquad (15.70)$$

and

$$\|f(x^*) - f(x)^*\|_B \le \phi(\underbrace{x, x, \ldots, x}_{n \ times}) \tag{15.71}$$

for all $\mu \in \mathbb{T}'$, *and* $x, y, x_1, x_2, \ldots, x_n \in A$. *If there exists an* $L = L(s) < 1$ *such that*

$$\phi(\underbrace{x, x, \ldots, x}_{n \ times}) \le \frac{1}{\lambda^s} L\phi(\underbrace{\lambda^s x, \lambda^s x, \ldots, \lambda^s x}_{n \ times}) \tag{15.72}$$

for all $x \in A$, *then there exists a unique* C^**-algebra homomorphism* $H : A \to B$ *such that*

$$\|f(x) - H(x)\|_B \le \frac{2L^{\frac{s+1}{2}}}{n\lambda^2(1 - L)}\phi(\underbrace{x, x, \ldots, x}_{n \ times}) \tag{15.73}$$

for all $x \in A$.

Proof. Consider the set $X = \{g : A \to B, \ g(0) = 0\}$, and introduce the generalized metric on X

$$d(g, h) = \inf\{C \in \mathbb{R}_+ : \|g(x) - h(x)\|_B \le C\phi(\underbrace{x, x, \ldots, x}_{n \ times}), \forall x \in A\}. \tag{15.74}$$

It is easy to show that (X, d) is complete.

Now, we consider the linear mapping $J : X \to X$ such that

$$J_g(x) = \lambda^s g\left(\frac{x}{\lambda^s}\right), \quad \text{for all } x \in A. \tag{15.75}$$

Then for all $g, h \in X$,

$$d(g, h) < C \Rightarrow \|g(x) - h(x)\|_B \le C\phi(x, x, \ldots, x), \quad \text{for all } x \in A$$

$$\Rightarrow \left\|\lambda^s g\left(\frac{x}{\lambda^s}\right) - \lambda^s h\left(\frac{x}{\lambda^s}\right)\right\|_B \le \lambda^s C\phi\left(\frac{x}{\lambda^s}, \frac{x}{\lambda^s}, \ldots, \frac{x}{\lambda^s}\right)$$

$$\Rightarrow \|J_g(x) - J_h(x)\|_B \le LC\phi(x, x, \ldots, x)$$

$$\Rightarrow d(J_g, J_h) \le LC.$$

Hence, we see that

$$d(J_g, J_h) \le Ld(g, h), \quad \text{for all } g, h \in X.$$

Letting $\mu = 1$, and $x_1 = x_2 = \cdots = x_n = x$ in the inequality (15.69), we obtain

$$\left\|\frac{n\lambda}{2}f(\lambda x) - \frac{n\lambda^2}{2}f(x)\right\|_B \le \phi(x, x, \ldots, x). \tag{15.76}$$

By using (15.72) with the case $s = -1$, we obtain

$$\left\| f(x) - \frac{1}{\lambda} f(\lambda x) \right\|_B \leq \frac{2}{n\lambda^2} \phi(x, x, \ldots, x)$$

for all $x \in A$, that is, $d(f, J_f) \leq \frac{2}{n\lambda^2} < \infty$.

Replace x by $\frac{x}{\lambda}$ in (15.76), and use (15.72) with the case $s = 1$, then we see that

$$\left\| f(x) - \lambda f\left(\frac{x}{\lambda}\right) \right\|_B \leq \frac{2}{n\lambda} \phi\left(\frac{x}{\lambda}, \frac{x}{\lambda}, \ldots, \frac{x}{\lambda}\right)$$

$$\leq \frac{2L}{n\lambda^2} \phi(x, x, \ldots, x)$$

for all $x \in A$, that is, $d(f, J_f) \leq \frac{2L}{n\lambda^2} < \infty$. By fundamental theorem of fixed point theory, there exists a mapping $H : A \to B$ such that the following conditions hold.

(1) H is a fixed point of J, that is,

$$H(\lambda x) = \lambda H(x) \tag{15.77}$$

for all $x \in A$. The mapping H is a unique fixed point of J in the set

$$Y = \{g \in X : d(f, g) < \infty\}.$$

This implies that H is a unique mapping satisfying (15.77) such that there exists $C \in (0, \infty)$ satisfying

$$\|H(x) - f(x)\|_B \leq C\phi(\underbrace{x, x, \ldots, x}_{n \text{ times}}) \quad \text{for all } x \in A.$$

(2) $d(J^n f, H) \to 0$ as $n \to \infty$. In both cases, this implies the equality

$$\lim_{m \to \infty} \lambda^{ms} f\left(\frac{x}{\lambda^{ms}}\right) = H(x), \quad \text{for all } x \in A. \tag{15.78}$$

(3) $d(f, g) \leq \frac{1}{1-L} d(f, J_f)$, which implies the inequality

$$d(f, H) \leq \frac{2L^{\frac{s+1}{2}}}{n\lambda^2(1 - L)}.$$

This implies that the inequality (15.73) holds.

It follows from (15.68), (15.69), and (15.78) that

$$\|D_H(x_1, x_2, \ldots, x_n)\|_B \leq \lim_{m \to \infty} \lambda^{ms} \left\| D_f\left(\frac{x_1}{\lambda^{ms}}, \frac{x_2}{\lambda^{ms}}, \ldots, \frac{x_n}{\lambda^{ms}}\right) \right\|_B$$

$$\leq \lim_{m \to \infty} \lambda^{ms} \phi\left(\frac{x_1}{\lambda^{ms}}, \frac{x_2}{\lambda^{ms}}, \ldots, \frac{x_n}{\lambda^{ms}}\right)$$

$$\leq \lim_{m \to \infty} \lambda^{2ms} \phi\left(\frac{x_1}{\lambda^{ms}}, \frac{x_2}{\lambda^{ms}}, \ldots, \frac{x_n}{\lambda^{ms}}\right) = 0.$$

By Theorem 1.1 of [276], the mapping $H : A \to B$ is Cauchy additive, that is, $H(x + y) = H(x) + H(y)$, for all $x, y \in A$.

Letting $x_i = x$, for all $i = 1, 2, \ldots, n$ in $D_\mu f(x_1, x_2, \ldots, x_n) = 0$, we obtain $\mu f(\lambda x) = f(\mu \lambda x)$ for all $\mu \in \mathbb{T}'$, and all $x \in A$. By a similar method to above we get $\mu H(\lambda x) = H(\mu \lambda x)$ for all $\mu \in \mathbb{T}'$, and all $x \in A$.

Thus, we can show that the mapping $H : A \to B$ is $\mathbb{C}$-linear.

It follows from (15.70) that

$$\|H(xy) - H(x)H(y)\|_B = \lim_{m \to \infty} \lambda^{2ms} \left\| f\left(\frac{xy}{\lambda^{2ms}}\right) - f\left(\frac{x}{\lambda^{ms}}\right) f\left(\frac{y}{\lambda^{ms}}\right) \right\|_B$$

$$\leq \lim_{m \to \infty} \lambda^{2ms} \phi\left(\frac{x}{\lambda^{ms}}, \frac{y}{\lambda^{ms}}, \underbrace{0, \ldots, 0}_{(n-2)\ \text{times}}\right) = 0$$

for all $x, y \in A$. So $H(xy) = H(x)H(y)$, for all $x, y \in A$.

It follows from (15.71) that

$$\|H(x^*) - H(x)^*\|_B = \lim_{m \to \infty} \lambda^{ms} \left\| f\left(\frac{x^*}{\lambda^{ms}}\right) - f\left(\frac{x}{\lambda^{ms}}\right)^* \right\|_B$$

$$\leq \lim_{m \to \infty} \lambda^{ms} \phi\left(\frac{x}{\lambda^{ms}}, \frac{x}{\lambda^{ms}}, \ldots, \frac{x}{\lambda^{ms}}\right) = 0$$

for all $x \in A$. So $H(x)^* = H(x)^*$ for all $x, y \in A$.

Thus $H : A \to B$ is a C^*-algebra homomorphism satisfying (15.73) as desired. This completes the proof of the theorem.   □

Corollary 15.1 ([291]). *Suppose that $s \in \{-1, 1\}$ be fixed, and $r > 0$ be given with $r \neq 2$ and let θ be non-negative real numbers, and let $f : A \to B$ be a mapping such that*

$$\|D_\mu f(x_1, x_2, \ldots, x_n)\|_B \leq \theta \sum_{i=1}^{n} \|x_i\|_A^r, \tag{15.79}$$

$$\|f(xy) - f(x)f(y)\|_B \leq \theta \left(\|x\|_A^r + \|y\|_A^r\right), \tag{15.80}$$

and

$$\|f(x^*) - f(x)^*\|_B \leq n\theta \|x\|_A^r \tag{15.81}$$

for all $\mu \in \mathbb{T}'$, and $x, y, x_1, x_2, \ldots, x_n \in A$. Further, assume that $f(0) = 0$ in (15.79) for the case $r > 2$. Then there exists a unique C^*-algebra homomorphism $H : A \to B$ such that

$$\|f(x) - H(x)\|_B \leq \frac{2s\theta}{\lambda^{r+1} - \lambda^2} \|x\|_A^r \tag{15.82}$$

for all $x \in A$, where $rs < 2s$.

Proof. The proof follows from Theorem 15.22 by taking $\phi(x_1, x_2, \ldots, x_n) = \theta \sum_{i=1}^{n} \|x_i\|_A^r$ for all $x_1, x_2, \ldots, x_n \in A$. Then taking $L = \lambda^{(1-r)s}$, we get the desired result. The proof is now complete. $\qquad\square$

15.4.2. *Stability of generalized derivations on C^*-algebras*

Throughout this section, A is a C^*-algebra with norm $\|\cdot\|_A$.

Definition 15.5 ([291]). A generalized derivation $\delta : A \to A$ is involutive $\mathbb{C}$-linear and fulfills

$$\delta(xyz) = \delta(xy)z - x\delta(y)z + x\delta(yz)$$

for all $x, y, z \in A$.

For a given mapping $f : A \to A$, we define

$$D_\mu f(x_1, x_2, \ldots, x_n) = \sum_{1 \leq i < j \leq n} f\left(\frac{\mu x_i + \mu x_j}{2} + \sum_{l=1, k_l \neq i,j}^{n-2} \mu x_{k_l}\right)$$
$$- \frac{(n-1)^2}{2} \sum_{i=1}^{n} \mu f(x_i)$$

for all $\mu \in \mathbb{T}' = \{\nu \in C : |\nu| = 1\}$, and all $x_1, x_2, \ldots, x_n \in A$, where $n \in \mathbb{N}$ is a fixed integer with $n \geq 3$.

In the following theorem, we obtain the Hyers–Ulam–Rassias stability of derivations on C^*-algebras for functional equation $D_\mu f(x_1, x_2, \ldots, x_n) = 0$, and we set $\lambda = n - 1$.

Theorem 15.23 ([291]). *Suppose that $s \in \{-1, 1\}$, and let $f : A \to A$ be a mapping satisfying $f(0) = 0$ for which there exists a function $\phi : A^n \to [0, \infty)$ such that*

$$\lim_{j \to \infty} \lambda^{3js} \phi\left(\frac{x_1}{\lambda^{js}}, \frac{x_2}{\lambda^{js}}, \ldots, \frac{x_n}{\lambda^{js}}\right) = 0, \tag{15.83}$$

$$\|D_\mu f(x_1, x_2, \ldots, x_n)\|_A \leq \phi(x_1, x_2, \ldots, x_n), \tag{15.84}$$

and

$$\|f(xyz) - f(xy)z + xf(y)z - xf(yz)\|_A \leq \phi(x, y, z, \underbrace{0, \ldots, 0}_{(n-3)\ times}) \tag{15.85}$$

for all $\mu \in \mathbb{T}'$, and $x, y, z, x_1, x_2, \ldots, x_n \in A$. If there exists an $L = L(s) < 1$ satisfying (15.72), then there exists a unique generalized derivation $\delta : A \to A$ such *that*

$$\|f(x) - \delta(x)\|_A \leq \frac{2L^{\frac{s+1}{2}}}{n\lambda^2(1-L)}\phi(\underbrace{x, x, \ldots, x}_{n \; times}) \tag{15.86}$$

for all $x \in A$.

Proof. By the same reasoning as the proof of Theorem 15.22, there exists a unique involutive $\mathbb{C}$-linear mapping $\delta : A \to A$ satisfying (15.86). The mapping $\delta : A \to A$ is given by

$$\delta(x) = \lim_{m\to\infty} \lambda^{ms} f\left(\frac{x}{\lambda^{ms}}\right) \tag{15.87}$$

for all $x \in A$.

It follows from (15.85) that

$$\|\delta(xyz) - \delta(xy)z + x\delta(y)z - x\delta(yz)\|_A$$

$$= \lim_{m\to\infty} \left\| f\left(\frac{xyz}{\lambda^{3ms}}\right) - f\left(\frac{xy}{\lambda^{2ms}}\right)\frac{z}{\lambda^{ms}} + \frac{x}{\lambda^{ms}}f\left(\frac{y}{\lambda^{ms}}\right)\frac{z}{\lambda^{ms}} - \frac{x}{\lambda^{ms}}f\left(\frac{yz}{\lambda^{2ms}}\right)\right\|_A$$

$$\leq \lambda^{3ms}\phi\left(\frac{x}{\lambda^{ms}}, \frac{y}{\lambda^{ms}}, \frac{z}{\lambda^{ms}}, \underbrace{0, \ldots, 0}_{(n-3) \; times}\right) = 0$$

for all $x, y, z \in A$. So $\delta(xyz) = \delta(xy)z - x\delta(y)z + x\delta(yz)$ for all $x, y, z \in A$. Thus $\delta : A \to A$ is a generalized derivation satisfying (15.86). The proof is now complete. $\square$

Corollary 15.2 ([291]). *Suppose that $s \in \{-1, 1\}$ be fixed, $r > 0$ be given with $r \neq 3$, θ be non-negative real numbers, and let $f : A \to A$ be a mapping such that*

$$\|D_\mu f(x_1, x_2, \ldots, x_n)\|_A \leq \theta \sum_{i=1}^{n} \|x_i\|_A^r, \tag{15.88}$$

and

$$\|f(xyz) - f(xy)z + xf(y)z - xf(yz)\|_A \leq \theta \sum_{i=1}^{3} \|x_i\|_A^r \tag{15.89}$$

for all $x, y, z, x_1, x_2, \ldots, x_n \in A$. Further, assume that $f(0) = 0$ in (15.79) for the case $r > 3$. Then there exists a unique generalized derivation $H : A \to A$ such that

$$\|f(x) - H(x)\|_A \leq \frac{2s\theta}{\lambda^{r+1} - \lambda^2}\|x\|_A^r \tag{15.90}$$

for all $x \in A$, where $rs < 3s$.

Proof. The proof follows from Theorem 15.23 by taking $\phi(x_1, x_2, \ldots, x_n) = \theta \sum_{i=1}^{n} \|x_i\|_A^r$ for all $x_1, x_2, \ldots, x_n \in A$. Then take $L = \lambda^{(1-r)s}$, and we get the desired result. This completes the proof. $\square$

15.4.3. *Stability of homomorphisms in Lie C^*-algebras*

Throughout this section, A is a Lie C^*-algebra with norm $\| \cdot \|_A$, and B is a Lie C^*-algebra with norm $\| \cdot \|_B$.

Definition 15.6 ([291]). A $\mathbb{C}$-linear mapping $H : A \to B$ is called a Lie C^*-algebra homomorphism if $H([x, y]) = [H(x), H(y)]$ for all $x, y \in A$.

For a given mapping $f : A \to B$, we define

$$D_\mu f(x_1, x_2, \ldots, x_n) = \sum_{1 \le i < j \le n} f \left(\frac{\mu x_i + \mu x_j}{2} + \sum_{l=1, k_l \ne i, j}^{n-2} \mu x_{k_l} \right)$$

$$- \frac{(n-1)^2}{2} \sum_{i=1}^{n} \mu f(x_i)$$

for all $\mu \in \mathbb{T}' = \{\nu \in C : |\nu| = 1\}$ and all $x_1, x_2, \ldots, x_n \in A$, where $n \in \mathbb{N}$ is a fixed integer with $n \ge 3$.

The following theorem presents the Hyers–Ulam–Rassias stability of homomorphisms in Lie C^*-algebras for the above functional equation $D_\mu f(x_1, x_2, \ldots, x_n) = 0$, and we set $\lambda = n - 1$.

Theorem 15.24 ([291]). *Suppose that $s \in \{-1, 1\}$, and let $f : A \to B$ be a mapping satisfying $f(0) = 0$ for which there exists a function $\phi : A^n \to [0, \infty)$ satisfying (15.68), and (15.69) such that*

$$\|f([x, y]) - [f(x), f(y)]\|_B \le \phi(x, y, \underbrace{0, \ldots, 0}_{(n-2) \ times}) \tag{15.91}$$

for all $x, y \in A$. If there exists an $L = L(s) < 1$ satisfying (15.72) then there exists a unique Lie C^*-algebra homomorphism $H : A \to B$ satisfying (15.73).

Proof. By the same reasoning as the proof of Theorem 15.22, there exists a unique $\mathbb{C}$-linear mapping $H : A \to B$ satisfying (15.73). The mapping $H : A \to B$ is given by (15.78).

It follows from (15.91) that

$$\|H([x, y]) - [H(x), H(y)]\|_B = \lim_{m \to \infty} \lambda^{2ms} \left\| f \left(\frac{[x, y]}{\lambda^{2ms}} \right) - \left[f \left(\frac{x}{\lambda^{ms}} \right), f \left(\frac{y}{\lambda^{ms}} \right) \right] \right\|_B$$

$$\le \lim_{m \to \infty} \lambda^{2ms} \phi \left(\frac{x}{\lambda^{ms}}, \frac{y}{\lambda^{ms}}, \underbrace{0, \ldots, 0}_{(n-2) \ times} \right) = 0$$

for all $x, y \in A$. So $H([x, y]) = [H(x), H(y)]$ for all $x, y \in A$.

Thus $H : A \to B$ is a Lie C^*-algebra homomorphism satisfying (15.73) as desired. The proof is now complete. $\square$

Corollary 15.3 ([291]). *Suppose that $s \in \{-1, 1\}$ be fixed and $r > 0$ be given with $r \neq 2$, and let θ be non-negative real numbers, and let $f : A \to B$ be a mapping satisfying (15.79) such that*

$$\|f([x, y]) - [f(x), f(y)]\|_B \leq \theta \left(\|x\|_A^r + \|y\|_A^r \right) \tag{15.92}$$

for all $x, y \in A$. Further, assume that $f(0) = 0$ in (15.79) for the case $r > 2$. Then there exists a unique Lie C^*-algebra homomorphism $H : A \to B$ satisfying (15.82).

Proof. The proof follows from Theorem 15.24 by taking $\phi(x_1, x_2, \ldots, x_n) = \theta \sum_{i=1}^n \|x_i\|_A^r$ for all $x_1, x_2, \ldots, x_n \in A$. Then take $L = \lambda^{(1-r)s}$, and we get the desired result. This completes the proof. $\qquad\square$

Chapter 16

Ulam Stability of Mixed Type Mappings on Restricted Domains

16.1. Introduction

In the previous chapter, we have discussed about the stability of Jensen equation in C^*-algebras, and Hyers–Ulam–Rassias stability of isomorphisms in C^*-algebras using direct method, and the generalized Hyers–Ulam stability of generalized additive functional equation in C^*-algebras, and Lie C^*-algebras using fixed point method. This chapter concerns the investigation of the Ulam stability of Jensen and Jensen type mappings, Ulam stability of quadratic functional equation, Ulam stability of a generalized quadratic functional equation, Ulam stability of quadratic functional equation with improved bound, Ulam stability of mixed type functional equation involving three variables, and Ulam stability of mixed type functional equation involving four variables on restricted domains. Also Ulam stability of quadratic functional equation in a set of Lebesgue measure zero is also studied in this chapter.

16.2. Ulam stability of Jensen and Jensen type mappings on restricted domains

This section deals with additive mappings of two forms: "Jensen" and "Jensen type", and investigate the Ulam stability of these mappings on restricted domains.

Throughout this section, X is a real normed space, and Y is a real Banach space in the case of functional inequalities, as well let X and Y be real linear spaces for functional equations.

Definition 16.1 ([272]). A mapping $A : X \to Y$ is called additive of the first form if A satisfies the additive functional equation

$$A(x_1 + x_2) + A(x_1 - x_2) = 2A(x_1) \qquad (16.1)$$

for all $x_1, x_2 \in X$. We note that (16.1) is equivalent to the Jensen equation

$$A\left(\frac{x+y}{2}\right) = \frac{1}{2}[A(x) + A(y)] \tag{16.2}$$

for $x = x_1 + x_2$, $y = x_1 - x_2$.

Definition 16.2 ([272]). A mapping $A : X \to Y$ is called additive of the second form if A satisfies the additive functional equation

$$A(x_1 + x_2) - A(x_1 - x_2) = 2A(x_2) \tag{16.3}$$

for all $x_1, x_2 \in X$. We note that (16.3) is equivalent to the Jensen type equation

$$A\left(\frac{x-y}{2}\right) = \frac{1}{2}[A(x) - A(y)] \tag{16.4}$$

for $x = x_1 + x_2$, $y = x_1 - x_2$.

Definition 16.3 ([272]). A mapping $f : X \to Y$ is called approximately odd if f satisfies the functional inequality

$$\|f(x) + f(-x)\| \leq \theta \tag{16.5}$$

for some fixed $\theta \geq 0$, and for all $x \in X$.

We present the following theorem which was proved by Rassias [263] in 1994.

Theorem 16.1 ([272]). *If a mapping $f : X \to Y$ satisfies the inequalities*

$$\|f(x_1 + x_2) + f(x_1 - x_2) - 2f(x_1)\| \leq \delta, \tag{16.6}$$

and

$$\|f(0)\| \leq \delta_0 \tag{16.7}$$

for some fixed $\delta, \delta_0 \geq 0$, and for all $x_1, x_2 \in X$, then there exists a unique additive mapping $A : X \to Y$ of the first form which satisfies the inequality

$$\|f(x) - A(x)\| \leq \delta + \delta_0 \tag{16.8}$$

for all $x \in X$. If moreover, f is measurable or $f(tx)$ is continuous in t for each fixed $x \in X$, then $A(Tx) = tA(x)$ for all $x \in X$, and $t \in \mathbb{R}$.

The last assertion holds according to Rassias [255] work in 1982.

16.2.1. *Ulam stability of functional equation (16.1) on a restricted domain*

Theorem 16.2 ([272]). *Let $d > 0$, and $\delta, \delta_0 \geq 0$ be fixed. If a mapping $f : X \to Y$ satisfies inequality (16.6) for all $x_1, x_2 \in X$, with $\|x_1\| + \|x_2\| \geq d$, and (16.7), then* there exists a unique additive mapping $A : X \to Y$ of the first form such that

$$\|f(x) - A(x)\| \le \frac{5}{2}\delta + \delta_0 \qquad (16.9)$$

for all $x \in X$. If, moreover, f is measurable or $f(tx)$ is continuous in t for each fixed $x \in X$, then $A(tx) = tA(x)$ for all $x \in X$, and $t \in \mathbb{R}$.

Proof. Assume $\|x_1\| + \|x_2\| < d$. If $x_1 = x_2 = 0$, then we choose $t \in X$ with $\|t\| = d$. Otherwise, let us choose

$$t = \left(a + \frac{d}{\|x_1\|}\right) x_1 \quad \text{if} \quad \|x_1\| \ge \|x_2\|, \quad t = \left(a + \frac{d}{\|x_2\|}\right) x_2 \quad \text{if} \quad \|x_1\| \le \|x_2\|.$$

We note that $\|t\| = \|x_1\| + d > d$ if $\|x_1\| \ge \|x_2\|$, $\|t = \|x_2\|\| + d > d$ if $\|x_1\| \le \|x_2\|$. Clearly, we see that

$$\|x_1 - t\| + \|x_2 + t\| \ge 2\|t\| - (\|x_1\| + \|x_2\|) \ge d, \quad \|x_1 - x_2\| + \|2t\| \ge d,$$
$$\|x_1 + t\| + \|-x_2 + t\| \ge 2\|t\| - (\|x_1\| + \|x_2\|) \ge d, \quad \|x_1\| + \|t\| \ge d. \qquad (16.10)$$

Inequalities (16.10) come from the corresponding substitutions attached between the right-hand sided parentheses of the following functional identity.

Therefore from (16.6), (16.10), the triangle inequality, and the functional identity

$$\begin{aligned}
2 & \left[f(x_1 + x_2) + f(x_1 - x_2) - 2f(x_1)\right] \\
&= [f(x_1 + x_2) + f(x_1 - x_2 - 2t) - 2f(x_1 - t)] \\
&\quad \text{(with } x_1 - t \text{ on } x_1 \text{ and } x_2 + t \text{ on } x_2) \\
&\quad - [f(x_1 - x_2 - 2t) + f(x_1 - x_2 + 2t) - 2f(x_1 - x_2)] \\
&\quad \text{(with } x_1 - x_2 \text{ on } x_1 \text{ and } 2t \text{ on } x_2) \\
&\quad + [f(x_1 - x_2 + 2t) + f(x_1 + x_2) - 2f(x_1 + t)] \\
&\quad \text{(with } x_1 + t \text{ on } x_1 \text{ and } -x_2 + t \text{ on } x_2) \\
&\quad + 2[f(x_1 + t) + f(x_1 - t) - 2f(x_1)] \\
&\quad \text{(with } x_1 \text{ on } x_1 \text{ and } t \text{ on } x_2),
\end{aligned}$$

we obtain

$$2\|f(x_1 + x_2) + f(x_1 - x_2) - 2f(x_1)\| \le \delta + \delta + \delta + 2\delta = 5\delta,$$

or

$$\|f(x_1 + x_2) + f(x_1 - x_2) - 2f(x_1)\| \le \frac{5}{2}\delta. \qquad (16.11)$$

Applying now Theorem 16.1, and inequality (16.11), one obtains that there exists a unique additive mapping $A : X \to Y$ of the first form that satisfies the additive equation (16.1), and inequality (16.9), such that $A(x) = \lim_{n \to \infty} 2^{-n} f(2^n x)$. Our last assertion is trivial according to Theorem 16.1. The proof is now complete. $\qquad \square$

Note. If we define $S_2 = \{(x_1, x_2) \in X^2 : \|x_i\| < d, i = 1, 2\}$ for some $d > 0$, then $\{(x_1, x_2) \in X^2 : \|x_1\| + \|x_2\| \geq 2d\} \subset X^2 \backslash S_2$.

Corollary 16.1 ([272]). *If we assume that a mapping $f : X \to Y$ satisfies inequalities (16.6), and (16.7) for some fixed $\delta, \delta_0 \geq 0$, and for all $(x_1, x_2) \in X^2 \backslash S_2$, then there exists a unique additive mapping $A : X \to Y$ of the first form, satisfying (16.9) for all $x \in X$. If, moreover, f is measurable or $f(tx)$ is continuous in t for each fixed $x \in X$, then $A(tx) = tA(x)$ for all $x \in X$, and all $t \in \mathbb{R}$.*

Proof. The proof follows immediately from Theorem 16.2, and the details are left to the reader. $\qquad\square$

16.2.2. *Stability of equation* (16.3)

Theorem 16.3 ([272]). *If a mapping $f : X \to Y$ satisfies the inequality*

$$\|f(x_1 + x_2) - f(x_1 - x_2) - 2f(x_2)\| \leq \delta \qquad (16.12)$$

for some $\delta \geq 0$, and for all $x_1, x_2 \in X$, then there exists a unique additive mapping $A : X \to Y$ of the second form which satisfies the inequality

$$\|f(x) - A(x)\| \leq \frac{3}{2}\delta \qquad (16.13)$$

for all $x \in X$. If, moreover, f is measurable or $f(tx)$ is continuous in t for each fixed $x \in X$, then $A(tx) = tA(x)$ for all $x \in X$, and $t \in \mathbb{R}$.

Proof. Replacing $x_1 = x_2 = 0$ in (16.12), we find

$$\|f(0)\| \leq \frac{\delta}{2}. \qquad (16.14)$$

Thus, substituting $x_1 = x_2 = x$ in (16.12), one obtains

$$\|f(2x) - f(0) - 2f(x)\| \leq \delta,$$

or

$$\left\|f(x) - 2^{-1}f(2x)\right\| \leq \frac{3}{2}\delta\left(1 - 2^{-1}\right) \qquad (16.15)$$

for all $x \in X$. Therefore from (16.15), with $2^i x$ in place of x ($i = 1, 2, \ldots, n - 1$), we obtain

$$\left\|f(x) - 2^{-n}f\left(2^n x\right)\right\| \leq \left\|f(x) - 2^{-1}f(2x)\right\| + \left\|2^{-1}f(2x) - 2^{-2}f\left(2^2 x\right)\right\| + \ldots$$

$$+ \left\|2^{-(n-1)}f(2^{n-1}x) - 2^{-n}f(2^n x)\right\|$$

$$\leq \frac{3}{2}\delta(1 + 2^{-1} + \cdots + 2^{-(n-1)})(1 - 2^{-1})$$

$$\leq \frac{3}{2}\delta\left(1 - 2^{-n}\right) \qquad (16.16)$$

for any $n \in \mathbb{N}$, and all $x \in X$.

We claim that

$$A(x) = 2^{-n} A (2^n x) \tag{16.17}$$

holds for any $n \in \mathbb{N}$, and all $x \in X$. In fact, replacing $x_1 = x_2 = 0$ in (16.3), one finds $A(0) = 0$. Thus substituting $x_1 = x_2 = x$ in (16.3), we get $A(2x) = 2A(x)$ for all $x \in X$. Therefore by induction on n, one obtains that

$$A \left(2^{n+1} x\right) = A \left(2 \cdot 2^n x\right) = 2A \left(2^n x\right) = 2 \cdot 2^n A(x) = 2^{n+1} A(x)$$

for all $x \in X$, completing the proof of (16.17).

By (16.16), for $n \geq m > 0$, and $h = 2^m x$, we have

$$\left\|2^{-n} f \left(2^n x\right) - 2^{-m} f \left(2^m x\right)\right\| = 2^{-m} \left\|2^{-(n-m)} f \left(2^{n-m} h\right) - f(h)\right\|$$

$$\leq 2^{-m} \left[\frac{3}{2} \delta \left(1 - 2^{-(n-m)}\right)\right]$$

$$\leq \frac{3}{2} \delta 2^{-m} \to 0 \quad \text{as } m \to \infty. \tag{16.18}$$

From (16.18), and the completeness of Y, we get that the Cauchy sequence $\{2^{-n} f \left(2^n x\right)\}$ converges. Therefore, we may apply a direct method to the definition of A, such that the formula

$$A(x) = \lim_{n \to \infty} 2^{-n} f \left(2^n x\right) \tag{16.19}$$

holds for all $x \in X$. From formula (16.19), and inequality (16.12), it follows that

$$\|A(x_1 + x_2) - A(x_1 - x_2) - 2A(x_2)\|$$
$$= \lim_{n \to \infty} 2^{-n} \|f(2^n x_1 + 2^n x_2) - f(2^n x_1 - 2^n x_2) - 2f(2^n x_2)\| \leq \lim_{n \to \infty} 2^{-n} \delta = 0$$

or equation (16.3) holds for all $x_1, x_2 \in X$. Thus $A : X \to Y$ is an additive mapping of the second form. According to inequality (16.16), and formula (16.19), one gets that inequality (16.13) holds.

Assume now that there is another additive mapping $A' : X \to Y$ of the second form which satisfies equation (16.3), formula (16.17), and inequality (16.13). Therefore

$$\|A(x) - A'(x)\| = 2^{-n} \|A \left(2^n x\right) - A' \left(2^n x\right)\|$$

$$\leq 2^{-n} \left(\|A \left(2^n x\right) - f \left(2^n x\right)\| + \|f \left(2^n x\right) - A' \left(2^n x\right)\|\right)$$

$$\leq 2^{-n} \left(\frac{3}{2} \delta + \frac{3}{2} \delta\right)$$

$$= 3\delta 2^{-n} \to 0 \quad \text{as } n \to \infty,$$

or

$$A(x) = A'(x) \tag{16.20}$$

for all $x \in X$, completing the proof of the first part of our theorem. The proof of the last assertion is obvious. This completes the proof. $\square$

16.2.3. *Ulam stability of functional equation (16.3) on a restricted domain*

We note that from (16.5), and $\|f(-2x) + f(2x)\| \leq \theta$ (from (16.5) with $2x$ on x), and (16.15) as well as $\|f(-2x) - 2f(-x)\| \leq \frac{3}{2}\delta$ (from (16.15) with $2x$ on x), and the triangle inequality, one gets

$$2\|f(-x) + f(x)\| \leq \|-[f(-2x) - 2f(-x)]\|$$

$$+ \|-[f(2x) - 2f(x)]\| + \|f(-2x) + f(2x)\|$$

$$\leq \frac{3}{2}\delta + \frac{3}{2}\delta + \theta = 3\delta + \theta,$$

or

$$\|f(-x) + f(x)\| \leq \frac{3}{2}\delta + \frac{\theta}{2} = \theta, \qquad \text{using (16.5).}$$

Therefore $\theta = 3\delta$, and (16.5) takes the independent of θ equivalent form

$$\|f(-x) + f(x)\| \leq 3\delta. \tag{16.21}$$

Theorem 16.4 ([272]). *Let $d > 0$, and $\delta \geq 0$ be fixed. If an approximately odd mapping $f : X \to Y$ satisfies inequality (16.12) for all $x_1, x_2 \in X$ with $\|x_1\| + \|x_2\| \geq d$, and inequality (16.21) for all $x \in X$ with $\|x\| \geq d$, then there exists a unique additive mapping $A : X \to Y$ of the second form such that*

$$\|f(x) - A(x)\| \leq \frac{33}{2}\delta \tag{16.22}$$

for all $x \in X$. If, moreover, f is measurable or $f(tx)$ is continuous in t for each fixed $x \in X$, then $A(tx) = tA(x)$ for all $x \in X$, and $t \in \mathbb{R}$.

Proof. Assume $\|x_1\| + \|x_2\| < d$. If $x_1 = x_2 = 0$, then we choose $t \in X$ with $\|t\| = d$. Otherwise, let us choose

$$t = \left(1 + \frac{d}{\|x_1\|}\right) x_1 \quad \text{if } \|x_1\| \geq \|x_2\|, \quad t = \left(1 + \frac{d}{\|x_2\|}\right) x_2 \quad \text{if } \|x_1\| \leq \|x_2\|.$$

We note that $\|t\| = \|x_1\| + d > d$ if $\|x_1\| \geq \|x_2\|$, $\|t\| = \|x_2\| + d > d$ if $\|x_1\| \leq \|x_2\|$. Clearly, we see that

$$\|x_1 - t\| + \|x_2 + t\| \geq 2\|t\| - (\|x_1\| + \|x_2\|) \geq d,$$

$$\|x_1 - t\| + \|x_2 - t\| \geq 2\|t\| - (\|x_1\| + \|x_2\|) \geq d,$$

$$\|x_1 - 2t\| + \|x_2\| \geq 2\|t\| - (\|x_1\| + \|x_2\|) \geq d, \quad \|t\| + \|x_2\| \geq d,$$

$$\|t - x_2\| \geq \|t\| - \|x_2\| = (\|x_2\| + d) - \|x_2\| = d, \quad \text{because } \|t\| = \|x_2\| + d.$$

$$\tag{16.23}$$

Therefore from (16.21), (16.12), (16.23), and the functional identity

$$
f(x_1 + x_2) - f(x_1 - x_2) - 2f(x_2)
$$
$$
= [f(x_1 + x_2) + f(x_1 - x_2 - 2t) - 2f(x_2 + t)]
$$
$$
\text{(with } x_1 - t \text{ on } x_1 \text{ and } x_2 + t \text{ on } x_2\text{)}
$$
$$
+ [f(x_1 + x_2 - 2t) - f(x_1 - x_2) - 2f(x_2 - t)]
$$
$$
\text{(with } x_1 - t \text{ on } x_1 \text{ and } x_2 - t \text{ on } x_2\text{)}
$$
$$
- [f(x_1 + x_2 - 2t) - f(x_1 - x_2 - 2t) - 2f(x_2)]
$$
$$
\text{(with } x_1 - 2t \text{ on } x_1 \text{ and } x_2 \text{ on } x_2\text{)}
$$
$$
+ 2[f(t + x_2) + f(t - x_2) - 2f(x_2)]
$$
$$
\text{(with } t \text{ on } x_1 \text{ and } x_2 \text{ on } x_2\text{)},
$$
$$
+ 2[f(t - x_2) + f(-(t - x_2))] \qquad \text{(with } t - x_2 \text{ on } x\text{)},
$$

we obtain

$$
\|f(x_1 + x_2) - f(x_1 - x_2) - 2f(x_2)\| \le \delta + \delta + \delta + 2\delta + 6\delta = 11\delta. \tag{16.24}
$$

Applying Theorem 16.3, and inequality (16.24), we prove that there exists a unique additive mapping $A : X \to Y$ of the second form that satisfies equation (16.3), and inequality (16.22), completing the proof of the theorem. $\square$

Note. If we define $S_1 = \{x \in X : \|x\| < d\}$, and $S_2 = \{(x_1, x_2) \in X^2 : \|x_i\| < d, i = 1, 2\}$ for some fixed $d > 0$, then $\{x \in X : \|x\| \ge 2d\} \subset X \backslash S_1$, and $\{(x_1, x_2) \in X^2 : \|x_1\| + \|x_2\| \ge 2d\} \subset X^2 \backslash S_2$.

Corollary 16.2 ([272]). *If we assume that a mapping $f : X \to Y$ satisfies inequality (16.12) for some fixed $\delta \ge 0$, and (16.21) for all $x \in X \backslash S_1$ and for all $(x_1, x_2) \in X^2 \backslash S_2$, then there exists a unique additive mapping $A : X \to Y$ of the second form, satisfying (16.22) for all $x \in X$. If, moreover, f is measurable or $f(tx)$ is continuous in t for each fixed $x \in X$, then $A(tx) = tA(x)$, for all $x \in X$, and $t \in \mathbb{R}$.*

Proof. The proof follows from Theorem 16.4, and the details are left to the reader. $\square$

16.3. Ulam stability of quadratic functional equation (4.1) on a restricted domain

This section deals with the Hyers–Ulam stability of the quadratic functional equation (4.1) on a restricted domain.

In 1992, Czerwik proved in [59] a Hyers–Ulam–Rassias stability theorem on the quadratic functional equation.

Lemma 16.1 ([59]). *Assume that there exist $\xi \geq 0$, $\eta \geq 0$, and $v \in \mathbb{R}$ such that a function $f : E_1 \to E_2$ between the normed spaces satisfies the inequality*

$$\|f(x+y) + f(x-y) - 2f(x) - 2f(y)\| \leq \xi + \eta \left(\|x\|^v + \|y\|^v\right) \tag{16.25}$$

for all $x, y \in E_1 \backslash \{0\}$. Then for $x \in E_1 \backslash \{0\}$, and $n \in \mathbb{N}$

$$\|f(2^n x) - 4^n f(x)\|$$
$$\leq 3^{-1}\left(4^n - 1\right)(\xi + c) + 2 \cdot 4^{n-1}\eta\|x\|^v \left(1 + a + \cdots + a^{n-1}\right), \tag{16.26}$$

and

$$\|f(x) - 4^n f\left(2^{-n}x\right)\|$$
$$\leq 3^{-1}\left(4^n - 1\right)(\xi + c) + 2^{1-v}\eta\|x\|^v \left(1 + b + \cdots + b^{n-1}\right), \tag{16.27}$$

where $a = 2^{v-2}$, $b = 2^{2-v}$, $c = \|f(0)\|$.

Proof. Put $x = y \neq 0$ in (16.25), we obtain

$$\|f(2x) - 4f(x)\| \leq \|f(0)\| + \xi + 2\eta \|x\|^v$$

which proves (16.26) for $n = 1$. Now, let us assume that (16.26) is true for $k \leq n$, and $x \in E_1 \backslash \{0\}$. Then for $n + 1$, we have

$$\|f\left(2^{n+1}x\right) - 4^{n+1}f(x)\|$$
$$\leq \|f\left(2 \cdot 2^n x\right) - 4f\left(2^n x\right)\| + 4\|f\left(2^n x\right) - 4^n f(x)\|$$
$$\leq (\xi + c) + 2\eta \|2^n x\|^v + \frac{4}{3}(\xi + c)\left(4^n - 1\right) + 2 \cdot 4^n \eta\|x\|^v \left(1 + a + \cdots + a^{n-1}\right)$$
$$= 3^{-1}\left(4^{n+1} - 1\right)(\xi + c) + 2 \cdot 4^n \eta\|x\|^v \left(1 + a + \cdots + a^n\right),$$

which proves the inequality (16.26) for all natural n.

Analogously, taking $x = y = \frac{t}{2}$, we can verify the inequality (16.27) for $n = 1$. Applying the induction principle, we get the result for all $n \in \mathbb{N}$, which completes the proof. $\qquad\qquad\square$

Using Lemma 16.1, we have the following theorem.

Theorem 16.5. *Let E_1 be a normed space, and E_2 be a Banach space, and let $f : E_1 \to E_2$ be a function satisfying the inequality (16.25) for all $x, y \in E_1 \backslash \{0\}$ and let $v < 2$. Then there exists exactly one quadratic mapping $g : E_1 \to E_2$ such that*

$$\|g(x) - f(x)\| \leq 3^{-1}(\xi + c) + 2\left(4 - 2^v\right)^{-1}\eta\|x\|^v, \tag{16.28}$$

for all $x \in E_1 \backslash \{0\}$. If, moreover, f is measurable (i.e., $f^{-1}(G)$ is a Borel set in E_1 for every open set G in E_2) or $\mathbb{R} \ni t \to f(tx)$ is continuous for each fixed $x \in E_1$, *then g satisfies the condition*

$$g(tx) = t^2 g(x) \tag{16.29}$$

for all $x \in E_1$, and $t \in \mathbb{R}$.

Proof. Define the sequence of functions $\{g_n\}$ by the formula

$$g_n(x) = 4^{-n} f(2^n x), \tag{16.30}$$

for all $x \in E_1$, $n \in \mathbb{N}$. Then $\{g_n\}$ is a Cauchy sequence for every $x \in E_1$. Obviously, for $x = 0$, it is trivial. Let $0 \neq x \in E_1$. We have for $n > m$,

$$\|g_n(x) - g_m(x)\|$$
$$= 4^{-n} \left\| f\left(2^{n-m} \cdot 2^m x\right) - 4^{n-m} f\left(2^m x\right) \right\|$$
$$\leq 4^{-n} \cdot 3^{-1} \left(4^{n-m} - 1\right) (\xi + c) + 2 \cdot 4^{-m-1} \eta \|2^m x\|^v \left(1 + a + \cdots + a^{n-m-1}\right)$$
$$\leq 3^{-1} \cdot 4^{-m} (\xi + c) + 2^{m(v-2)-1} (1-a)^{-1} \eta \|x\|^v.$$

Since $v < 2$, we guess that $\{g_n(x)\}$ is a Cauchy sequence. Define

$$g(x) = \lim_{n \to \infty} g_n(x), \qquad x \in E_1.$$

It is easy to verify that g is a quadratic function. If $x = y = 0$, since $g(0) = 0$, it is clear. For $y = 0$, $x \neq 0$, we have

$$g(x + 0) + g(x - 0) - 2g(x) - 2g(0) = 0.$$

Let us now consider the case $x, y \in E_1 \backslash \{0\}$. We get the following estimations

$$\|g_n(x + y) + g_n(x - y) - 2g_n(x) - 2g_n(y)\|$$
$$= 4^{-n} \left\| f\left(2^n(x + y)\right) + f\left(2^n(x - y)\right) - 2f\left(2^n x\right) - 2f\left(2^n y\right) \right\|$$
$$\leq 4^{-n} \xi + 2^{n(v-2)} \eta \left(\|x\|^v + \|y\|^v\right).$$

For $n \to \infty$ we get the equality

$$g(x + y) + g(x - y) - 2g(x) - 2g(y) = 0.$$

Considering the case $x = 0$, $y \neq 0$, we put into that equation $x = y$ getting $g(2x) = 4g(x)$ for $x \in E_1$. Moreover, setting $y = -x$, we obtain $g(-x) = g(x)$ for $x \in E_1$. Therefore,

$$g(y) + g(-y) - 2g(0) - 2g(y) = 0,$$

that is, g is a quadratic function.

The estimation (16.28) may be obtained directly from the inequality (16.26).

To prove the uniqueness assume that there exist two quadratic functions $g_i :$ $E_1 \to E_2$, $i = 1, 2$ such that

$$\|g_i(x) - f(x)\| \le c_i + b_i \|x\|^v,$$

for all $x \in E_1 \setminus \{0\}$, $i = 1, 2$, where $c_i, b_i, i = 1, 2$ are given non-negative constants. Then we have

$$g_i(2^n x) = 4^n g_i(x),$$

for all $x \in E_1$, $n \in \mathbb{N}$, $i = 1, 2$. Now, we obtain for every $x \in E_1 (g_1(0) = g_2(0) = 0)$

$$\|g_1(x) - g_2(x)\| \le 4^{-n} (g_1(2^n x) - f(2^n x)) + \|g_2(2^n x) - f(2^n x)\|$$
$$\le 4^{-n}(c_1 + c_2) + 2^{n(v-2)}(b_1 + b_2)\|x\|^v,$$

whence, if $n \to \infty$, we get $g_1(x) = g_2(x)$ for all $x \in E_1$.

Let L be any continuous linear functional defined on the space E_2. Consider the mapping $\varphi : \mathbb{R} \to \mathbb{R}$ defined as follows:

$$\varphi(t) = L[g(tx)] \quad \text{for all } t \in \mathbb{R} \text{ and } x \in E_1, x \text{ fixed.}$$

It is easy to see that φ is a quadratic function. If we assume that f is measurable, then φ as the pointwise limit of the sequence of measurable functions

$$\varphi_n(t) = 4^{-n} L[f(2^n tx)], \quad n \in \mathbb{N}, \ t \in \mathbb{R}$$

is measurable. Therefore φ as a measurable quadratic function is continuous, so has the form

$$\varphi(t) = t^2 \varphi(1) \quad \text{for } t \in \mathbb{R}.$$

Consequently

$$L[g(tx)] = \varphi(t) = t^2 \varphi(1) = t^2 L[g(x)] = L[t^2 g(x)],$$

whence taking into account that L is any continuous linear functional, we get the property (16.29). This completes the proof of the theorem. □

Theorem 16.6 ([59]). *Let $\delta \ge 0$ be fixed. If a mapping $f : X \to Y$ satisfies the inequality*

$$\|f(x + y) + f(x - y) - 2f(x) - 2f(y)\| \le \delta \tag{16.31}$$

for all $x, y \in X$, then there exists a unique quadratic mapping $Q : X \to Y$ such that

$$\|f(x) - Q(x)\| \le \frac{\delta}{2}$$

for all $x \in X$. Moreover, if f is measurable or if $f(tx)$ is continuous in t for each fixed $x \in X$, then $Q(tx) = t^2 Q(x)$ for all $x \in X$, and $t \in \mathbb{R}$.

Proof. By taking $\xi = \frac{\delta}{2}$, $\eta = \frac{\delta}{2}$, $\upsilon = 0$, and $f(0) = 0$ in Theorem 16.5, the proof is complete. $\qquad\square$

Theorem 16.7 ([146]). *Let $d > 0$, and $\delta \geq 0$ be given. Assume that a mapping $f : X \to Y$ satisfies the inequality (16.31) for all $x, y \in X$ with $\|x\| + \|y\| \geq d$. Then there exists a unique quadratic mapping $Q : X \to Y$ such that*

$$\|f(x) - Q(x)\| \leq \frac{7}{2}\delta \qquad (16.32)$$

for all $x \in X$. If, moreover, f is measurable or $f(tx)$ is continuous in t for each fixed $x \in X$, then $Q(tx) = t^2 Q(x)$ for all $x \in X$, and all $t \in \mathbb{R}$.

Proof. Assume $\|x\| + \|y\| < d$. If $x = y = 0$, then we choose a $z \in X$ with $\|z\| = d$. Otherwise, let $z = (1 + \frac{d}{\|x\|})x$, for $\|x\| \geq \|y\|$ or $z = (1 + \frac{d}{\|y\|})y$, for $\|x\| < \|y\|$. Clearly, we see

$$\|x - z\| + \|y + z\| \geq d, \qquad \|x + z\| + \|y + z\| \geq d,$$
$$\|y + z\| + \|z\| \geq d, \qquad \|x\| + \|y + 2z\| \geq d, \qquad \|x\| + \|z\| \geq d. \qquad (16.33)$$

From (16.31), (16.33), and the relation

$$\begin{aligned}
f(x + y) &+ f(x - y) - 2f(x) - 2f(y) \\
&= f(x + y) + f(x - y - 2z) - 2f(x - z) - 2f(y + z) \\
&\quad + f(x + y + 2z) + f(x - y) - 2f(x + z) - 2f(y + z) \\
&\quad - 2f(y + 2z) - 2f(y) + 4f(y + z) + 4f(z) \\
&\quad - f(x + y + 2z) - f(x - y - 2z) + 2f(x) + 2f(y + 2z) \\
&\quad + 2f(x + z) + 2f(x - z) - 4f(x) - 4f(z),
\end{aligned}$$

we obtain

$$\|f(x + y) + f(x - y) - 2f(x) - 2f(y)\| \leq 7\delta. \qquad (16.34)$$

Obviously, the inequality (16.34) holds true for all $x, y \in X$. According to (16.34) and Theorem 16.5, there exists a unique quadratic mapping $Q : X \to Y$ which satisfies the inequality (16.32) for all $x \in X$. Our last assertion is trivial in view of Theorem 16.5. The proof is now complete. $\qquad\square$

Note. If we define $B = \{(x, y) \in X^2 : \|x\| < d, \|y\| < d\}$ for some $d > 0$, then $\{(x, y) \in X^2 : \|x\| + \|y\| \geq 2d\} \subset X^2 \backslash B$.

Corollary 16.3 ([146]). *Assume that a mapping $f : X \to Y$ satisfies the inequality (16.31) for some $\delta \geq 0$, and for all $(x, y) \in X^2 \backslash B$. Then there exists a unique quadratic mapping $Q : X \to Y$ satisfying the inequality (16.32) for all $x \in X$. If, moreover, f is measurable or $f(tx)$ is continuous in t for each fixed $x \in X$, then $Q(tx) = t^2 Q(x)$ for all $x \in X$, and all $t \in \mathbb{R}$.*

Proof. The proof of this corollary is obtained by similar arguments as in Theorem 16.7. The details are left to the reader. □

16.4. Ulam stability of a generalized quadratic functional equation (16.35) on a restricted domain

This section deals with the Hyers–Ulam stability of a generalized quadratic functional equation

$$f(rx + sy) + rsf(x - y) = rf(x) + sf(y) \tag{16.35}$$

where r, s are non-zero real numbers with $r + s = 1$ on a restricted domain.

Theorem 16.8 ([207]). *Let $d > 0$, and $\delta \geq 0$ be given. Assume that an even mapping $f : X \to Y$ satisfies the inequality*

$$\|f(rx + sy) + rsf(x - y) - rf(x) - sf(y)\| \leq \delta \tag{16.36}$$

for all $x, y \in X$ with $\|x\| + \|y\| \geq d$. Then there exists $K > 0$ such that f satisfies

$$\|f(x + y) + f(x - y) - 2f(x) - 2f(y)\| \leq \frac{4(2 + |r| + |s|)}{|rs|}\delta \tag{16.37}$$

for all $x, y \in X$ with $\|x\| + \|y\| \geq K$.

Proof. Let $x, y \in X$ with $\|x\| + \|y\| \geq 2d$. Then, since $\|x + y\| + \|y\| \geq \max\{\|x\|\, 2\|y\| - \|x\|\}$, we obtain $\|x + y\| + \|y\| \geq d$. So it follows from (16.36) that

$$\|f(rx + y) + rsf(x) - rf(x + y) - sf(y)\| \leq \delta \tag{16.38}$$

for all $x, y \in X$ with $\|x\| + \|y\| \geq 2d$. So

$$\|f(ry + x) + rsf(y) - rf(x + y) - sf(x)\| \leq \delta \tag{16.39}$$

for all $x, y \in X$ with $\|x\| + \|y\| \geq 2d$.

Let $x, y \in X$ with $\|x\| + \|y\| \geq 4d(\frac{1}{|r|} + |1 - \frac{1}{|r|}|)$. We have two cases.

Case 1. $\|y\| > \frac{2d}{|r|}$. Then $\|x\| + \|x + ry\| \geq |r|\,\|y\| \geq 2d$.

Case 2. $\|y\| \leq \frac{2d}{|r|}$. Then we have $\|x\| \geq 2d(\frac{1}{|r|} + 2|1 - \frac{1}{|r|}|)$. So

$$\|x\| + \|x + ry\| \geq 2\|x\| - |r|\,\|y\| \geq 2d\left(\frac{2}{|r|} + 4\left|1 - \frac{1}{|r|}\right| - 1\right) \geq 2d. \tag{16.40}$$

Therefore we get that $\|x\| + \|x + ry\| \geq 2d$ from Cases 1 and 2. Hence by (16.38), we have

$$\|f(r(x+y)+x)+rsf(x)-rf(2x+ry)-sf(x+ry)\| \leq \delta \qquad (16.41)$$

for all $x, y \in X$ with $\|x\| + \|y\| \geq 4d(\frac{1}{|r|} + |1 - \frac{1}{|r|}|)$. Set $M = 4d\left(\frac{1}{|r|} + \left|1 - \frac{1}{|r|}\right|\right)$. Then

$$\|x+y\| + \|x\| \geq \frac{M}{2} \geq 2d, \qquad \text{and} \qquad \|2x\| + \|y\| \geq M \geq 4d \qquad (16.42)$$

for all $x, y \in X$ with $\|x\| + \|y\| \geq M$. From (16.38) and (16.39), we get the following inequalities:

$$\|f(r(x+y)+x)+rsf(x+y)-rf(2x+y)-sf(x)\| \leq \delta,$$

$$\left\|rf(ry+2x)+r^2sf(y)-r^2f(2x+y)-rsf(2x)\right\| \leq \delta|r|,$$

and

$$\left\|sf(ry+x)+rs^2f(y)-rsf(x+y)-s^2f(x)\right\| \leq \delta|s|. \qquad (16.43)$$

Using (16.41), and the above inequalities, we have

$$\|f(2x+y)+2f(x)+f(y)-2f(x+y)-f(2x)\| \leq \frac{2+|r|+|s|}{|rs|}\delta \qquad (16.44)$$

for all $x, y \in X$ with $\|x\| + \|y\| \geq M$. If $x, y \in X$ with $\|x\| + \|y\| \geq 2M$, then $\|x\| + \|y - x\| \geq M$. So it follows from (16.44) that

$$\|f(x+y)+2f(x)+f(y-x)-2f(y)-f(2x)\| \leq \frac{2+|r|+|s|}{|rs|}\delta. \qquad (16.45)$$

Letting $y = 0$ in (16.45), we obtain

$$\|4f(x)-f(2x)-2f(0)\| \leq \frac{2+|r|+|s|}{|rs|}\delta \qquad (16.46)$$

for all $x, y \in X$ with $\|x\| \geq 2M$. Letting $x = 0$ (and $y \in X$ with $\|y\| \geq 2M$) in (16.45), we get $\|f(0)\| \leq (\frac{2+|r|+|s|}{|rs|})\delta$. Therefore it follows from (16.45) and (16.46) that

$$\|f(x+y)+f(y-x)-2f(x)-2f(y)\|$$

$$\leq \|f(x+y)+2f(x)+f(y-x)-2f(y)-f(2x)\|$$

$$+ \|4f(x) - f(2x) - 2f(0)\| + 2\|f(0)\|$$

$$\leq \frac{4(2 + |r| + |s|)}{|rs|}\delta \tag{16.47}$$

for all $x, y \in X$ with $\|x\| \geq 2M$. Since f is even, the inequality (16.47) holds for all $x, y \in X$ with $\|y\| \geq 2M$. Therefore

$$\|f(x + y) + f(x - y) - 2f(x) - 2f(y)\| \leq \frac{4(2 + |r| + |s|)}{|rs|}\delta \tag{16.48}$$

for all $x, y \in X$ with $\|x\| + \|y\| \geq 4M$. This completes the proof by letting $K = 4M$. The proof is now complete. $\qquad \square$

Theorem 16.9 ([207]). *Let $d > 0$, and $\delta \geq 0$ be given. Assume that an even mapping $f : X \to Y$ satisfies the inequality (16.36) for all $x, y \in X$ with $\|x\| + \|y\| \geq d$. Then f satisfies*

$$\|f(x + y) + f(x - y) - 2f(x) - 2f(y)\| \leq \frac{19(2 + |r| + |s|)}{|rs|}\delta \tag{16.49}$$

for all $x, y \in X$.

Proof. By Theorem 16.8, there exists $K > 0$ such that f satisfies (16.37) for all $x, y \in X$ with $\|x\| + \|y\| \geq K$ and $\|f(0)\| \leq (\frac{2 + |r| + |s|}{|rs|})\delta$ (see the proof of Theorem 16.8). Using Theorem 2 of [271], we get that

$$\|f(x + y) + f(x - y) - 2f(x) - 2f(y)\| \leq \frac{18(2 + |r| + |s|)}{|rs|}\delta + \|f(0)\|$$

$$\leq \frac{19(2 + |r| + |s|)}{|rs|}\delta \tag{16.50}$$

for all $x, y \in X$. $\qquad \square$

Theorem 16.10 ([207]). *Let $d > 0$, and $\delta \geq 0$ be given. Assume that an even mapping $f : X \to Y$ satisfies the inequality (16.36) for all $x, y \in X$ with $\|x\| + \|y\| \geq d$. Then there exists a unique quadratic mapping $Q : X \to Y$ such that $Q(x) = \lim_{n \to \infty} 4^{-n} f(2^n x)$, and*

$$\|f(x) - Q(x)\| \leq \frac{19(2 + |r| + |s|)}{2|rs|}\delta \tag{16.51}$$

for all $x \in X$.

Proof. The result follows from Theorems 16.8 and 16.9. $\qquad \square$

16.5. Hyers–Ulam stability of quadratic functional equation (4.1) on a restricted domain with improved bound

In this section, the Hyers–Ulam stability result obtained in Section 16.2 is improved with better upper bound. Throughout this section, X is a real normed space, and Y is a real Banach space in the case of functional inequalities, as well as let X and Y be real linear spaces for functional equations.

Theorem 16.11 ([271]). *Let $d > 0$, and $\delta \geq 0$ be fixed. If a mapping $f : X \to Y$ satisfies the quadratic inequality (16.31) for all $x, y \in X$, with $\|x\| + \|y\| \geq d$, then there exists a unique quadratic mapping $Q : X \to Y$ such that*

$$\|f(x) - Q(x)\| \leq \frac{5}{2}\delta \tag{16.52}$$

for all $x \in X$.

Proof. Assume $\|x\| + \|y\| < d$. If $x = y = 0$, then we choose a $t \in X$ with $\|t\| = d$. Otherwise, let

$$t = \left(1 + \frac{d}{\|x\|}\right) x, \qquad \text{if } \|x\| \geq \|y\|;$$

and

$$t = \left(1 + \frac{d}{\|y\|}\right) y, \qquad \text{if } \|x\| \leq \|y\|.$$

We note that $\|t\| = \|x\| + d > d$, if $\|x\| \geq \|y\|$; $\|t\| = \|y\| + d > d$, if $\|x\| \leq \|y\|$. Clearly, we see

$$\|x - t\| + \|y + t\| \geq 2\|t\| - (\|x\| + \|y\|) \geq d,$$
$$\|x - y\| + \|2t\| \geq \|x - y\| + 2d \geq d,$$
$$\|x + t\| + \|-y + t\| \geq d, \tag{16.53}$$
$$\|x\| + \|t\| \geq d, \quad \|t\| + \|y\| \geq d, \quad \|t\| + \|t\| \geq d.$$

These inequalities (16.53) come from the corresponding substitutions attached between the right-hand sided parentheses of the following functional identity.

Besides from (16.31) with $x = y = 0$ we get that $\|f(0)\| \leq \frac{\delta}{2}$. Therefore from (16.31), (16.53), and the new functional identity

$$2[f(x + y) + f(x - y) - 2f(x) - 2f(y) - f(0)]$$
$$= [f(x + y) + f(x - y - 2t) - 2f(x - t) - 2f(y + t)]$$
$$\text{(with } x - t \text{ on } x \text{ and } y + t \text{ on } y)$$
$$- [f(x - y - 2t) + f(x - y + 2t) - 2f(x - y) - 2f(2t)]$$
$$\text{(with } x - y \text{ on } x \text{ and } 2t \text{ on } y)$$

$$+ [f(x - y + 2t) + f(x + y) - 2f(x + t) - 2f(-y + t)]$$

(with $x + t$ on x and $-y + t$ on y)

$$+ 2[f(x + t) + f(x - t) - 2f(x) - 2f(t)]$$

(with x on x and t on y)

$$+ 2[f(t + y) + f(t - y) - 2f(t) - 2f(y)]$$

(with t on x and y on y)

$$- 2[f(2t) + f(0) - 2f(t) - 2f(t)] \qquad \text{(with } t \text{ on } x \text{ and } t \text{ on } y)$$

we obtain

$$2 \, \| f(x + y) + f(x - y) - 2f(x) - 2f(y) - f(0) \|$$
$$\leq \delta + \delta + \delta + 2\delta + 2\delta + 2\delta = 9\delta,$$

or

$$\| f(x + y) + f(x - y) - 2f(x) - 2f(y) \| \leq \frac{9}{2}\delta + \| f(0) \| \leq 5\delta. \qquad (16.54)$$

Applying now Theorem 16.5, and the above inequality (16.54), one gets that there exists a unique quadratic mapping $Q : X \to Y$ that satisfies the quadratic equation (16.35) and the inequality (16.52), such that $Q(x) = \lim_{n \to \infty} 2^{-2n} f(2^n x)$, completing the proof of Theorem 16.11. $\qquad \square$

Note. If we define $S_2 = \{(x, y) \in X^2 : \|x\|, \|y\| < d\}$ for some $d > 0$, then $\{(x, y) \in X^2 : \|x\| + \|y\| \geq 2d\} \subset X^2 \backslash S_2$.

Remark 16.1. The inequalities (16.52) and (16.54) are sharper than the corresponding inequalities obtained in Section 16.2, where the right-hand sides are equal to $\frac{7}{2}\delta$ and 7δ, respectively.

Corollary 16.4 ([271]). *If we assume that a mapping $f : X \to Y$ satisfies the quadratic inequality (16.31) for some fixed $\delta \geq 0$, and for all $(x, y) \in X^2 \backslash S_2$, then there exists a unique quadratic mapping $Q : X \to Y$ satisfying (16.52) for all $x \in X$.*

Proof. The proof is similar to that of Theorem 16.11. The details are left to the reader. $\qquad \square$

16.6. Hyers–Ulam stability of mixed type functional equation involving three variables in a restricted domain

This section deals with the Hyers–Ulam stability of a mixed type functional equation involving three variables in a restricted domain. Throughout this section, X is a real normed space, and Y is a real Banach space in the case of functional inequalities, as well as let X and Y be real linear spaces for functional equations.

Definition 16.4 ([271]). A mapping $M : X \to Y$ is called additive (respectively, quadratic) in X if M satisfies the functional equation of two types

$$M\left(\sum_{i=1}^{3} x_i\right) + \sum_{i=1}^{3} M(x_i) = \sum_{1 \le i < j \le 3} M(x_i + x_j) \qquad (16.55)$$

for all $x_i \in X$ $(i = 1, 2, 3)$.

We note that all the real mappings $M : \mathbb{R} \to \mathbb{R}$ of the two types: $M(x) = \alpha x$ or $M(x) = \beta x^2$ satisfy (16.55) for all $x \in \mathbb{R}$, and all arbitrary but fixed $\alpha, \beta \in \mathbb{R}$. We also note that the mapping $M : X \to Y$ may be called mixed type as it is either additive or quadratic.

16.6.1. *Hyers–Ulam stability of functional equation* (16.55) *of two types*

Lemma 16.2 ([271]). *Assume that a mapping* $f : X \to Y$ *satisfies the inequality*

$$\left\| f\left(\sum_{i=1}^{3} x_i\right) - f(x_1 + x_2) - f(x_1 + x_3) - f(x_2 + x_3) + \sum_{i=1}^{3} f(x_i) \right\| \le \delta \qquad (16.56)$$

for some fixed $\delta \ge 0$, *and for all* $x_i \in X$ $(i = 1, 2, 3)$. *It then holds that*

$$\left\| f(x) - \frac{2^n + 1}{2^{2n+1}} f(2^n x) + \frac{2^n - 1}{2^{2n+1}} f(-2^n x) \right\| \le 3\delta \sum_{i=1}^{n} 2^{-i} \left(= 3\delta \left(1 - 2^{-n}\right)\right) \qquad (16.57)$$

for all $x \in X$, *and* $n \in \mathbb{N}$.

Proof. Let us denote

$$a_i = \frac{2^i + 1}{2^{2i+1}}, \quad A_i(x) = 3f\left(2^{i-1} x\right) + f\left(-2^{i-1} x\right) - f(2^i x),$$

$$b_i = -\frac{2^i - 1}{2^{2i+1}}, \quad B_i(x) = 3f\left(-2^{i-1} x\right) + f\left(2^{i-1} x\right) - f(-2^i x),$$

$$T_i(x) = a_i f\left(2^i x\right) + b_i f(-2^i x), \quad S_n(x) = T_0(x) - T_n(x),$$

such that $T_0(x) = f(x)$, for all $x \in X$, $i \in \mathbb{N}$.

We note that

$$a_{i-1} = 3a_i + b_i, \quad b_{i-1} = a_i + 3b_i, \quad \text{hold for any } i \in \mathbb{N}.$$

From these identities, we obtain that

$$
\begin{aligned}
T_{i-1}(x) &= a_{i-1}f\left(2^{i-1}x\right) + b_{i-1}f\left(-2^{i-1}x\right) - T_i(x) \\
&= (3a_i + b_i)f\left(2^{i-1}x\right) + (a_i + 3b_i)f\left(-2^{i-1}x\right) - a_i f\left(2^i x\right) - b_i f\left(-2^i x\right) \\
&= a_i\left[3f\left(2^{i-1}x\right) + f\left(-2^{i-1}x\right) - f\left(2^i x\right)\right] \\
&\quad + b_i[3f\left(-2^{i-1}x\right) + f\left(2^{i-1}x\right) - f\left(-2^i x\right)],
\end{aligned}
$$

or the formula

$$T_{i-1}(x) - T_i(x) = a_i A_i(x) + b_i B_i(x) \tag{16.58}$$

holds for any $i \in \mathbb{N}$.

We note that

$$S_n(x) = T_0(x) - T_n(x) = \sum_{i=1}^{n}[T_{i-1}(x) - T_i(x)].$$

Therefore from this formula, and (16.58), one obtains the new formula

$$S_n(x) = \sum_{i=1}^{n}\left[a_i A_i(x) + b_i B_i(x)\right]. \tag{16.59}$$

Replacing $x_i = 0$ $(i = 1, 2, 3)$ in (16.56), one gets

$$\|f(0)\| \leq \delta. \tag{16.60}$$

Setting $x_1 = x$, $x_2 = x$, $x_3 = x$ in (16.56), we find from (16.60) that

$$\|3f(x) + f(-x) - f(2x) - 2f(0)\| \leq \delta,$$

or

$$\|3f(x) + f(-x) - f(2x)\| \leq 3\delta \tag{16.61}$$

holds for all $x \in X$.

Substituting $-x$ for x in (16.61), one obtains

$$\|3f(-x) + f(x) - f(-2x)\| \leq 3\delta. \tag{16.62}$$

Placing $2^{i-1}x$ on x in (16.61) and (16.62) we get

$$\|A_i(x)\| \leq 3\delta, \quad \text{and} \quad \|B_i(x)\| \leq 3\delta \tag{16.63}$$

for all $i \in \mathbb{N}$. Thus from the formula (16.59), the inequalities (16.63), and the triangle inequality we prove

$$\|S_n(x)\| \leq \sum_{i=1}^{n} [|a_i|\,\|A_i(x)\| + |b_i|\,\|B_i(x)\|]$$

$$\leq 3\delta \sum_{i=1}^{n} \left[\frac{2^i + 1}{2^{2i+1}} + \frac{2^i - 1}{2^{2i+1}}\right] = 3\delta \sum_{i=1}^{n} 2^{-i} = 3\delta(1 - 2^{-n}), \tag{16.64}$$

for all $x \in X$, and $n \in \mathbb{N}$, completing the proof of this lemma. $\qquad\square$

16.6.2. Hyers–Ulam stability of functional equation (16.55) on a restricted domain

Theorem 16.12 ([271]). *Let $d > 0$, and $\delta \geq 0$ be fixed. If an approximately even mapping $f : X \to Y$ satisfies the quadratic inequality (16.56) for all $x_i \in X$, $(i = 1,2,3)$ with $\sum_{i=1}^{3} \|x_i\| \geq d$, then there exists a unique quadratic mapping $Q : X \to Y$, such that*

$$\|f(x) - Q(x)\| \leq 15\delta \tag{16.65}$$

for all $x \in X$.

Proof. Assume $\sum_{i=1}^{3} \|x_i\| < d$. If $x_i = 0$ $(i = 1,2,3)$, then we choose a $t \in X$ with $\|t\| \geq 2d$. Otherwise, choose a $t \in X$ with $\|t\| \geq d$; clearly

$$\|x_1 - t\| + \|x_2\| + \|x_3 + t\| \geq 2\|t\| - \sum_{i=1}^{3} \|x_i\| \geq d,$$

$$\|x_1\| + \|x_2\| + \|-t\| \geq d, \quad \|x_2\| + \|x_3\| + \|t\| \geq d, \tag{16.66}$$

$$\|x_2\| + \|-t\| + \|t\| = 2\|t\| + \|x_2\| \geq d.$$

Besides from (16.56) with $x_i = 0$ $(i = 1,2,3)$, we obtain that $\|f(0)\| \leq \delta$.

Therefore from (16.56), (16.66), and the new functional identity

$$f\left(\sum_{i=1}^{3} x_i\right) - f(x_1 + x_2) - f(x_1 + x_3) - f(x_2 + x_3) + \sum_{i=1}^{3} f(x_i) + f(0)$$

$$= [f(x_1 + x_2 + x_3) - f(x_1 + x_2 - t) - f(x_1 + x_3) - f(x_2 + x_3 + t)$$

$$\quad + f(x_1 - t) + f(x_2) + f(x_3 + t)]$$

$$\quad \text{(with } x_1 - t \text{ on } x_1,\ x_2 \text{ on } x_2 \text{ and } x_3 + t \text{ on } x_3)$$

$$\quad + [f(x_1 + x_2 - t) - f(x_1 + x_2) - f(x_1 - t) - f(x_2 - t) + f(x_1)$$

$$\quad + f(x_2) + f(-t)]$$

(with x_1 on x_1, x_2 on x_2 and $-t$ on x_3)

$$+ [f(x_2 + x_3 + t) - f(x_2 + x_3) - f(x_2 + t) - f(x_3 + t) + f(x_2)$$
$$+ f(x_3) + f(t)]$$

(with x_2 on x_1, x_3 on x_2 and t on x_3)

$$- [f(x_2) - f(x_2 - t) - f(x_2 + t) - f(0) + f(x_2) + f(-t) + f(t)]$$

(with x_2 on x_1, $-t$ on x_2 and t on x_3),

we obtain

$$\left\| f\left(\sum_{i=1}^{3} x_i\right) - f(x_1 + x_2) - f(x_1 + x_3) - f(x_2 + x_3) \right.$$
$$\left. + \sum_{i=1}^{3} f(x_i) + f(0) \right\| \le \delta + \delta + \delta + \delta = 4\delta,$$

or

$$\left\| f\left(\sum_{i=1}^{3} x_i\right) - f(x_1 + x_2) - f(x_1 + x_3) - f(x_2 + x_3) + \sum_{i=1}^{3} f(x_i) \right\|$$

$$\le 4\delta + \|f(0)\| \le 5\delta. \tag{16.67}$$

Hence there exists a unique quadratic mapping $Q : X \to Y$ that satisfies the quadratic equation (16.55), and the inequality (16.65), completing the proof of the theorem. □

Note. If we define $S_3 = \{(x_1, x_2, x_3) \in X^3 : \|x_i\| < d, i = 1, 2, 3\}$ for some fixed $d > 0$, then $\{(x_1, x_2, x_3) \in X^3 : \sum_{i=1}^{3} \|x_i\| \ge 3d\} \subset X^3 \backslash S_3$.

Corollary 16.5 ([271]). *If we assume that an approximately even mapping $f : X \to Y$ satisfies the inequality (16.56) for some fixed $\delta \ge 0$, and for all $(x_1, x_2, x_3) \in X^3 \backslash S_3$, then there exists a unique quadratic mapping $Q : X \to Y$ satisfying (16.65) for all $x \in X$.*

Proof. The proof is similar to that of Theorem 16.12, and the details are left to the reader. □

Theorem 16.13 ([271]). *Let $d > 0$, and $\delta \ge 0$ be fixed. If an approximately odd mapping $f : X \to Y$ satisfies the additive inequality (16.56) for all $x_i \in X$ ($i = 1, 2, 3$) with $\sum_{i=1}^{3} \|x_i\| \ge d$, then there exists a unique additive mapping $A : X \to Y$, such that*

$$\|f(x) - A(x)\| \le 15\delta \tag{16.68}$$

for all $x \in X$.

Proof. The proof is obtained by similar arguments as in Theorem 16.12, and the details are left to the reader. □

Remark 16.2. The inequalities (16.67) and (16.68) are sharper than the corresponding inequalities obtained in Section 16.2, where the right-hand sides are equal to 7δ and 21δ, respectively.

Corollary 16.6 ([271]). *If we assume that an approximately odd mapping $f : X \to Y$ satisfies the inequality (16.56) for some fixed $\delta \geq 0$, and for all $(x_1, x_2, x_3) \in X^3 \backslash S_3$, then there exists a unique additive mapping $A : X \to Y$ satisfying (16.68) for all $x \in X$.*

Proof. The proof is similar to that of Theorem 16.13, and the details are left to the reader. □

16.7. Ulam stability of mixed type functional equation involving four variables in a restricted domain

This section deals with the Hyers–Ulam stability of a mixed type functional equation involving four variables in a restricted domain. Throughout this section, X is a real normed space, and Y is a real Banach space in the case of functional inequalities, as well as X and Y are real linear spaces for functional equations. In the following definition, the functional equation (16.55) is generalized.

Definition 16.5 ([271]). A mapping $M : X \to Y$ is called additive (respectively quadratic) in $\mathbb{R}^4$ if M satisfies the functional equation of two types

$$M\left(\sum_{i=1}^{4} x_i\right) + \sum_{1 \leq i < j \leq 4} M(x_i + x_j) - \sum_{i=1}^{4} M(x_i) + \sum_{1 \leq i < j < k \leq 4} M(x_i + x_j + x_k) \quad (16.69)$$

for all $x_i \in X$ $(i = 1, 2, 3, 4)$.

16.7.1. *Ulam stability of functional equation (16.69)*

Theorem 16.14 ([271]). *Assume an approximately even mapping $f : X \to Y$ satisfies the following quadratic inequality*

$$\left\| f\left(\sum_{i=1}^{4} x_i\right) + \sum_{1 \leq i < j \leq 4} f(x_i + x_j) - \sum_{i=1}^{4} f(x_i) \right.$$

$$\left. - \sum_{1 \leq i < j < k \leq 4} f(x_i + x_j + x_k) \right\| \leq \delta, \quad (16.70)$$

for some fixed $\delta \geq 0$, and $\theta \geq 0$ and for all $x_i \in X$ $(i = 1, 2, 3, 4)$. Then there exist a unique quadratic mapping $Q : X \to Y$ which satisfies the quadratic equation

(16.69), *and the inequality*

$$\|f(x) - Q(x)\| \leq \delta + \frac{5}{6}\theta \tag{16.71}$$

for all $x \in X$.

Proof. Replacing $x_i = 0$ $(i = 1, 2, 3, 4)$ in (16.70), we find $\|f(0)\| \leq \delta$. Thus, substituting $x_i = x$, $(i = 1, 2)$, and $x_j = -x$, $j = 3, 4$ in (16.70), we obtain that

$$\|4f(x) + 4f(-x) - f(2x) - f(-2x)\| \leq 6\delta \tag{16.72}$$

for all $x \in X$. Therefore from (16.72), (16.5) for approximately even mappings, the quadratic inequality (16.70), and the triangle inequality, we obtain that

$$2\|4f(x) - f(2x)\| \leq \|4f(x) + 4f(-x) - f(2x) - f(-2x)\|$$
$$+ \|-4\left[f(-x) - f(x)\right]\| + \|f(-2x) - f(2x)\|$$
$$\leq 6\delta + 4\theta + \theta = 6\delta + 5\theta,$$

or

$$\|f(x) - 2^{-2}f(2x)\| \leq \frac{3}{4}\delta + \frac{5}{8}\theta \left(= \left(\delta + \frac{5}{6}\theta\right)\left(1 - 2^{-2}\right)\right).$$

Using mathematical induction on a positive integer n, we can prove that

$$\|f(x) - 2^{-2n}f\left(2^n x\right)\| \leq \left(\delta + \frac{5}{6}\theta\right)\left(1 - 2^{-2n}\right), \tag{16.73}$$

holds for all $x \in X$. Similarly from (16.69), we get, by induction on n, that

$$Q(x) = 2^{-2n}Q\left(2^n x\right), \tag{16.74}$$

holds for any $n \in \mathbb{N}$, and all $x \in X$.

By (16.73), for $n \geq m > 0$, and $h = 2^m x$, we have

$$\|2^{-2n}f\left(2^n x\right) - 2^{-2m}f\left(2^m x\right)\| = 2^{-2m}\|2^{-2(n-m)}f(2^{n-m}.2^m x) - f(2^m x)\|$$
$$= 2^{-2m}\|2^{-2(n-m)}f(2^{n-m}h) - f(h)\|$$
$$\leq \left(\delta + \frac{5}{6}\theta\right)2^{-2m} \to 0, \qquad \text{as } m \to \infty. \tag{16.75}$$

From (16.75), and the completeness of Y, we get that the Cauchy sequence $\{2^{-2n}f\left(2^n x\right)\}$ converges. Therefore, applying direct method to the definition of Q such that $Q(x) = \lim_{n \to \infty} 2^{-2n}f\left(2^n x\right)$ holds for all $x \in X$. From the quadratic inequality (16.70), it follows that

$$\left\| Q\left(\sum_{i=1}^{4} x_i\right) + \sum_{1 \leq i < j \leq 4} Q(x_i + x_j) - \sum_{i=1}^{4} Q(x_i) \right.$$

$$\left. - \sum_{1 \leq i < j < k \leq 4} Q(x_i + x_j + x_k) \right\| \leq 2^{-2n} \delta \to 0, \quad \text{as } n \to \infty,$$

for all $x_i \in X$ $(i = 1, 2, 3, 4)$. Thus, it is obvious that Q satisfies the quadratic equation (16.69). Analogously, by (16.5), we can show that $Q(0) = 0$ (with $x_i = 0, i = 1, 2, 3, 4$) in (16.69), and that Q is even from (16.5) with $2^n x$ in place of x, $\|Q(x) - Q(-x)\| \leq 2^{-2n}\theta \to 0$, as $n \to \infty$, or $Q(-x) = Q(x)$.

According to (16.73), one gets that the inequality (16.71) holds. Assume now that there is another quadratic mapping $Q' : X \to Y$ which satisfies the quadratic equation (16.69), the formula (16.74), and the inequality (16.71). Therefore

$$\|Q(x) - Q'(x)\| = 2^{-2n} \|Q(2^n x) - Q'(2^n x)\|$$

$$\leq 2^{-2n} \left(\|Q(2^n x) - f(2^n x)\| + \|f(2^n x) - Q'(2^n x)\| \right)$$

$$\leq 2\left(\delta + \frac{5}{6}\theta\right) 2^{-2n} \to 0, \quad \text{as } n \to \infty,$$

or

$$Q'(x) = Q(x),$$

for all $x_i \in X$, completing the proof of the theorem. $\qquad \square$

16.7.2. *Hyers–Ulam stability of functional equation* (16.69) *on a restricted domain*

Next, we deal with the Hyers–Ulam stability for more general equations of two types on a restricted domain.

Theorem 16.15 ([271]). *Let $d > 0$, $\delta \geq 0$, and $\theta \geq 0$ be fixed. If an approximately even mapping $f : X \to Y$ satisfies the quadratic inequality (16.70) for all $x_i \in X$, $(i = 1, 2, 3, 4)$ with $\sum_{i=1}^{4} \|x_i\| \geq d$, then there exists a unique quadratic mapping $Q : X \to Y$ such that*

$$\|f(x) - Q(x)\| \leq 5\left(\delta + \frac{\theta}{6}\right) \tag{16.76}$$

for all $x \in X$.

Proof. Assume $\sum_{i=1}^{4} \|x_i\| < d$. We choose a $t \in X$ with $\|t\| \geq 2d$. Clearly, we see

$$\|x_1 - t\| + \|x_2\| + \|x_3 + t\| + \|x_4\| \geq 2\|t\| - \sum_{i=1}^{4} \|x - i\| \geq d,$$

$$\|x_1\| + \|x_2\| + \|x_4\| + \|-t\| = \|t\| + (\|x_1\| + \|x_2\| + \|x_4\|) \geq d,$$
$$\|x_2\| + \|x_3\| + \|x_4\| + \|t\| \geq d,$$
$$\|x_2\| + \|x_4\| + \|t\| + \|-t\| \geq d. \tag{16.77}$$

Besides from (16.70) with $x_i = 0$ ($i = 1, 2, 3, 4$), we get that $\|f(0)\| \leq \delta$. Therefore from (16.70), (16.77), and the following new functional identity

$$f\left(\sum_{i=1}^{4} x_i\right) - f(x_1 + x_2 + x_3) - f(x_1 + x_2 + x_4) - f(x_1 + x_3 + x_4)$$

$$- f(x_2 + x_3 + x_4) + f(x_1 + x_2) + f(x_1 + x_3) + f(x_1 + x_4)$$

$$+ f(x_2 + x_3) + f(x_2 + x_4) + f(x_3 + x_4) - \sum_{i=1}^{4} f(x_i) - f(0)$$

$$= f\left(\sum_{i=1}^{4} x_i\right) - f(x_1 + x_2 + x_3) - f(x_1 + x_2 + x_4 - t)$$

$$- f(x_1 + x_3 + x_4) - f(x_2 + x_3 + x_4 + t) + f(x_1 + x_2 - t)$$

$$+ f(x_1 + x_3) + f(x_1 + x_4 - t) + f(x_2 + x_3 + t) + f(x_2 + x_4)$$

$$+ f(x_3 + x_4 + t) - f(x_1 - t) - f(x_2) - f(x_3 + t) - f(x_4)$$

(with $x_1 - t$ on x_1, x_2 on x_2, $x_3 + t$ on x_3 and x_4 on x_4)

$$+ [f(x_1 + x_2 + x_4 - t) - f(x_1 + x_2 + x_4) - f(x_1 + x_2 - t)$$

$$- f(x_1 + x_4 - t) - f(x_2 + x_4 - t) + f(x_1 + x_2 0 + f(x_1 + x_4)$$

$$+ f(x_1 - t) + f(x_2 + x_4) + f(x_2 - t) + f(x_4 - t) - f(x_1)$$

$$- f(x_2) - f(x_4) - f(-t)]$$

(with x_1 on x_1, x_2 on x_2, x_4 on x_3 and $-t$ on x_4)

$$+ [f(x_2 + x_3 + x_4 + t) - f(x_2 + x_3 + x_4) - f(x_2 + x_3 + t)$$

$$- f(x_2 + x_4 + t) - f(x_3 + x_4 + t) + f(x_2 + x_3) + f(x_2 + x_4)$$

$$+ f(x_2 + t) + f(x_3 + x_4) + f(x_3 + t) + f(x_4 + t) - f(x_2)$$

$$- f(x_3) - f(x_4) - f(t)]$$

(with x_2 on x_1, x_3 on x_2, x_4 on x_3 and t on x_4)

$$- [f(x_2 + x_4) - f(x_2 + x_4 + t) - f(x_2 + x_4 - t) - f(x_2) - f(x_4)$$

$$+ f(x_2 + x_4) + f(x_2 + t) + f(x_2 - t) + f(x_4 + t) + f(x_4 - t)$$

$$+ f(0) - f(x_2) - f(x_4) - f(t) - f(-t)]$$

(with x_2 on x_1, x_4 on x_2, t on x_3 and $-t$ on x_4), we obtain

$$\left\| f\left(\sum_{i=1}^{4} x_i\right) + \sum_{1\le i<j\le 4} f(x_i + x_j) - \sum_{i=1}^{4} f(x_i) \right.$$

$$\left. - \sum_{1\le i<j<k\le 4} f(x_i + x_j + x_k) \right\| \le 4\delta + \|f(0)\| \le 5\delta. \tag{16.78}$$

Applying Theorem 16.14, and the inequality (16.78), we prove that there exists a unique quadratic mapping $Q : X \to Y$ that satisfies the quadratic equation (16.69), and the inequality (16.76), completing the proof of the theorem. $\square$

Note. If we define $S_4 = \{(x_1, x_2, x_3, x_4) \in X^4 : \|x_i\| < d, i = 1, 2, 3, 4\}$ for some fixed $d > 0$, then $\{(x_1, x_2, x_3, x_4) \in X^4 : \sum_{i=1}^{4} \|x_i\| \ge 4d\} \subset X^4 \backslash S_4$.

Corollary 16.7 ([271]). *If we assume that an approximately even mapping $f : X \to Y$ satisfies the inequality (16.70) for some fixed $\delta \ge 0$, and $\theta \ge 0$, and for all $(x_1, x_2, x_3, x_4) \in X^4 \backslash S_4$, then there exists a unique quadratic mapping $Q : X \to Y$ satisfying (16.76) for all $x \in X$.*

Proof. The proof is similar to that of Theorem 16.15, and the details are left to the reader. $\square$

16.8. Hyers–Ulam stability of quadratic functional equation (4.1) in a set of Lebesgue measure zero

Let $\mathbb{R}$ be the set of real numbers, Y be a Banach space, and $f : \mathbb{R} \to Y$. The following theorem proves the Hyers–Ulam stability for the quadratic functional inequality

$$\|f(x + y) + f(x - y) - 2f(x) - 2f(y)\| \le \epsilon$$

for all $(x, y) \in \Omega$, where $\Omega \subset \mathbb{R}^2$ is of Lebesgue measure 0.

Throughout this section, X and Y is a real normed space, and a real Banach space, respectively.

Theorem 16.16 ([56]). *Let $d > 0$. Suppose that $f : X \to Y$ satisfies the functional inequality*

$$\|f(x + y) + f(x - y) - 2f(x) - 2f(y)\| \le \epsilon \tag{16.79}$$

for all $x, y \in X$ with $\|x\| + \|y\| \ge d$. Then there exists a unique quadratic mapping $Q : X \to Y$ such that

$$\|f(x) - Q(x)\| \le 3\epsilon \tag{16.80}$$

for all $x \in X$.

We prove that the result of Theorem 16.16 holds by constructing a subset Ω of $\mathbb{R}^2$ of Lebesgue measure $m(\Omega) = 0$ based on the Baire category theorem.

Let $\Omega \subset X^2$. In the following theorem, we assume that Ω satisfies the condition: For given $x, y \in X$, there exists $t \in X$ such that

$$(C1) \qquad \{(t,t), (t,y), (x,t), (x+t, -y+t), (x-y, 2t), (x-t, y+t)\} \subset \Omega.$$

Theorem 16.17 ([56]). *Let $\epsilon \geq 0$ be fixed. Suppose that $f : X \to Y$ satisfies the functional inequality*

$$\|f(x+y) + f(x-y) - 2f(x) - 2f(y)\| \leq \epsilon \qquad (16.81)$$

for all $(x,y) \in \Omega$. Then there exists a unique quadratic mapping $Q : X \to Y$ such that

$$\|f(x) - Q(x)\| \leq 3\epsilon \qquad (16.82)$$

for all $x \in X$.

Proof. Let $D(x,y) = f(x+y) + f(x-y) - 2f9x) - 2f(y)$. Since Ω satisfies $(C1)$, for given $x, y \in X$, there exists $t \in X$ such that

$$\|D(t,t)\| \leq \epsilon, \quad \|D(t,y)\| \leq \epsilon, \quad \|D(x,t)\| \leq \epsilon, \quad \|D(x+t, -y+t)\| \leq \epsilon$$
$$\|D(x-y, 2t)\| \leq \epsilon, \quad \|D(x-t, y+t)\| \leq \epsilon.$$

Thus, using the triangle inequality we have

$$\|f(x+y) + f(x-y) - 2f(x) - 2f(y) - f(0)\|$$
$$= \left\| -D(t,t) + D(t,y) + D(x,t) + \frac{1}{2}D(x+t, -y+t) \right.$$
$$\left. - \frac{1}{2}D(x-y, 2t) + \frac{1}{2}D(x-t, y+t) \right\|$$
$$\leq \epsilon + \epsilon + \epsilon + \frac{1}{2}\epsilon + \frac{1}{2}\epsilon + \frac{1}{2}\epsilon = \frac{9}{2}\epsilon \qquad (16.83)$$

for all $x, y \in X$. Putting $x = 0$ in (16.83), we have $\|f(0)\| \leq \frac{3}{2}\epsilon$. Using the triangle inequality with (16.83), we obtain

$$\|f(x+y) + f(x-y) - 2f(x) - 2f(y)\| \leq \|f(0)\| + \frac{9}{2}\epsilon \leq 6\epsilon \qquad (16.84)$$

for all $x, y \in X$. Hence, there exists a unique quadratic mapping $Q : X \to Y$ such that

$$\|f(x) - Q(x)\| \leq 3\epsilon \qquad (16.85)$$

for all $x \in X$. This completes the proof. $\qquad \square$

Corollary 16.8 ([56]). *Suppose that* $f : X \to Y$ *satisfy the functional equation*

$$f(x + y) + f(x - y) = 2f(x) + 2f(y) \qquad (16.86)$$

for all $(x, y) \in \Omega$. *Then equation* (16.86) *holds for all* $x, y \in X$.

Proof. The proof is obtained by taking $\epsilon = 0$ in Theorem 16.17. $\qquad \square$

Now, we construct a set Ω of measure zero satisfying the condition $(C1)$ when $X = \mathbb{R}$. From now on, we identify $\mathbb{R}^2$ with $\mathbb{C}$. The following lemma is a crucial key of our construction.

Lemma 16.3 ([213]). *The set of* $\mathbb{R}$ *of real numbers can be partitioned as*

$$\mathbb{R} = F \cup K,$$

where F *is of first Baire category, that is,* F *is a countable union of nowhere dense subsets of* $\mathbb{R}$, *and* K *is of Lebesgue measure* 0.

Proof. Let $a_1, a_2, \ldots$ be an enumeration of the set of rational numbers (or of any countable dense subset of the line). Let I_{ij} be the open interval with center a_i, and length $\frac{1}{2^{i+j}}$. Let $G_j = \bigcup_{i=1}^{\infty} I_{ij} (j = 1, 2, \ldots)$, and $K = \bigcap_{j=1}^{\infty} G_j$. For any $\epsilon > 0$ we can choose j so that $\frac{1}{2^j} < \epsilon$. Then $K \subset \bigcup_i I_{ij}$, and $\sum_i |I_{ij}| = \sum_i \frac{1}{2^{i+j}} = \frac{1}{2^j} < \epsilon$. Hence K is a null set. On the other hand, G_j is a dense open subset of $\mathbb{R}$, since it is the union of a sequence of open intervals and it includes all rational points. Therefore its complement G'_j is nowhere dense, and $F = K' = \bigcup_j G'_j$ is of first category. This completes the proof. $\qquad \square$

Lemma 16.4 ([56]). *Let* K *be a subset of* $\mathbb{R}$ *of measure* 0 *such that* $K^c = R \backslash K$ *is of first Baire category. Then, for any countable subsets* $U \subset \mathbb{R}$, $V \subset \mathbb{R} \backslash \{0\}$, *and* $M > 0$, *there exists* $\lambda \geq M$ *such that*

$$U + \lambda V = \{u + \lambda v : u \in U, v \in V\} \subset K. \qquad (16.87)$$

Proof. Let $U = \{u_1, u_2, u_3, \ldots\}$ and $V = \{v_1, v_2, v_3, \ldots\}$, and $K^c_{m,n} = v_m^{-1}(K^c - u_n)$, $m, n = 1, 2, 3, \ldots$. Then, since K^c is of first Baire category, $K^c_{m,n}$ are also of first Baire category for all $m, n = 1, 2, 3, \ldots$. Since each $K^c_{m,n}$ consists of a countable union of nowhere dense subsets, by the Baire category theorem, countable many of them cannot cover $[M, \infty)$, i.e.,

$$\bigcap_{m,n=1}^{\infty} K^c_{m,n} \not\supset [M, \infty).$$

Thus, there exists $\lambda \geq d$ such that $\lambda \notin K^c_{m,n}$ for all $m, n = 1, 2, 3, \ldots$. This means that $u_n + v_m \lambda \in K$ for all $m, n = 1, 2, 3, \ldots$ This completes the proof. $\qquad \square$

Theorem 16.18 ([56]). *Let* $\Omega = e^{-\frac{\pi}{6}i}(K \times K)$ *be the rotation of* $K \times K$ *by* $-\frac{\pi}{6}$. *Then* Ω *satisfies the condition* $(C1)$ *which has two-dimensional Lebesgue measure* 0.

Proof. Let $x, y, t \in \mathbb{R}$ and let

$$P_{x,y,t} = \{(t,t), (t,y), (x,t), (x+t, -y+t), (x-y, 2t), (x-t, y+t)\}. \tag{16.88}$$

Then by the construction of Ω, the condition $(C1)$ is equivalent to the condition that for every $x, y \in \mathbb{R}$, there exists $t \in \mathbb{R}$ such that

$$e^{\frac{\pi}{6}i} P_{x,y,t} \subset K \times K. \tag{16.89}$$

The inclusion (16.89) is equivalent to

$$S_{x,y,t} = \left\{ \frac{\sqrt{3}}{2} u - \frac{1}{2} v, \frac{1}{2} u + \frac{\sqrt{3}}{2} v : (u,v) \in P_{x,y,t} \right\} \subset K. \tag{16.90}$$

It is easy to check that the set $S_{x,y,t}$ is contained in a set of form $U + tV$, where

$$U = \left\{ 0, \frac{\sqrt{3}}{2} x, \frac{1}{2} x, \frac{\sqrt{3}}{2} y, -\frac{1}{2} y, \frac{\sqrt{3}}{2} x \pm \frac{1}{2} y, \frac{1}{2} x \pm \frac{\sqrt{3}}{2} y, \frac{1}{2}(x-y), \frac{\sqrt{3}}{2}(x-y) \right\},$$

$$V = \left\{ \frac{\sqrt{3}}{2}, \pm\frac{1}{2}, -1, \sqrt{3}, \pm\frac{\sqrt{3}+1}{2}, \frac{\sqrt{3}-1}{2} \right\}.$$

By Lemma 16.4, for given $x, y \in \mathbb{R}$ and $M > 0$ there exists $t \geq M$ such that

$$S_{x,y,t} \subset U + tV \subset K. \tag{16.91}$$

Thus, Ω satisfies $(C1)$. This completes the proof. $\qquad\square$

Corollary 16.9 ([56]). *Let $d > 0$, and $\Omega_d = \{(p,q) \in \Omega : |p| + |q| \geq d\}$. Then Ω_d satisfies the condition $(C1)$.*

Proof. In view of the proof of Theorem 16.18, the inclusion (16.91) implies that for every $x, y \in \mathbb{R}$, and $M > 0$ there exists $t \geq M$ such that

$$P_{x,y,t} \subset \Omega. \tag{16.92}$$

For given $x, y \in \mathbb{R}$ if we take $M = d + |x| + |y|$, and if $t \geq M$, then we have

$$P_{x,y,t} \subset \{(p,q) : |p| + |q| \geq d\}. \tag{16.93}$$

It follows from (16.92) and (16.93) that for every $x, y \in \mathbb{R}$ there exists $t \in \mathbb{R}$ such that

$$P_{x,y,t} \subset \Omega_d. \tag{16.94}$$

Thus, Ω_d satisfies $(C1)$. This completes the proof. $\qquad\square$

Remark 16.3. As a consequence of Theorem 16.17 and Corollary 16.9, we obtain the asymptotic behavior of f satisfying

$$\|f(x+y) + f(x-y) - 2f(x) - 2f(y)\| \to 0 \tag{16.95}$$

as $(x, y) \in \Omega$, $|x| + |y| \to \infty$.

Corollary 16.10 ([56]). *Suppose that $f : \mathbb{R} \to \mathbb{R}$ satisfy the condition (16.95). Then f is a quadratic mapping.*

Proof. The condition (16.95) implies that for each $n \in \mathbb{N}$, there exists $d_n > 0$ such that

$$\|f(x+y) + f(x-y) - 2f(x) - 2f(y)\| \leq \frac{1}{n} \tag{16.96}$$

for all $(x, y) \in \Omega_{d_n}$. By Corollary 16.9, $\Omega_{d_n} = \{(x, y) \in \Omega : |x| + |y| \geq d_n\}$ satisfies the condition $(C1)$. Thus, by Theorem 16.17, there exists a unique quadratic mapping $Q_n : \mathbb{R} \to \mathbb{R}$ such that

$$\|f(x) - Q_n(x)\| \leq \frac{3}{n} \tag{16.97}$$

for all $x \in \mathbb{R}$. Replacing $n \in \mathbb{N}$ by $m \in \mathbb{N}$ in (16.97), and using the triangle inequality, we have

$$\|Q_n(x) - Q_m(x)\| \leq \frac{3}{n} + \frac{3}{m} \leq 6 \tag{16.98}$$

for all $m, n \in \mathbb{N}$, and $x \in \mathbb{R}$. For every $x \in \mathbb{R}$, and $k \in \mathbb{N}$, we have

$$\|Q_n(x) - Q_m(x)\| = \frac{1}{k^2} \|Q_n(kx) - Q_m(kx)\| \leq \frac{6}{k^2}. \tag{16.99}$$

Letting $k \to \infty$ in (16.99), we have $Q_n = Q_m$. Now, letting $n \to \infty$ in (16.97), we get the result. This completes the proof. $\qquad\square$

Chapter 17

Related Topics on Distributions and Hyperfunctions and Jordan Lie Homomorphisms

17.1. Introduction

In the previous chapter, we have discussed about stability results of Jensen, and Jensen type mappings, quadratic functional equation, a generalized quadratic functional equation, quadratic functional equation with improved bound, mixed type functional equation involving three variables, mixed type functional equation involving four variables on restricted domains, and Ulam stability of quadratic functional equation in a set of Lebesgue measure zero. This chapter contains the study of Hyers–Ulam stability of a generalized quadratic functional equation in the spaces of distributions of Schwartz and hyperfunctions of Gelfand modulo bounded distributions, stability of generalized quadratic-additive functional equation and stability of n-Lie homomorphisms and Jordan n-Lie homomorphisms on n-Lie algebras. Using the fixed point method, Hyers–Ulam–Rassias stability of the generalized Cauchy–Jensen additive functional equation in n-Lie Banach algebras is also studied in this chapter.

17.2. Hyers–Ulam stability on a generalized quadratic functional equation in distributions and hyperfunctions

Consider the following stability problem of the quadratic functional equation.

Let f be a map from a vector space (or a commutative group) G to a Banach space B satisfying the inequality

$$\|f(x+y) + f(x-y) - 2f(x) - 2f(y)\| \leq \epsilon \tag{17.1}$$

for all $x, y \in G$. Then there exists a unique map $q : G \to E$ satisfying the *quadratic functional equation*,

$$q(x+y) + q(x-y) - 2q(x) - 2q(y) = 0,$$

such that

$$\|f(x) - q(x)\| \leq \frac{\epsilon}{2} \tag{17.2}$$

for all $x \in G$. The above stability problem (17.1), and its Pexiderized version,

$$\|f_1(x+y) + f_2(x-y) - 2f_3(x) - 2f_4(y)\| \leq \epsilon, \tag{17.3}$$

have been investigated in various spaces of generalized functions such as the spaces $\mathcal{S}'(\mathbb{R}^n)$, $\mathcal{D}'(\mathbb{R}^n)$ of tempered distributions, distributions of Schwartz, respectively, and the spaces $\mathcal{F}'(\mathbb{R}^n)$, $\mathcal{G}'(\mathbb{R}^n)$ of Fourier hyperfunctions, Gelfand generalized functions, respectively. A distributional version of the inequality (17.2) has been reformulated as

$$u_1 \circ A + u_2 \circ B - 2u_2 \circ P_1 - 2u_4 \circ P_2 \in L_\epsilon^\infty(\mathbb{R}^{2n}), \tag{17.4}$$

where $u_j, j = 1, 2, 3, 4$, are generalized functions, $\circ$ denotes the pullback, $A(x+y) = x+y$, $B(x,y) = x-y$, $P_1(x,y) = x$, $P_2(x,y) = y$, and $L_\epsilon^\infty(\mathbb{R}^{2n})$ denotes the space of bounded measurable functions f on $\mathbb{R}^{2n}$ satisfying $\|f\| \leq \epsilon$. The formulation (17.3) is, however, regarded as an incomplete generalization of the inequality (17.2) in the sense of distributions since the left-hand side of (17.3) is a generalized function, in general. Due to Schwartz, the space L^∞ has been generalized to the space $\mathcal{D}'_{L^\infty}$ of bounded distributions as a subspace of distributions and later the space $\mathcal{D}'_{L^\infty}$ is further generalized to the space $\mathcal{A}'_{L^\infty}$ of bounded hyperfunctions as a subspace of Sato hyperfunctions which is a generalization of Schwartz distributions. In the space of bounded distributions and hyperfunctions, however, the validity of the bound $\epsilon > 0$ is deprived. Thus, it is natural to consider

$$u_1 \circ A + u_2 \circ B - 2u_2 \circ P_1 - 2u_4 \circ P_2 \in \mathcal{D}'_{L^\infty}(\mathbb{R}^{2n}) \quad \left[\mathcal{A}'_{L^\infty}(\mathbb{R}^{2n})\right]. \tag{17.5}$$

The main tool of treating (17.4) is *the heat kernel method* initiated by Matsuzawa which represents the generalized functions as the initial values of solutions of the heat equation with some growth conditions. Making use of the heat kernel method we can convert (17.4) to the Hyers–Ulam stability problems of *quadratic-additive functional equation*. There exists $C > 0$ (for every $\epsilon > 0$; there exists $C_\epsilon > 0$) such that

$$|f_1(x+y, t+s) + f_2(x-y, t+s) - 2f_3(x,t) - 2f_4(y,s)|$$

$$\leq C\left(\frac{1}{t} + \frac{1}{s}\right) \quad [C_\epsilon e^{\epsilon(1/t + 1/s)}] \tag{17.6}$$

for all $x, y \in \mathbb{R}^n$, $t, s > 0$, where $f_j : \mathbb{R}^n \times (0, \infty) \to \mathbb{C}$, $j = 1, 2, 3, 4$, are solutions of the heat equation.

In the next section, the Hyers–Ulam stability of a generalization of the inequality (17.5) is obtained and combined with the heat kernel method, the Hyers–Ulam-type stability problem of (17.4) is presented. As a result, we have the following.

Let $u_j \in \mathcal{G}'(\mathbb{R}^n)$, $j = 1, 2, 3, 4$, satisfy (17.4). Then there exist a unique quadratic function $q(x) = \sum_{1 \le j \le k \le n} a_{jk} x_j x_k$, $a_{jk} \in \mathbb{C}$, $j = 1, \ldots, n$, a unique $a \in \mathbb{C}^n$, a unique $b \in \mathbb{C}^n$, and $r_1, r_2, r_3, r_4 \in \mathcal{D}'_{L^\infty}(\mathbb{R}^n)[\mathcal{A}'_{L^\infty}(\mathbb{R}^n)]$, not necessary unique, such that

$$u_1 = q(x) + (a + b) \cdot x + r_1, \quad u_2 = q(x) + (a - b) \cdot x + r_2,$$
$$u_3 = q(x) + a \cdot x + r_3, \quad u_4 = q(x) + b \cdot x + r_4.$$

17.2.1. *Bounded distributions and hyperfunctions*

We first introduce the space $\mathcal{S}'$ of Schwartz tempered distributions, and $\mathcal{G}'$ of Gelfand generalized functions. We use the notations $|\alpha| = \alpha_1 + \cdots + \alpha_n$, $\alpha! = \alpha_1! \ldots \alpha_n!$, $|x| = \sqrt{x_1^2 + \cdots + x_n^2}$, $x^\alpha x_1^{\alpha_1} \ldots x_n^{\alpha_n}$, and $\partial^\alpha = \partial_1^{\alpha_1} \ldots \partial_n^{\alpha_n}$ for $x = (x_1, \ldots, x_n) \in \mathbb{R}^n$, $\alpha = (\alpha_1, \ldots, \alpha_n) \in \mathbb{N}_0^n$, where $\mathbb{N}_0$ is the set of non-negative integers and $\partial_j = \partial/\partial x_j$.

Definition 17.1 ([53]). We denote by $\mathcal{S}$ or $\mathcal{S}(\mathbb{R}^n)$ the Schwartz space of all infinitely differentiable functions φ in $\mathbb{R}^n$, such that

$$\|\varphi\|_{\alpha,\beta} = \sup_x |x^\alpha \partial^\beta \varphi(x)| < \infty \tag{17.7}$$

for all $\alpha, \beta \in \mathbb{N}_0^n$, equipped with the topology defined by the seminorms $\|\cdot\|_{\alpha,\beta}$. The elements of $\mathcal{S}$ are called rapidly decreasing functions and the elements of the dual space $\mathcal{S}'$ are called tempered distributions.

Definition 17.2 ([53]). We denote by $\mathcal{G}$ or $\mathcal{G}(\mathbb{R}^n)$ the Gelfand space of all infinitely differentiable functions φ in $\mathbb{R}^n$, such that

$$\|\varphi\|_{h,k} = \sup_{x \in \mathbb{R}^n; \alpha, \beta \in \mathbb{N}_0^n} \frac{|x^\alpha \partial^\beta \varphi(x)|}{h^{|\alpha|} k^{|\beta|} \alpha!^{1/2} \beta!^{1/2}} < \infty$$

for some $h, k > 0$. We say that $\varphi_j \to 0$ as $j \to \infty$ if $\|\varphi\|_{h,k} \to \infty$ as $j \to \infty$ for some h, k, and denote by $\mathcal{G}'$ the dual space of $\mathcal{G}$ and call its elements Gelfand generalized functions.

As a generalization of the space L^∞ of bounded measurable functions, Schwartz introduced the space $\mathcal{D}'_{L^\infty}$ of bounded distributions as a subspace of tempered distributions.

Definition 17.3 ([53]). We denote by $\mathcal{D}_{L^1}(\mathbb{R}^n)$ the space of smooth functions on $\mathbb{R}^n$, such that $\partial^\alpha \varphi \in L^1(\mathbb{R}^n)$ for all $\alpha \in \mathbb{N}_0^n$ equipped with the topology defined by the countable family of seminorms,

$$\|\varphi\|_m = \sum_{|\alpha| \le m} \|\partial^\alpha \varphi\|_{L^1}, \quad m \in \mathbb{N}_0.$$

We denote by $\mathcal{D}'_{L^\infty}$ the dual space of $\mathcal{D}'_L$ and call its elements bounded distributions.

Generalizing bounded distributions the space $\mathcal{A}'_{L\infty}$ of bounded hyperfunctions has been introduced as a subspace of Gelfand generalized functions $\mathcal{S}'$.

Definition 17.4 ([53]). We denote by $\mathcal{A}_{L^1}$ the space of smooth functions on $\mathbb{R}^n$ satisfying

$$\|\varphi\|_h = \sup_\alpha \frac{\|\partial^\alpha \varphi\|_{L^1}}{h^{|\alpha|}\alpha!} < \infty$$

for some constant $h > 0$. We say that $\varphi_j \to 0$ in $\mathcal{A}_{L^1}$ as $j \to \infty$ if there is a positive constant h, such that

$$\sup_\alpha \frac{\|\partial^\alpha \varphi_j\|_{L^1}}{h^{|\alpha|}\alpha!} \to 0 \qquad \text{as } j \to \infty.$$

We denote by $\mathcal{A}'_{L\infty}$ the dual space of $\mathcal{A}_{L^1}$.

It is well known that the following topological inclusions hold:

$$\mathcal{G} \hookrightarrow \mathcal{S} \hookrightarrow \mathcal{D}_{L^1}, \qquad \mathcal{D}'_{L\infty} \hookrightarrow \mathcal{S}' \hookrightarrow \mathcal{G}',$$

$$\mathcal{G} \hookrightarrow \mathcal{A}_{L^1} \hookrightarrow \mathcal{D}_{L^1}, \qquad \mathcal{D}'_{L\infty} \hookrightarrow \mathcal{A}'_{L\infty} \hookrightarrow \mathcal{G}'.$$

It is known that the space $\mathcal{G}(\mathbb{R}^n)$ consists of all infinitely differentiable functions $\varphi(x)$ on $\mathbb{R}^n$ which can be extended to an entire function on $\mathbb{C}^n$ satisfying

$$|\varphi(x + iy)| \leq C \exp\left(-a|x|^2 + b|y|^2\right) \tag{17.8}$$

for some $a, b, C > 0$. It is easy to see that the n-dimensional heat kernel $E_l(x)$ is given by

$$E_l(x) = (4\pi t)^{-\pi/2}\exp\left(-|x|^2/4t\right), \qquad t > 0$$

belongs to the Gelfand space $\mathcal{G}(\mathbb{R}^n)$ for each $t > 0$.

Definition 17.5 ([53]). Let $u_j \in \mathcal{G}'(\mathbb{R}^n_j)$ for $j = 1, 2$, with $n_1 \geq n_2$, and let $\lambda : \mathbb{R}^{n_1} \to \mathbb{R}^{n_2}$ be a smooth function such that for each $x \in \mathbb{R}^{n_1}$, the Jacobian matrix $\nabla\lambda(x)$ of λ at x has rank n_2. Then there exists a unique continuous linear map $\lambda^* : \mathcal{G}(\mathbb{R}^{n_2}) \to \mathcal{G}'(\mathbb{R}^{n_1})$, such that $\lambda^* u = u \circ \lambda$ when u is a continuous function. We call $\lambda^* u$ the pullback of u by λ and often denoted by $u \circ \lambda$.

In particular, let $A, B, P_1, P_2 : \mathbb{R}^{2n} \to \mathbb{R}^n$ defined by $A(x, y) = x + y$, $B(x, y) = x - y$, $P_1(x, y) = x$, $P_2(x, y) = y$, $x, y \in \mathbb{R}^n$. Then we have

$$\langle u \circ A, \varphi(x, y)\rangle = \left\langle u, \int \varphi(x - y, y)dy\right\rangle, \tag{17.9}$$

$$\langle u \circ B, \varphi(x,y)\rangle = \left\langle u, \int \varpi(x+y,y)dy \right\rangle, \tag{17.10}$$

$$\langle u \circ P_1, \varphi(x,y)\rangle = \left\langle u, \int \varphi(x,y)dy \right\rangle, \tag{17.11}$$

$$\langle u \circ P_2, \varphi(x,y)\rangle = \left\langle u, \int \varphi(y,x)dy \right\rangle. \tag{17.12}$$

17.2.2. *Stability of generalized quadratic-additive functional equation*

Let G and S be an Abelian group and a semigroup divisible by 2, respectively, and B be a Banach space. Let $f_j : G \times S \to B$, $j = 1, 2, 3, 4$. This section deals with the following stability problem:

$$\|f_1(x+y,t+s) + f_2(x-y,t+s) - 2f_3(x,t) - 2f_4(y,s)\| \le \psi(t,s) \tag{17.13}$$

for all $x, y \in G$, $t, s \in S$, where $\psi : S \times S \to [0, \infty)$.

Theorem 17.1 ([53]). *Let $f_j : G \times S \to B$, $j = 1, 2, 3, 4$, satisfy the inequality (17.13). Then there exist a unique quadratic function q and unique additive functions a and b, such that*

$$\|f_1(x,t) - q(x) - a(x) - b(x) - f_1(0,t)\| \le \frac{20}{3}\psi\left(\frac{t}{2},\frac{t}{2}\right),$$

$$\|f_2(x,t) - q(x) - a(x) + b(x) - f_2(0,t)\| \le \frac{20}{3}\psi\left(\frac{t}{2},\frac{t}{2}\right),$$

$$\|f_3(x,t) - q(x) - a(x) - f_3(0,t)\| \le \frac{10}{3}\psi\left(\frac{t}{2},\frac{t}{2}\right),$$

and

$$\|f_4(x,t) - q(x) - b(x) - f_4(0,t)\| \le \frac{10}{3}\psi\left(\frac{t}{2},\frac{t}{2}\right).$$

Proof. For each $j = 1, 2, 3, 4$ we denote by $f_j^e(\cdot, t)$ the even parts of $f_j(\cdot, t)$, that is,

$$f_j^e(x,t) = \frac{1}{2}[f_j(x,t) + f_j(-x,t)], \quad j = 1, 2, 3, 4.$$

Then we have

$$\|f_1^e(x+y,t+s) + f_2^e(x-y,t+s) - 2f_3^e(x,t) - 2f_4^e(y,s)\| \le \psi(t,s). \tag{17.14}$$

Let $F^e(x, y, t, s)$ be the difference

$$F^e(x,y,t,s) = f_1^e(x+y,t+s) + f_2^e(x-y,t+s) - 2f_3^e(x,t) - 2f_4^e(y,s). \tag{17.15}$$

Then we have

$$\frac{1}{2}F^e(x, y, 2t, s) + \frac{1}{2}F^e(x, -y, 2t, 2s) - \frac{1}{2}F^e(x + y, 0, t + s, t + s)$$

$$- \frac{1}{2}F^e(x - y, 0, t + s, t + s) + F^e(y, 0, 2s, 2s) - F^e(0, y, 2s, 2s)$$

$$= f_3^e(x + y, t + s) + f_3^e(x - y, t + s) - 2f_3^e(x, 2t)$$

$$- 2f_3^e(y, 2s) + 2f_3(0, 2s) + 2f_4(0, t + s) - 2f_4(0, 2s). \tag{17.16}$$

Thus the equality (17.16) gives

$$\|f_3^e(x + y, t + s) + f_3^e(x - y, t + s) - 2f_3^e(x, 2t)$$

$$- 2f_3^e(y, 2s) + 2f_3(0, 2s) + 2f_4(0, t + s) - 2f_4(0, 2s)\|$$

$$\leq \psi(2t, 2s) + \psi(t + s, t + s) + 2\psi(2s, 2s). \tag{17.17}$$

In (17.17), replacing y by x, both t and s by $\frac{t}{2}$, and dividing by 4, we obtain

$$\left\| f_3^e(x, t) - \frac{1}{4}f_3^e(2x, t) - \frac{3}{4}f_3(0, t) \right\| \leq \psi(t, t). \tag{17.18}$$

Making the induction hypothesis, we have for all $k = 1, 2, 3, \ldots,$

$$\left\| f_3^e(x, t) - 4^{-k}f_3^e\left(2^k x, t\right) - f_3(0, t) + 4^{-k}f_3(0, t) \right\| \leq \frac{4}{3}\psi(t, t). \tag{17.19}$$

From (17.19), it is easy to see that

$$Q(x, t) = \lim_{k \to \infty} 4^{-k}f_3^e\left(2^k x, t\right)$$

exists, and satisfies the functional equation

$$Q(x + y, t + s) + Q(x - y, t + s) - 2Q(x, 2t) - 2Q(y, 2s) = 0. \tag{17.20}$$

It follows from (17.20) that

$$Q(0, t) = 0,$$

and

$$Q(x, t) = Q(x, s)$$

for all $x \in G$, and $t, s \in S$. Thus $Q(x, t)$ is independent of $t \in S$, and we may write $Q(x, t) = q(x)$ for all $x \in G$, $t \in S$ and obtain the equation

$$q(x + y) + q(x - y) - 2q(x) - 2q(y) = 0. \tag{17.21}$$

Letting $k \to \infty$ in (17.19), we have

$$\|f_3^e(x, t) - f_3(0, t) - q(x)\| \leq \frac{4}{3}\psi(t, t). \tag{17.22}$$

On the other hand, for each $j = 1, 2, 3, 4$, we denote by $f_j^o(\cdot, t)$ the odd parts of $f_j(\cdot, t)$, that is,

$$f_j^o(x, t) = \frac{1}{2}[f_j(x, t) - f_j(-x, t)], \qquad j = 1, 2, 3, 4.$$

Then we have

$$\|f_1^o(x + y, t + s) + f_2^o(x - y, y + s) - 2f_3^o(x, t) - 2f_4^o(y, s)\| \leq \psi(t, s). \qquad (17.23)$$

Similarly as in the even part of f, we denote by $F^o(x, y, t, s)$ the difference

$$F^o(x, y, t, s) = f_1^o(x + y, t + s) + f_2^o(x - y, t + s) - 2f_3^o(x, t) - 2f_4^o(y, s). \qquad (17.24)$$

Then we have

$$\frac{1}{2}F^o(x, y, 2t, 2s) + \frac{1}{2}F^o(x, -y, 2t, 2s)$$

$$- \frac{1}{2}F^o(x + y, 0, t + s, t + s) - \frac{1}{2}F^o(x - y, 0, t + s + t + s)$$

$$= f_3^o(x + y, t + s) + f_3^o(x - y, t + s) - 2f_3^o(x, 2t).$$

Thus we obtain

$$\|f_3^o(x + y, t + s) + f_3^o(x - y, t + s) - 2f_3^o(x, 2t)\|$$

$$\leq \psi(2t, 2s) + \psi(t + s, t + s). \qquad (17.25)$$

In (17.25), replacing y by x, both t and s by $\frac{t}{2}$, and dividing by 2, we obtain

$$\left\| f_3^o(x, t) - \frac{1}{2}f_3^o(2x, t) \right\| \leq \psi(t, t). \qquad (17.26)$$

Making the induction hypothesis, we have for all $k = 1, 2, 3, \ldots$,

$$\left\| f_3^o(x, t) - 2^{-k}f_3^o\left(2^k x, t\right) \right\| \leq 2\psi(t, t). \qquad (17.27)$$

Thus

$$A(x, t) = \lim_{k \to \infty} 2^{-k} f_3\left(2^k x, t\right)$$

exists, and satisfies the functional equation

$$A(x + y, t + s) + A(x - y, t + s) - 2A(x, 2t) - 0. \qquad (17.28)$$

It follows from (17.28) that

$$A(0, t) = 0,$$

and

$$A(x, t) = A(x, s)$$

for all $x \in G$, and $t, s, \in S$. Thus $A(x, t)$ is independent of $t \in S$, and we may write $A(x, t) = a(x)$ for all $x \in G$, $t \in S$, and obtain the equation

$$a(x + y) + a(x - y) - 2a(x) = 0, \qquad a(0) = 0,$$

which is equivalent to

$$a(x + y) - a(x) - a(y) = 0. \tag{17.29}$$

Letting $k \to \infty$ in (17.27), we have

$$\|f_3^o(x, t) - a(x)\| \leq 2\psi(t, t). \tag{17.30}$$

From (17.22) and (17.30), we have

$$\|f_3(x, t) - q(x) - a(x) - f_3(0, t)\| \leq \frac{10}{3}\psi(t, t). \tag{17.31}$$

Changing the roles of f_3^e and f_4^e in (17.14), in view of (17.22), we have

$$\|f_4^e(x, t) - q^*(x) - f_4(0, t)\| \leq \frac{4}{3}\psi(t, t) \tag{17.32}$$

for some quadratic function q^*. Now we show that $q = q^*$. In view of (17.14) and (17.15), we have

$$\|f_3^e(x, t) - f_3^e(0, t) - f_4^e(x, t) + f_4^e(0, t)\|$$

$$= \left\|\frac{1}{2}F^e(0, x, t, t) - \frac{1}{2}F^e(x, 0, t, t)\right\|$$

$$\leq \frac{1}{2}\|F^e(0, x, t, t)\| + \frac{1}{2}\|F^e(x, 0, t, t)\| \leq \psi(t, t). \tag{17.33}$$

Thus it follows from (17.22), (17.32), and (17.33) that

$$\|q(x) - q^*(x)\| \leq \frac{11}{3}\psi(t, t). \tag{17.34}$$

Since $q(kx) = k^2 q(x)$, $q^*(kx) = k^2 q^*(x)$ for all $x \in G$, and all positive integer k, we have

$$\|q(x) - q^*(x)\| \leq \frac{11}{3k^2}\psi(t, t) \tag{17.35}$$

for all $x \in G$, and all positive integer k. Letting $k \to \infty$ in (17.35) we have $q = q^*$.

On the other hand, changing the roles of f_3^o and f_4^o in (17.23), in view of (17.30), we have

$$\|f_4^o(x, t) - b(x)\| \leq 2\psi(t, t) \tag{17.36}$$

for some additive function b. In view of (17.32) and (17.36), we have

$$\|f_4(x,t) - q(x) - b(x) - f_4(0,t)\| \le \frac{10}{3}\psi(t,t). \tag{17.37}$$

Replacing x and y by $\frac{x}{2}$, t and s by $\frac{t}{2}$ in (17.13), and using (17.31), (17.37), and the triangle inequality, we have

$$\left\| f_1(x,t) + f_2(0,t) - q(x) - a(x) - b(x) - 2f_3\left(0,\frac{t}{2}\right) - 2f_4\left(0,\frac{t}{2}\right) \right\|$$

$$\le \frac{43}{3}\psi\left(\frac{t}{2},\frac{t}{2}\right). \tag{17.38}$$

Similarly we have

$$\left\| f_2(x,t) + f_1(0,t) - q(x) - a(x) + b(x) - 2f_3\left(0,\frac{t}{2}\right) - 2f_4\left(0,\frac{t}{2}\right) \right\|$$

$$\le \frac{43}{3}\psi\left(\frac{t}{2},\frac{t}{2}\right). \tag{17.39}$$

On the other hand, in view of (17.14), and (17.15), we have

$$\|f_1^e(x+y,t) + f_1^e(x-y,t) - 2f_1^e(x,t) - 2f_1^e(y,t) + 2f_1(0,t)\|$$

$$\le \left\| F^e\left(\frac{x+y}{2},\frac{x+y}{2},\frac{t}{2},\frac{t}{2}\right) + F^e\left(\frac{x-y}{2},\frac{x-y}{2},\frac{t}{2},\frac{t}{2}\right) \right\|$$

$$+ \left\| -F^e\left(\frac{x+y}{2},\frac{x-y}{2},\frac{t}{2},\frac{t}{2}\right) - F^e\left(\frac{x-y}{2},\frac{x+y}{2},\frac{t}{2},\frac{t}{2}\right) \right\|$$

$$+ \left\| -2F^e\left(\frac{y}{2},\frac{-y}{2},\frac{t}{2},\frac{t}{2}\right) \right\|$$

$$\le 8\psi\left(\frac{t}{2},\frac{t}{2}\right). \tag{17.40}$$

We obtain that for each $t \in S$, there exists a quadratic function $q_1(\cdot,t)$, such that

$$\|f_1^e(x,t) - q_1(x,t) - f_1(0,t)\| \le \frac{8}{3}\psi\left(\frac{t}{2},\frac{t}{2}\right). \tag{17.41}$$

Now for the odd part, in view of (17.23), and (17.24) we have

$$\|2f_1^o(x+y,t) - f_1^o(2x,t) - f_1^o(2y,t)\|$$

$$= \left\| F^o\left(x,y,\frac{t}{2},\frac{t}{2}\right) + F^o\left(y,x,\frac{t}{2},\frac{t}{2}\right) - F^o\left(x,x,\frac{t}{2},\frac{t}{2}\right) - F^o\left(y,y,\frac{t}{2},\frac{t}{2}\right) \right\|$$

$$\le 4\psi\left(\frac{t}{2},\frac{t}{2}\right). \tag{17.42}$$

Putting $y = 0$ in (17.42), and dividing by 2 we have

$$\left\| f_1^o(x,t) - \frac{1}{2}f_1(2x,t) \right\| \le 2\psi\left(\frac{t}{2},\frac{t}{2}\right). \tag{17.43}$$

Using the induction argument, we can see that

$$a_1(x,t) = \lim_{k\to\infty} 2^{-k} f_1^o\left(2^k x, t\right) \tag{17.44}$$

exists, and satisfies the functional equation

$$2a_1(x+y,t) - a_1(2x,t) - a_1(2y,t) = 0, \tag{17.45}$$

and the inequality

$$\|f_1^o(x,t) - a_1(x,t)\| \le 4\psi\left(\frac{t}{2},\frac{t}{2}\right). \tag{17.46}$$

In view of (17.44), for each $t \in S$, $a_1(\cdot,t)$ is an odd function. Putting $y = 0$ in (17.45), we have $a_1(2x,t) = 2a_1(x,t)$ for all $x \in G$ and $t \in S$, which implies that $a_1(\cdot,t)$ is additive for each $t \in S$. From (17.41) and (17.46), we have

$$\|f_1(x,t) - q_1(x,t) - a_1(x,t) - f_1(0,t)\| \le \frac{20}{3}\psi\left(\frac{t}{2},\frac{t}{2}\right). \tag{17.47}$$

Replacing $-y$ by y in (17.13), and changing the roles of f_1 and f_2 we have

$$\|f_2(x,t) - q_2(x,t) - a_2(x,t) - f_2(0,t)\| \le \frac{20}{3}\psi\left(\frac{t}{2},\frac{t}{2}\right), \tag{17.48}$$

where $q_2(\cdot,t)$ is quadratic function and $a_2(\cdot,t)$ is an additive function for each $t \in S$. Using the fact that

$$q_1(kx,t) = k^2 q_1(x,t), \quad q(kx) = k^2 q(x), \quad a_1(kx,t) = ka_1(x,t),$$

$$a(kx) = ka(x), \quad b(kx) = kb(x),$$

for all rational numbers k, it follows from (17.38), (17.47), and the triangle inequality that for each $t \in S$, there exists a positive constant $M(t)$, such that

$$\left\| [q_1(x,t) - q(x)] k^2 + [a_1(x,t) - a(x) - b(x)] k \right\| \le M(t). \tag{17.49}$$

Letting $k \to \infty$ in (17.49), it follows that $q_1(x,t) = q(x)$, and $a_1(x,t) = a(x) + b(x)$ for all $x \in G$, $t \in S$. Similarly, using the inequalities (17.39), and (17.48), we can show that $q_2(x,t) = q(x)$, and $a_2(x,t) = a(x) - b(x)$ for all $x \in G$, $t \in S$. This completes the proof. $\square$

Remark 17.1. In particular, if $f_j : \mathbb{R}^n \times (0, \infty) \to \mathbb{C}$, $j = 1, 2, 3, 4$, are continuous functions, the quadratic function $q(x)$ is given by

$$q(x) = \sum_{1 \leq j \leq k \leq n} a_{jk} x_j x_k, \qquad a_{jk} \in \mathbb{C}, \qquad j, k = 1, \ldots, n,$$

since $q(x)$ is a Lebesgue measurable (in fact, a continuous) solution of the quadratic functional equation

$$q(x + y) + q(x - y) - 2q(x) - 2q(y) = 0.$$

Indeed, as we see in the proof of Theorem 17.1, the function $q(x)$ is given by a limit of a sequence of Lebesgue measurable functions (in fact, a uniform limit of a sequence of continuous functions),

$$q(x) = \lim_{k \to \infty} 4^{-k} f_3^e \left(2^k x, 1 \right).$$

Thus it follows that $q(x)$ is Lebesgue measurable functions (in fact, a continuous function). Similarly, the additive functions $a(x)$ and $b(x)$ are also given by $a(x) = a \cdot x$, $b(x) = b \cdot x$ for some $a, b \in \mathbb{C}^n$. Thus we have the following corollary.

Corollary 17.1 ([53]). *Let* $f_j : \mathbb{R}^n \times (0, \infty) \to \mathbb{C}$, $j = 1, 2, 3, 4$, *be continuous functions satisfying*

$$|f_1(x + y) + f_2(x - y) - 2f_3(x) - 2f_4(y)| \leq \psi(t, s)$$

for all $x, y \in \mathbb{R}^n$, $t, s > 0$. *Then there exist a unique quadratic function* $q = \sum_{1 \leq j \leq k \leq n} a_{jk} x_j x_k$, $a_{jk} \in \mathbb{C}$, $j, k = 1, \ldots, n$, *and unique* $a, b \in \mathbb{C}^n$, *such that*

$$\|f_1(x, t) - q(x) - (a + b) \cdot x - f_1(0, t)\| \leq \frac{20}{3} \psi \left(\frac{t}{2}, \frac{t}{2} \right),$$

$$\|f_2(x, t) - q(x) - (a - b) \cdot x - f_2(0, t)\| \leq \frac{20}{3} \psi \left(\frac{t}{2}, \frac{t}{2} \right),$$

$$\|f_3(x, t) - q(x) - a \cdot x - f_3(0, t)\| \leq \frac{10}{3} \psi \left(\frac{t}{2}, \frac{t}{2} \right),$$

and

$$\|f_4(x, t) - q(x) - b \cdot x - f_4(0, t)\| \leq \frac{10}{3} \psi \left(\frac{t}{2}, \frac{t}{2} \right).$$

17.3. Stability of n-Lie homomorphisms and Jordan n-Lie homomorphisms on n-Lie algebras

This section deals with the Hyers–Ulam stability problems of some kinds of functional equations to the classes of n-Lie homomorphisms and n-Lie algebras by using the structures of n-Lie homomorphisms and n-Lie algebras. Also, this section deals with the generalized Hyers–Ulam–Rassias stability of n-Lie homomorphisms and Jordan n-Lie homomorphisms on n-Lie algebras associated to the generalized Cauchy–Jensen–Rassias additive functional equation are investigated using the fixed point methods.

The theory of finite-dimensional complex Lie algebras is an important part of Lie theory. It has several applications in physics and connections with other parts of mathematics. With an increasing amount of theory and applications concerning lie algebras of various dimensions, it has become necessary to ascertain which tools are applicable for handling them. The miscellaneous characteristics of Lie algebras constitute such tools and have also found applications.

In the year 1985, the notion of an n-Lie algebras was introduced by Filippov [76], which resulted a natural generalization of a Lie algebra for n-ary algebra for each $n \geq 3$. The Lie product is taken between n elements of the algebra instead of two. This new bracket is n-linear, skew-symmetric and satisfies a generalization of the Jacobi identity. For $n = 3$, this product is a special case of the Nambu bracket, well known in physics, which was introduced by Nambu [212] in 1973, as a generalization of the Poisson bracket in Hamiltonian mechanics.

An n-Lie algebra is a natural generalization of a Lie algebra. A vector spce V together with a multilinear, antisymmetric n-ary operation $[\,] : \Lambda^n V \to V$ is called an n-*Lie algebra*, $n \geq 3$, if the n-ary bracket is a derivation with respect to itself, that is,

$$[[x_1, \ldots, x_n], x_{n+1}, \ldots, x_{2n-1}]$$

$$= \sum_{i=1}^{n} [x_1, \ldots, x_{i-1}, [x_i, x_{n+1}, \ldots, x_{2n-1}], x_{i+1}, \ldots, x_n], \qquad (17.50)$$

where $x_1, x_2, \ldots, x_{2n-1} \in V$. Equation (17.50) is called the *generalized Jacobi identity*. The meaning of this identity is similar to that of the usual Jacobi identity for a Lie algebra (which is a 2-Lie algebra).

Let $\mathcal{A}$ be an n-Lie algebras over the field of complex numbers $\mathbb{C}$. An n-Lie algebra $\mathcal{A}$ is a normed n-Lie algebra if there exists a norm $\|\cdot\|_{\mathcal{A}}$ on $\mathcal{A}$ such that $\|[x_1, \ldots, x_n]\|_{\mathcal{A}} \leq \|x_1\|_{\mathcal{A}} \cdots \|x_n\|_{\mathcal{A}}$ for all $x_1, \ldots, x_n \in \mathcal{A}$. A normed n-Lie algebra $\mathcal{A}$ is called a *Banach n-Lie algebra* if $(\mathcal{A}, \|\cdot\|_{\mathcal{A}})$ is a Banach space. Let $(\mathcal{A}, [\,]_{\mathcal{A}})$ and $(\mathcal{B}, [\,]_{\mathcal{B}})$ be two Banach n-Lie algebras. A $\mathbb{C}$-linear mapping $H : (\mathcal{A}, [\,]_{\mathcal{A}}) \to (\mathcal{B}, [\,]_{\mathcal{B}})$ is called an n-Lie homomorphism if

$$H([x_1, \ldots, x_n]_{\mathcal{A}}) = [H(x_1), \ldots, H(x_n)]_{\mathcal{B}},$$

and H is called a *Jordan n-Lie Homomorphism* if

$$H([x, \ldots, x]_{\mathcal{A}}) = [H(x), \ldots, H(x)]_{\mathcal{B}},$$

for all $x \in \mathcal{B}$.

Consider a mapping $f : X \to Y$ satisfying the following functional equation

$$\sum_{1 \leq i < j \leq n} f\left(\frac{x_i + x_j}{2} + \sum_{l=1, k_l \neq i,j}^{n-2} x_{k_l}\right) = \frac{(n-1)^2}{2} \sum_{i=1}^{n} f(x_i), \tag{17.51}$$

for all $x_i, x_j \in X$, where $n \in \mathbb{N}$ is a fixed integer with $n \geq 2$. The functional equation (17.51) was introduced by Rassias [276] in the year 2009. We observe that the case $n = 2$ of (17.51) yields the Cauchy–Jensen additive functional equation $2f\left(\frac{x+y}{2}\right) = f(x) + f(y)$ and there are many interesting results concerning the stability problems of the Cauchy–Jensen equation. Therefore, the functional equation (17.51) is a generalized form of the Cauchy–Jensen additive equation.

17.3.1. *Hyers–Ulam–Rassias stability of the generalized Cauchy–Jensen additive functional equation using the fixed point method*

Assume that $(\mathcal{A}, [\,]_{\mathcal{A}})$ and $(\mathcal{B}, [\,]_{\mathcal{B}})$ are two n-Lie Banach algebras and for a fixed positive integer $n_0 \in \mathbb{N}$, let $T^1_{1/n_0} = \{e^{i\theta} : 0 \leq \theta \leq \frac{2\pi}{n_0}\}$. For any mapping $f : \mathcal{A} \to \mathcal{B}$, we define

$$\Delta_\mu f(x_1, \ldots, x_n) = \sum_{1 \leq i < j \leq n} f\left(\frac{\mu x_i + \mu x_j}{2} + \sum_{l=1, k_l \neq i,j}^{n-2} \mu x_{k_l}\right) - \frac{(n-1)^2}{2} \sum_{i=1}^{n} \mu f(x_i)$$

for all $x_1, \ldots, x_n \in \mathcal{A}$ $(n \geq 3)$ and $\mu \in T^1_{1/n_0} \cup \{1\}$.

The following lemmas are needed to prove the main results.

Lemma 17.1 ([276]). *Let X, Y be linear spaces and $n \geq 3$ be a fixed positive integer. A mapping $f : X \to Y$ satisfies the functional equation (17.51) if and only if f is additive.*

Proof. Putting $x_1 = \cdots = x_n = 0$ in (17.51), we obtain $f(0) = 0$. If we put $x_1 = x$, $x_2 = y$, $x_3 = \cdots = x_n = 0$ in (17.51), then we have

$$f\left(\frac{x+y}{2}\right) + (n-2)f\left(\frac{x}{2} + y\right) + (n-2)f\left(\frac{y}{2} + x\right)$$
$$+ \frac{(n-2)(n-3)}{2} f(x + y) = \frac{(n-1)^2}{2} [f(x) + f(y)] \tag{17.52}$$

for all $x, y \in X$. For $y = 0$ it follows from (17.52) that $f(2x) = 2f(x)$ for all $x \in X$. Therefore equation (17.52) may be rewritten in the form

$$f(x + y) + (n-2)f(x + 2y) + (n-2)f(3x + y)$$
$$+ (n-2)(n-3)f(x + y) = (n-1)^2[f(x) + f(y)] \tag{17.53}$$

for all $x, y \in X$. For $y = -x$ in (17.53), we get $f(-x) = -f(x)$, for all $x \in X$. By virtue of (17.51), we also obtain

$$f\left(\frac{x+y}{2} + z\right) + (n-3)f\left(\frac{x}{2} + y + z\right) + f\left(\frac{x+z}{2} + y\right) + (n-3)f\left(\frac{y}{2} + x + z\right)$$
$$+ f\left(\frac{y+z}{2} + x\right) + (n-3)f\left(\frac{z}{2} + x + y\right) + \frac{(n-3)(n-4)}{2}f(x+y+z)$$
$$= \frac{(n-1)^2}{2}[f(x) + f(y) + f(z)]$$

for all $x, y, z \in X$. Putting here $z = -x - y$ and using properties $f(0) = 0$, $f(2x) = 2f(x)$ and $f(-x) = -f(x)$, we infer that f is an additive function. The converse implication is obvious. This completes the proof. $\qquad\square$

Lemma 17.2 ([110]). *Let $f : \mathcal{A} \to \mathcal{B}$ be an additive function. Then f is linear if and only if $f(\mu x) = \mu f(x)$ for all $\mu \in T^1_{1/n_0}$ and $x \in \mathcal{A}$.*

Proof. Suppose that f is additive and $f(\mu x) = \mu f(x)$ for all $x \in \mathcal{A}$ and $\mu \in T^1_{1/n_0}$. Let μ be in T^1, then $\mu = e^{i\theta}$ for $0 \le \theta \le 2\pi$. We set $\mu_1 = e^{\frac{i\theta}{n_0}}$, thus μ_1 is in T^1_{1/n_0} and $f(\mu x) = f\left(\mu_1^{n_0} x\right) = \mu_1^{n_0} f(x) = \mu f(x)$, for all $x \in \mathcal{A}$. If μ belongs to $nT^1 = \{nz : z \in T^1\}$, then by additivity of f, $f(\mu x) = \mu f(x)$, for all $x \in \mathcal{A}$ and $\mu \in nT^1$. If $t \in (0, \infty)$, then by Archimedean property, there exists a natural number n such that the point $(t, 0)$ lies in the interior of circle with center at origin and radius n.

Let $t_1 = t + \sqrt{n^2 - t^2}i \in nT^1$ and $t_2 = t - \sqrt{n^2 - t^2}i \in nT^1$.
We have $t = \frac{t_1 + t_2}{2}$ and $f(tx) = f\left(\frac{t_1 + t_2}{2}x\right) = \frac{t_1 + t_2}{2}f(x) = tf(x)$, for all $x \in \mathcal{A}$.

If $\mu \in \mathbb{C}$, then $\mu = |\mu|e^{i\mu_1}$. So $f(\mu x) = f\left(|\mu|e^{i\mu_1}x\right) = |\mu|e^{i\mu_1}f(x) = \mu f(x)$ for all $x \in \mathcal{A}$.

The converse is trivial. This completes the proof. $\qquad\square$

Theorem 17.2 ([164]). *Let $n_0 \in \mathbb{N}$ be a fixed positive integer number. Let $f : \mathcal{A} \to \mathcal{B}$ be a mapping for which there exists a function $\phi : \mathcal{A}^n \to [0, \infty)$ such that*

$$\|\Delta_\mu f(x_1, \ldots, x_n)\|_\mathcal{B} \le \phi(x_1, \ldots, x_n), \tag{17.54}$$

and

$$\|f([x_1, \ldots, x_n]) - [f(x_1), \ldots, f(x_n)]_\mathcal{B}\|_\mathcal{B} \le \phi(x_1, \ldots, x_n), \tag{17.55}$$

for all $x_1, \ldots, x_n \in \mathcal{A}$ ($n \ge 3$), and $\mu \in T^1_{1/n_0} = \{e^{i\theta} : 0 \le \theta \le 2\pi/n_0\}$. If there exists $L < 1$ such that

$$\phi(x_1, \ldots, x_n) \le nL\phi\left(\frac{x_1}{n}, \ldots, \frac{x_n}{n}\right), \tag{17.56}$$

for all $x_1, \ldots, x_n \in \mathcal{A}$, then there exists a unique n-Lie homomorphism $H : \mathcal{A} \to \mathcal{B}$ such that

$$\|f(x) - H(x)\|_{\mathcal{B}} \leq \frac{2}{n(n-1)^2(1-L)}\phi(x, \ldots, x), \tag{17.57}$$

for all $x \in \mathcal{A}$.

Proof. Let Ω be the set of all functions from $\mathcal{A}$ into $\mathcal{B}$ and a generalized metric d on Ω as follows:

$$d(g, h) = \inf\{C \in [0, \infty) : \|g(x) - h(x)\|_{\mathcal{B}} \leq C\phi(x, \ldots, x) \text{ for all } x \in \mathcal{A}\}.$$

It is easy to show that (Ω, d) is a generalized complete metric space. Now, we consider the mapping $T : \Omega \to \Omega$ defined by $(Tg)(x) = \frac{1}{n}g(nx)$ for all $g \in \Omega$, $x \in \mathcal{A}$ and $n \in \mathbb{N}$. Let $g, h \in \Omega$ and $C \in [0, \infty)$ be an arbitrary constant with $d(g, h) < C$. By the definition of d, we have

$$\|(Tg)(x) - (Th)(x)\|_{\mathcal{B}} \leq LC\phi(x, \ldots, x),$$

for all $x \in \mathcal{A}$, which means that

$$d(Tg, Th) \leq Ld(g, h),$$

for all $g, h \in \Omega$. Thus T is strictly contractive self-mapping on Ω with the Lipschitz constant L. Putting $\mu = 1$, $x_1 = x_2 = \cdots = x_n = x$ in (17.54), we have

$$\left\| \frac{1}{\lambda}f(\lambda x) - f(x) \right\|_{\mathcal{B}} \leq \frac{2}{n\lambda^2}\phi(x, \ldots, x), \tag{17.58}$$

for all $x \in \mathcal{A}$, and $n \in \mathbb{N}$ with $n \geq 3$, where $\lambda = n - 1$. Thus, we have $d(Tf, f) \leq \frac{2}{n\lambda^2} < \infty$. By the fundamental theorem of fixed point theory, there exists a unique mapping $H \in \{g : \Omega : d(f, g) < \infty\}$ which is a unique fixed point of T and so let

$$H(x) = \lim_{m \to \infty} \frac{1}{\lambda^m}f(\lambda^m x), \tag{17.59}$$

for all $x \in \mathcal{A}$. Again, by the fundamental theorem of fixed point theory, we have

$$d(f, H) \leq \frac{1}{1-L}d(Tf, f) \leq \frac{2}{n\lambda^2(1-L)} \tag{17.60}$$

and so (17.57) holds for all $x \in \mathcal{A}$. It follows from (17.56) that

$$\lim_{m \to \infty} \frac{1}{\lambda^m}\phi(\lambda^m x_1, \lambda^m x_2, \ldots, \lambda^m x_n) = 0, \tag{17.61}$$

for all $x_1, \ldots, x_n \in \mathcal{A}$ $(n \geq 3)$, where $\lambda = n - 1$.

Also, from (17.59), and (17.61), we have

$$H([x_1, \ldots, x_n]_\mathcal{A}) - [H(x_1), \ldots, H(x_n)]_\mathcal{B}$$

$$= \lim_{m \to \infty} \frac{1}{\lambda^{nm}} \|f([\lambda^m x_1, \ldots, \lambda^m x_n]_\mathcal{A}) - [f(\lambda^m x_1), \ldots, f(\lambda^m x_n)]_\mathcal{B}\|_\mathcal{B}$$

$$\leq \lim_{m \to \infty} \frac{1}{\lambda^{nm}} \phi(\lambda^m x_1, \ldots, \lambda^m x_n) = 0,$$

which gives

$$H([x_1, \ldots, x_n]_\mathcal{A}) = [H(x_1), \ldots, H(x_n)]_\mathcal{B}, \tag{17.62}$$

for all $x_1, \ldots, x_n \in \mathcal{A}$. Then H is an n-Lie homomorphism on $\mathcal{A}$. On the other hand, it follows from (17.54), (17.59), and (17.61) that

$$\|\Delta_\mu H(x_1, \ldots, x_n)\|_\mathcal{B} \leq \lim_{m \to \infty} \|\Delta_\mu f(\lambda^m x_1, \ldots, \lambda^m x_n)\|_\mathcal{B}$$

$$\leq \lim_{m \to \infty} \frac{1}{\lambda^m} \phi(\lambda^m x_1, \ldots, \lambda^m x_n) = 0, \tag{17.63}$$

for all $x_1, \ldots, x_n \in \mathcal{A}$. If we put $\mu = 1$ in $\Delta_\mu H(x_1, \ldots, x_n) = 0$, then H is additive by Lemma 17.1. Also, letting $x_1 = x_2 = \cdots = x_n = x$, it follows that $H(\mu x) = \mu H(x)$ and so $H \in \Omega$ is $\mathbb{C}$-linear by Lemma 17.2. Therefore, $H : \mathcal{A} \to \mathcal{B}$ is an n-Lie homomorphism satisfying (17.57). This completes the proof of the theorem. $\square$

Theorem 17.3 ([164]). *Let $f : \mathcal{A} \to \mathcal{B}$ be a mapping for which there exists a function $\phi : \mathcal{A}^n \to [0, \infty)$ satisfying (17.54), and (17.55). If there exists $L < 1$ such that*

$$\phi(nx_1, \ldots, nx_n) \leq \frac{L}{n} \phi(x_1, \ldots, x_n), \tag{17.64}$$

for all $x_1, \ldots, x_n \in \mathcal{A}$, then there exists a unique n-Lie homomorphism $H : \mathcal{A} \to \mathcal{B}$ such that

$$\|f(x) - H(x)\|_\mathcal{B} \leq \frac{2L}{n(n-1)^2(1-L)} \phi(x, \ldots, x), \tag{17.65}$$

for all $x \in \mathcal{A}$.

Proof. Let Ω and d be as in the proof of Theorem 17.2. Then (Ω, d) becomes a generalized complete metric space and the mapping $T : \Omega \to \Omega$ defined by $(Tg)(x) = ng\left(\frac{x}{n}\right)$ for all $g \in \Omega$, $x \in \mathcal{A}$ and $n \in \mathbb{N}$. Then $d(Tg, Th) \leq Ld(g, h)$ for all $g, h \in \Omega$. It follows from (17.54) that

$$\left\|\lambda f\left(\frac{x}{\lambda}\right) - f(x)\right\| \leq \frac{2L}{n\lambda^2} \phi(x, \ldots, x),$$

for all $x \in \mathcal{A}$ and $n \geq 3$, where $\lambda = n - 1$. Then we obtain $d(Tf, f) \leq \frac{2L}{n(n-1)^2}$. The remaining assertion is similar to the corresponding part of Theorem 17.2. This completes the proof. $\square$

The following corollary deals with the product stability of the functional equation (17.51) using Theorems 17.2 and 17.3.

Corollary 17.2 ([164]). *Let $\ell \in \{-1, 1\}$, $r_i \in \mathbb{R}$ with $r = \sum_{i=1}^{n} r_i \neq 1$ for each $n \geq 3$ and θ be non-negative real numbers. Suppose that a mapping $f : \mathcal{A} \to \mathcal{B}$ satisfies*

$$\|\Delta_\mu f(x_1, \ldots, x_n)\|_{\mathcal{B}} \leq \prod_{i=1}^{n} \|x_i\|_{\mathcal{A}}^{r_i},$$

and

$$\|f([x_1, \ldots, x_n]_{\mathcal{A}}) - [f(x_1), \ldots, f(x_n)]_{\mathcal{B}}\|_{\mathcal{B}} \leq \prod_{i=1}^{n} \|x_i\|_{\mathcal{A}}^{r_i}, \tag{17.66}$$

for all $x_1, \ldots, x_n \in \mathcal{A}$ ($\mathcal{A}\backslash\{0\}$ if $r \leq 0$) and $\mu \in T_{1/n_0}^1 = \{e^{i\theta} : 0 \leq \theta \leq 2\pi/n_0\}$. Then there exists a unique n-Lie homomorphism $H : \mathcal{A} \to \mathcal{B}$ such that, if $\ell r < \ell$,

$$\|f(x) - H(x)\|_{\mathcal{B}} \leq \frac{2\theta \|x\|_{\mathcal{A}}^r}{n\ell(\lambda^2 - \lambda^{r+1})}, \tag{17.67}$$

for all $x \in \mathcal{A}$ ($\mathcal{A}\backslash\{0\}$ if $r \leq 0$), where $\lambda = n - 1$.

Proof. Putting $\phi(x_1, \ldots, x_n) = \theta \prod_{i=1}^{n} \|x_i\|_{\mathcal{A}}^{r_i}$ for all $x_1, \ldots, x_n \in \mathcal{A}$, and $L = \lambda^{\ell(r-1)}$, the desired result is obtained. $\square$

The following theorem presents the stability of Jordan n-Lie homomorphisms on n-Lie Banach algebras associated to the functional equation (17.51).

Theorem 17.4 ([164]). *Let $n_0 \in \mathbb{N}$ be a fixed positive integer number. Let $f : \mathcal{A} \to \mathcal{B}$ be a mapping for which there exists a function $\phi : A^n \to [0, \infty)$ such that*

$$\|\Delta_\mu f(x_1, \ldots, x_n)\|_{\mathcal{B}} \leq \phi(x, \ldots, x), \tag{17.68}$$

$$\|f([x, \ldots, x]_{\mathcal{A}}) - [f(x), \ldots, f(x)]_{\mathcal{B}}\|_{\mathcal{B}} \leq \phi(x, \ldots, x), \tag{17.69}$$

for all $x_1, \ldots, x_n, x \in \mathcal{A}$ ($n \geq 3$), and $\mu \in T_{1/n_0}^1 = \{e^{i\theta} : 0 \leq \theta \leq 2\pi/n_0\}$. If there exists $L < 1$ satisfying (17.56), then there exists a unique Jordan n-Lie homomorphism $H : \mathcal{A} \to \mathcal{B}$ such that

$$\|f(x) - H(x)\|_{\mathcal{B}} \leq \frac{2}{n(n-1)^2(1-L)}\phi(x, \ldots, x), \tag{17.70}$$

for all $x \in \mathcal{A}$.

Proof. By the same reasoning as in the proof of Theorem 17.2, we obtain a $\mathbb{C}$-linear mapping $H : \mathcal{A} \to \mathcal{B}$ satisfying (17.57). The mapping is given by

$$H(x) = \lim_{m \to \infty} \frac{1}{\lambda^m} f(\lambda^m x),$$

for all $x \in \mathcal{A}$, and $n \geq 3$, where $\lambda = n - 1$. It follows from (17.68) that

$$\|H([x,\ldots,x]_{\mathcal{A}}) - [H(x),\ldots,H(x)]_{\mathcal{B}}\|_{\mathcal{B}}$$

$$= \lim_{m\to\infty} \frac{1}{\lambda^{nm}} \|f([\lambda^m x,\ldots,\lambda^m x]_{\mathcal{A}}) - [f(\lambda^m x),\ldots,f(\lambda^m x)]_{\mathcal{B}}\|_{\mathcal{B}}$$

$$\leq \lim_{m\to\infty} \frac{1}{\lambda^{nm}} \phi(\lambda^m x,\ldots,\lambda^m x) = 0,$$

which gives $H([x,\ldots,x]_{\mathcal{A}}) = [H(x),\ldots,H(x)]_{\mathcal{B}}$ for all $x \in \mathcal{A}$. Therefore, $H : \mathcal{A} \to \mathcal{B}$ is a Jordan n-Lie homomorphism satisfying (17.70). This completes the proof. □

Theorem 17.5 ([164]). *Let $f : \mathcal{A} \to \mathcal{B}$ be a mapping for which there exists a function $\phi : \mathcal{A}^n \to [0,\infty)$ satisfying (17.68), and (17.69). If there exists an $L < 1$ satisfying (17.64), then there exists a unique Jordan n-Lie homomorphism $H : \mathcal{A} \to \mathcal{B}$ such that*

$$\|f(x) - H(x)\|_{\mathcal{B}} \leq \frac{2L}{n(n-1)^2(1-L)} \phi(x,\ldots,x), \tag{17.71}$$

for all $x \in \mathcal{A}$.

Proof. The proof is similar to that of Theorem 17.4, and the details are left to the reader. □

The following corollary deals with the sum stability of the functional equation (17.51) from Theorems 17.4 and 17.5.

Corollary 17.3 ([164]). *Let $\ell \in \{-1,1\}$, θ, $p \in \mathbb{R}$ be non-negative real numbers and $n \geq 3$. Suppose that a mapping $f : \mathcal{A} \to \mathcal{B}$ satisfies*

$$\|\Delta_\mu f(x-1,\ldots,x_n)\|_{\mathcal{B}} \leq \theta \sum_{i=1}^n \|x_i\|_{\mathcal{A}}^p,$$

and

$$\|f([x,\ldots,x]_{\mathcal{A}}) - [f(x),\ldots,f(x)]_{\mathcal{B}}\|_{\mathcal{B}} \leq \theta \sum_{i=1}^n \|x\|_{\mathcal{A}}^p, \tag{17.72}$$

for all $x_1,\ldots,x_n, x \in \mathcal{A}$ ($\mathcal{A}\backslash\{0\}$ if $r \leq 0$) for each $n \geq 3$ and $\mu \in T_{1/n_0}^1 = \{e^{i\theta} : 0 \leq \theta \leq 2\pi/n_0\}$. Then there exists a unique Jordan n-Lie homomorphism $H : \mathcal{A} \to \mathcal{B}$ such that, if $\ell p < \ell$,

$$\|f(x) - H(x)\|_{\mathcal{B}} \leq \frac{2\theta \|x\|_{\mathcal{A}}^p}{\ell(\lambda^2 - \lambda^{p+1})}, \tag{17.73}$$

for all $x \in \mathcal{A}$ ($\mathcal{A}\backslash\{0\}$ if $r \leq 0$), where $\lambda = n - 1$.

Proof. Putting $\phi(x_1,\ldots,x_n) = \theta \sum_{i=1}^n \|x_i\|_{\mathcal{A}}^p$ for all $x_1,\ldots,x_n \in \mathcal{A}$, and $L = \lambda^{\ell(p-1)}$, the desired result is obtained. This completes the proof. □

Chapter 18

Exercises and Open Problems

18.1. Problems of Chapter 3

1. Find all functions $f : \mathbb{Q} \to \mathbb{Q}$ such that $f(1) = 2$ and $f(xy) = f(x)f(y) - f(x + y) + 1$.

2. Given a function $f : \mathbb{R} \to \mathbb{R}$, if for every two real numbers x and y the equality $f(xy + x + y) = f(xy) + f(x) + f(y)$ holds, prove that $f(x + y) = f(x) + f(y)$ for every two real numbers x and y.

3. Find all functions $f : \mathbb{R}^+ \to \mathbb{R}^+$ such that

$$f(x + y) = \frac{f(x)f(y)}{f(x) + f(y)} \qquad \forall x, y \in \mathbb{R}^+.$$

4. Find all functions $f : \mathbb{R}^+ \to \mathbb{R}^+$ such that

$$f\left(\frac{x + y}{2}\right) = \frac{2f(x)f(y)}{f(x) + f(y)} \qquad \forall x, y \in \mathbb{R}^+.$$

5. Find all functions $f : \mathbb{R} \to \mathbb{R}$ which satisfy the equation $f\left((x - y)^2\right) = f(x)^2 - 2xf(y) + y^2$.

6. Determine all functions $f : \mathbb{R} \setminus \{0, 1\} \to \mathbb{R}$, which satisfy the equation

$$f(x) + f\left(\frac{1}{1 - x}\right) = \frac{2(1 - 2x)}{x(1 - x)},$$

valid for all $x \neq 0$ and $x \neq 1$.

7. Suppose $f : \mathbb{R} \to \mathbb{R}$ is a function such that

$$f\left(\frac{x + y}{x - y}\right) = \frac{f(x) + f(y)}{f(x) - f(y)},$$

for all $x \neq y$. Prove that $f(x) = x$ for all $x \in \mathbb{R}$.

8. Find all functions $f : \mathbb{R} \to \mathbb{R}$, which satisfy the functional equation

$$f(xf(x) + f(y)) = f(x)^2 + y, \qquad \text{for all } x, y \in \mathbb{R}.$$

9. Find all functions $f : \mathbb{R} \to \mathbb{R}$, which satisfy the equation

$$f\left(x^2 + y + f(y)\right) = 2y + f(x)^2,$$

for all real numbers x, y.

10. Find all $f : \mathbb{R} \to \mathbb{R}$ such that

$$f(f(x) + y) = f(x^2 - y) + 4f(x)y, \qquad \text{for all } x, y \in \mathbb{R}.$$

11. Find all functions $f : \mathbb{R} \to \mathbb{R}$ such that

$$f(f(x - y)) = f(x) - f(y) + f(x) + f(y) - xy,$$

holds for all reals x, y.

12. Find all functions $f : \mathbb{R} \to \mathbb{R}$, which satisfy the equation

$$f(x^2 + f(y)) = f(x)^2 + y, \qquad \text{for all } x, y \in \mathbb{R}.$$

13. Let $f : \mathbb{R} \to \mathbb{R}$ be a function such that

(a) $f(x + y) = f(x) + f(y)$, for all real numbers x, y;
(b) $f(xy) = f(x)f(y)$, for all real numbers x, y.

Show that $f(x) = 0$ for all reals x or $f(x) = x$ for all reals x.

18.2. Problems of Chapter 4

1. Prove that the functional equation

$$f(x - y - z) + f(x) + f(y) + f(z) = f(x - y) + f(y + z) + f(z - x)$$

is a quadratic functional equation.

2. Show the functional equation

$$f(x + 3y) + f(y + 3z) + f(z + 3x) - 3f(x + y + z) = 7(f(x) + f(y) + f(z))$$

is a quadratic functional equation.

3. Find the general solution of the mixed additive-quadratic functional equation

$$f(x + y + z) + f(x) + f(y) + f(z) = f(x + y) + f(y + z) + f(z + x).$$

4. If r is a non-zero rational number, then find the general solution of the quadratic functional equation

$$r^2 f\left(\frac{x + y}{4}\right) + r^2 f\left(\frac{x - y}{r}\right) = 2f(x) + 2f(y).$$

5. If a and b are fixed integers with $a, b \neq 0$ and $b \neq \pm a, -3a$, then prove that the functional equation

$$f(ax + by) + f(ax - by) = \frac{b(a+b)}{2}f(x+y) + \frac{b(a+b)}{2}f(x-y)$$
$$+ (2a^2 - ab - b^2)f(x) + (b^2 - ab)f(y)$$

is a quadratic functional equation.

6. Find the general solution of the functional equation

$$f(ax + by) + f(ax - by) = 2a^2 f(x) + 2b^2 f(y)$$

where a and b are fixed integers with $a, b \neq 0$ and $b \neq \pm a$.

7. Show that the functional equation

$$f(2x - y) + f(2y - z) + f(2z - x) + 2f(x + y + z) = 7(f(x) + f(y) + f(z))$$

is a quadratic functional equation.

8. Find the general solution of the following equation

$$f(2x + y) + f(2x - y) = 2f(x + y) + 2f(x - y) + 4f(x) - 2f(y).$$

9. Prove that the functional equation

$$f(2x + y) + f(x + 2y) = 4f(x + y) + f(x) + f(y)$$

is a quadratic functional equation.

10. Show that the functional equation

$$f(x + y + z) + f(x - y) + f(x - z) = f(x - y - z) + f(x + y) + f(x + z)$$

is a mixed additive-quadratic functional equation.

11. Find the general solution of the quadratic functional equation

$$f(nx + y) + f(nx - y) = f(x + y) + f(x - y) + 2(n^2 - 1)f(x)$$

where n is a positive integer.

12. Investigate the general solution of the quadratic functional equation

$$f(2x + y) + f(2x - y) = 8f(x) + 2f(y).$$

18.3. Problems of Chapter 5

1. Show that the functional equation

$$f(x + 3y) - 3f(x + y) + 3f(x - y) - f(x - 3y) = 48f(y)$$

is a cubic functional equation.

2. Find the general solution of the functional equation

$$f(x + 2y) + f(x - 2y) + f(2x) = 2f(x) + 4f(x + y) + 4f(x - y).$$

3. If $k \in \mathbb{N} - \{1\}$, then prove that the functional equation

$$f(x + ky) - kf(x + y) + kf(x - y) - f(x - ky) = 2k(k^2 - 1)f(y)$$

is a cubic functional equation.

4. If a is an integer with $a \neq 0, \pm 1$, then show that the functional equation

$$f(ax + y) + f(x + ay) = (a + 1)(a - 1)^2[f(x) + f(y)] + a(a + 1)f(x + y)$$

is a cubic functional equation.

18.4. Problems of Chapter 6

1. Show that the functional equation

$$f(x + 2y) + f(x - 2y) + 6f(x) = f[f(x + y) + f(x - y) + 6f(y)]$$

is a quartic functional equation.

2. Show that the functional equation

$$f(x + 3y) + f(x - 3y) + f(x + 2y) + f(x - 2y) + 22f(x)$$
$$= 13(f(x + y) + f(x - y)) + 168f(y)$$

is a quartic functional equation.

3. If m is a fixed positive integer, then show that the functional equation

$$f(x + my) + f(x - my) = 2(7m - 9)(m - 1)f(x) + 2m^2(m^2 - 1)f(y)$$
$$- (m - 1)^2 f(2x) + m^2[f(x + y) + f(x - y)]$$

is a quartic functional equation.

4. If $a \neq 0$, $b \neq 0$, $a \pm b \neq 0$, then find the kind (or type) of the following (a, b)-functional equation

$$f(ax + by) + f(bx + ay) + \frac{1}{2}ab(a - b)^2 f(x - y)$$
$$= (a^2 - b^2)[f(x) + f(y)] + \frac{1}{2}ab(a + b)^2 f(x + y).$$

18.5. Problems of Chapter 7

1. Show that the functional equation

$$f(x + y) + f(x - y) = 2f(x) + 20f(\sqrt[5]{x^3y^2}) + 10f(\sqrt[5]{xy^4})$$

is a quintic functional equation.

2. Show that the functional equation

$$f(x+y) + f(x-y) = 2f(x) + 30f(\sqrt[6]{x^4y^2}) + 30f(\sqrt[6]{x^2y^4}) + 2f(y)$$

is a sextic functional equation.

3. If n is a positive integer, then show that the functional equation

$$f(nx+y) + f(nx-y) + f(x+ny) + f(x-ny)$$
$$= (n^4 + n^2)[f(x+y) + f(x-y)] + 2(n^6 - n^4 - n^2 + 1)[f(x) + f(y)]$$

is a sextic functional equation.

18.6. Problems of Chapter 8

1. Show that the functional equation

$$f(x+2y) + f(x-2y) = f(x+y) + f(x-y) + 3f(2y) - 6f(y)$$

is a mixed type additive-quadratic equation.

2. Show that the functional equation

$$9f\left(\frac{x+y+z}{3}\right) + 4\left[f\left(\frac{x-y}{2}\right) + f\left(\frac{y-z}{2}\right) + f\left(\frac{z-x}{2}\right)\right]$$
$$= 3[f(x) + f(y) + f(z)]$$

is a mixed type additive-quadratic equation.

3. If $a, b \geq 2$, then show that the functional equation

$$f\left(\frac{x+y}{a} + \frac{z+w}{b}\right) + f\left(\frac{x+y}{a} - \frac{z+w}{b}\right)$$
$$= \frac{1}{a^2}[(1+a)f(x+y) + (1-a)f(-x-y)]$$
$$+ \frac{1}{b^2}[f(z+w) + f(-z-w)]$$

is a mixed type additive-quadratic equation.

4. Show that the Drygas functional equation

$$f(x+y) + f(x-y) - 2f(x) + f(y) + f(-y)$$

is a mixed type additive-quadratic equation.

5. Show that the functional equation

$$2f(x+y) + f(x-y) + f(y-x) = f(2x) + f(2y)$$

is a mixed type additive-quadratic equation.

6. If k is a positive integer, then show that

$$f(kx + y) + f(kx - y) = kf(x + y) + kf(x - y) + 2f(kx) - 2kf(x)$$

is a mixed type additive-cubic equation.

7. Show that the functional equation

$$f(2x + y) + f(2x - y) = 2f(x + y) + 2f(x - y) + 2f(2x) - 4f(x)$$

is a mixed type additive-cubic equation.

8. Show that the functional equation

$$f(2x + y) + f(2x - y) = 4f(x + y) + 4f(x - y) + 10f(x)$$
$$+ 14f(-x) - 3f(y) - 3f(-y)$$

is a mixed type additive-quartic equation.

9. Show that the functional equation

$$6f(x + y) - 6f(x - y) + 4f(3y) = 3f(x + 2y) - 3f(x - 2y) + 9f(2y)$$

is a mixed type quadratic-cubic equation.

10. If n is a fixed integer with $n \neq 0, \pm 1$, then show that the functional equation

$$f(nx + y) + f(nx - y) = n^2[f(x + y) + f(x - y)]$$
$$+ 2\left(f(nx) - n^2 f(x)\right) - 2(n^2 - 1)f(y)$$

is a mixed type quadratic-quartic equation.

11. If m is a fixed integer with $m \neq 0, \pm 1$, then show that the functional equation

$$m[f(x + my) + f(x - my)] = m^3[f(x + y) + f(x - y)]$$
$$- 2m^3(m + 1)f(y) - 2m(m^2 - 1)f(x)$$
$$+ 2(m + 1)f(my)$$

is a mixed type cubic-quartic equation.

12. If k is a fixed integer with $k \neq 0, \pm 1$, then show that the functional equation

$$f(x + ky) + f(x - ky) = k^2(f(x + y) + f(x - y)) + 2(1 - k^2)f(x)$$
$$+ \left(\frac{k^4 - k^2}{4}\right)(f(2y) - 8f(y))$$
$$+ \tilde{f}(2x) - 16\tilde{f}(x)$$

is a mixed type cubic-quartic functional equation where $\tilde{f} = f(x) + f(-x)$.

13. Show that the functional equation

$$f(x + 3y) - 3f(x + y) + 3f(x - y) - f(x - 3y) = 48f(y)$$

is a mixed type cubic-quartic equation.

14. Show that the functional equation

$$f(3x + y) + f(x + 3y) = 64f(x) + 64f(y) + 24f(x + y) - 6f(x - y)$$

is a mixed type cubic-quartic equation.

18.7. Problems of Chapter 9

1. Show that the functional equation

$$f(x + 3y) + f(x - 3y) = 9(f(x + y) + f(x - y)) - 16f(x)$$

is a mixed type additive-quadratic-cubic equation.

2. If $k \in \mathbb{Z} - \{0, \pm 1\}$, then show that the functional equation

$$f(x + ky) + f(x - ky) = k^2 f(x + y) + k^2 f(x - y)$$
$$+ (k^2 - 1) \left(k^2 f(y) + k^2 f(-y) - 2f(x) \right)$$

is a mixed type additive-quadratic-cubic equation.

3. Show that the functional equation

$$f(x + 2y) + f(x - 2y) = 2f(x + y) - 2f(-x - y) + 2f(x - y)$$
$$- 2f(y - x) + f(2y) + f(-2y) + 4f(-x) - 2f(x)$$

is a mixed type additive-quadratic-cubic equation.

4. If $k \in \mathbb{Z} - \{0, \pm 1\}$, then show that the functional equation

$$f(x + ky) + f(x - ky) = k^2 f(x + y) + k^2 f(x - y)$$
$$+ (k^2 - 1) \left(k^2 f(y) + k^2 f(-y) - 2f(x) \right)$$

is a mixed type additive-cubic-quartic equation.

5. Show that the functional equation

$$3(f(x + 2y) + f(x - 2y)) = 12(f(x + y) + f(x - y))$$
$$+ 4f(3y) - 18f(2y) + 36f(y) - 18f(x)$$

is a mixed type quadratic-cubic-quartic equation.

6. Show that the functional equation

$$f(x + 2y) + f(x - 2y) = f(x + y) + 4f(x - y)$$
$$- 6f(x) + f(2y) + f(-2y) - 4f(y) - 4f(-y)$$

is a mixed type additive-quadratic-cubic-quartic equation.

18.8. Problems of Chapter 10

1. Find the general solution of the following mixed type two-variable additive-quadratic functional equation

$$f(x + 2u, y + 2v) + 2f(x - u, y - v) = f(x - 2u, y - 2v) + 2f(x + u, y + v).$$

2. Obtain the general solution of the following two-variable additive-cubic functional equation

$$f(2x+y, 2z+w) - f(2x-y, 2z-w) = 4[f(x+y, z+w) - f(x-y, z-w)] - 6f(y, w).$$

3. Derive the general solution of the mixed type two-variable additive-quadratic functional equation

$$f(x + y, z + w) + f(x + y, z - w) = 2f(x, z) + 2f(x, w) + 2f(y, z) + 2f(y, w).$$

4. Find the general solution of the bi-Jensen functional equation

$$4f\left(\frac{x+y}{2}, \frac{z+w}{2}\right) = f(x, z) + f(x, w) + f(y, z) + f(y, w).$$

5. Solve the following bi-additive functional equation

$$f(x + y, z - w) + f(x - y, z + w) = 2f(x, z) - 2f(y, w).$$

18.9. Problems of Chapter 12

1. Solve the generalized Hyers–Ulam stability of the quadratic functional equation in Banach spaces:

$$f(x + y) + f(x - y) = 2f(x) + 2f(y).$$

Hence obtain Hyers–Ulam stability, Hyers–Ulam–Rassias stability and Ulam–Gavruta–Rassias stability.

2. Investigate the generalized Hyers–Ulam stability of the quadratic functional equation in Banach spaces:

$$f(2x + y) + f(2x - y) = f(x + y) + f(x - y) + 6f(x).$$

Hence obtain Hyers–Ulam stability, Hyers–Ulam–Rassias stability and Ulam–Gavruta–Rassias stability.

3. Obtain the generalized Hyers–Ulam stability of the quadratic functional equation in Banach spaces:

$$f(2x + y) + f(x + 2y) = 4f(x + y) + f(x) + f(y).$$

Hence obtain Hyers–Ulam stability, Hyers–Ulam–Rassias stability and Ulam–Gavruta–Rassias stability.

4. Find the generalized Hyers–Ulam stability of the cubic functional equation in Banach spaces:

$$f(x + 2y) + f(x - 2y) + 6f(x) = 4f(x + y) + 4f(x - y).$$

Hence obtain Hyers–Ulam stability, Hyers–Ulam–Rassias stability and Ulam–Gavruta–Rassias stability.

5. Establish the generalized Hyers–Ulam stability of the quartic functional equation in Banach spaces:

$$f(2x + y) + f(2x - y) = 4f(x + y) + 4f(x - y) + 24f(x) - 6f(y).$$

Hence obtain Hyers–Ulam stability, Hyers–Ulam–Rassias stability and Ulam–Gavruta–Rassias stability as well as J.M. Rassias mixed product-sum stability.

6. Solve the Hyers–Ulam–Rassias stability of the quadratic functional equation in paranormed spaces:

$$f(x + y) + f(x - y) = 2f(x) + 2f(y).$$

7. Find the generalized Hyers–Ulam stability of the quadratic functional equation in fuzzy Banach spaces:

$$f(x + y) + f(x - y) = 2f(x) + 2f(y).$$

8. Investigate the generalized Ulam–Hyers stability of the quadratic functional equation in generalized quasi-Banach spaces:

$$f(x + y) + f(x - y) = 2f(x) + 2f(y).$$

9. Obtain the generalized Hyers–Ulam stability of the mixed type additive-quadratic functional equation

$$9f\left(\frac{x + y + z}{3}\right) + 4\left[f\left(\frac{x - y}{2}\right) + f\left(\frac{y - z}{2}\right) + f\left(\frac{z - x}{2}\right)\right]$$
$$= 3[f(x) + f(y) + f(z)]$$

in Banach spaces.

10. Find the generalized Hyers–Ulam stability of the mixed type additive-cubic functional equation

$$f(2x + y) + f(2x - y) = 2f(x + y) + 2f(x - y) + 2f(2x) - 4f(x)$$

in non-Archimedean Banach spaces.

11. Investigate the generalized Hyers–Ulam stability of the mixed type additive-cubic-quartic functional equation

$$11[f(x+2y) + f(x-2y)]$$
$$= 44[f(x+y) + f(x-y)] + 12f(3y) - 48f(2y) + 60f(y) - 66f(x)$$

in Banach spaces.

12. Obtain the generalized Hyers–Ulam stability of the mixed type additive-quadratic-cubic functional equation

$$f(x+2y) - f(x-2y)$$
$$= 2[f(x+y) - f(x-y)] + 2f(3y) - 6f(2y) + 6f(y)$$

in Banach spaces.

13. Establish the generalized Hyers–Ulam stability of the quadratic functional equation

$$f(2x-y) + f(2y-z) + f(2z-x) + 2f(x+y+z) = 7f(x) + 7f(y) + 7f(z)$$

in Banach spaces.

14. Find the generalized Hyers–Ulam stability of the cubic functional equation

$$2f(x+2y) + f(2x-y) = 5f(x+y) + 5f(x-y) + 15f(y)$$

in Banach spaces.

15. Obtain the generalized Hyers–Ulam stability of the quintic functional equation

$$f(x+3y) - 5f(x+2y) + 10f(x+y) - 10f(x) + 5f(x-y) - f(x-2y) = 120f(y)$$

in Banach spaces.

16. Investigate the generalized Hyers–Ulam stability of the sextic functional equation

$$f(x+3y) - 6f(x+2y) + 15f(x+y) - 20f(x)$$
$$+ 15f(x-y) - 6f(x-2y) + f(x-3y) = 720f(y).$$

in Banach spaces.

18.10. Open problems

1. Discuss the effect of taking $p = 1$ in the stability result of the functional equation (11.53) in Theorem 11.7 of Chapter 11.
2. What is the effect of the stability result in Theorems 12.9 and 12.10 when $r = 2$ in Chapter 12?
3. What is the situation in Theorems 12.1 and 12.2 of Chapter 12 when $r = 1$?

4. What happens to the Hyers–Ulam–Rassias stability of the quadratic functional equation $f(x + y) + f(x - y) = 2f(x) + 2f(y)$ in Theorems 12.3 and 12.4 of Chapter 12 when $r = 2$?

5. Do the stability results hold good for the quadratic functional equation $f(x + y) + f(x - y) = 2f(x) + 2f(y)$ when $r = 2$ in Corollaries 12.3 and 12.4 of Chapter 12?

6. In Chapter 12, can we investigate the stability results in Corollaries 12.8 and 12.9 when $r = 1$?

7. Examine the stability of the functional equation (13.17) when $p = 1$ in Theorem 13.3 of Chapter 13.

8. Inspect the stability result of the functional equation (14.2) when $p = 3$ and $q = 3$ in Corollaries 14.1 and 14.2 of Chapter 14.

Bibliography

[1] Aczél, J.(1966). *Lectures on Functional Equations and Their Applications*, Academic Press, New York.

[2] Aczél, J. and Dhombres, J. (1989). *Functional Equations in Several Variables*, Cambridge University Press, London.

[3] Adam, M. (2011). On the stability of some quadratic functional equation, *J. Nonlinear Sci. Appl.*, **4(1)**, pp. 50–59.

[4] Alotaibi, A., Mursaleen, M., Dutta, H. and Mohiuddine, S.A. (2014). On the Ulam stability of Cauchy functional equation in IFN-spaces, *Appl. Math. Inf. Sci.*, **8(3)**, pp. 1135–1143.

[5] Alsina, C. and Garcia-Roig, J.L. (1994). On a conditional Cauchy equation on rhombuses, in *Functional Analysis, Approximation Theory and Numerical Analysis*, J.M. Rassias (Ed.), World Scientific.

[6] Alzer, H. (1999). Remark on the stability of the gamma functional equation, *Results Math.*, **35**, pp. 199–200.

[7] Amir, D. (1986). *Characterizations of Inner Product Spaces*, Birkhäuser, Basel.

[8] Aoki, T. (1950). On the stability of the linear transformation in Banach spaces, *J. Math. Soc. Japan*, **2**, pp. 64–66.

[9] Ara, P. and Mathieu, M. (2003). *Local Multipliers of C^*-algebras*, Springer, London.

[10] Arunkumar, M. and Ramamoorthi, S. (2011). Penta and hexa functional equations, *JP J. Math. Sci.*, **2(1&2)**, pp. 49–63.

[11] Arunkumar, M. (2013). Generalized Ulam–Hyers stability of derivations of an AQ-functional equation, *CUBO A Math. J.*, **15(1)**, pp. 159–169.

[12] Ashish and Chugh, R. (2012). Hyers–Ulam–Rassias stability of orthogonally cubic and quartic functional equations, *Int. J. Pure Appl. Math.*, **81(1)**, pp. 9–20.

[13] de Azcarraga, J.A. and Izquierdo, J.M. (2010). n-Ary algebras: A review with applications, *J. Phys. A*, **43**, pp. 293001-1–203001-117.

[14] Baak, C. and Moslehian, M.S. (2005). On the stability of J^*-homomorphisms, *Nonlinear Anal.*, **63**, pp. 42–48.

[15] Baak, C. (2006). Cauchy–Rassias stability of Cauchy–Jensen additive mappings in Banach spaces, *Acta Math. Sin. (Engl. Ser.)*, **22(6)**, pp. 1789–1796.

[16] Bae, J.H. (2000). On the stability of 3-dimensional quadratic functional equation, *Bull. Korean Math. Soc.*, **37(3)**, pp. 477–486.

[17] Bae, J.H. and Park, W.G. (2006). On the solution of a bi-Jensen functional equation and its stability, *Bull. Korean Math. Soc.*, **43(3)**, pp. 499–507.

[18] Bae, J.H. and Park, W.G. (2007). A functional equation originating from quadratic forms, *J. Math. Anal. Appl.*, **326**, pp. 1142–1148.

[19] Bagger, J. and Lambert, N. (2008). Comments on Multiple M2-branes, *J. High Energy Phys.*, **0802**, p. 105.

[20] Bahyrycz, A. and Brzdek, J. (2013). On solutions of the d'Alembert equation on a restricted domain, *Aequationes Math.*, **85**, pp. 169–183.

[21] Baker, J., Lawrence, J. and Zorzitto, F. (1979). The stability of the equation $f(x + y) = f(x)f(y)$, *Proc. Amer. Math. Soc.*, **74**, pp. 242–246.

[22] Baker, J. (1980). The stability of the cosine equation, *Proc. Amer. Math. Soc.*, **80**, pp. 411–416.

[23] Baker, J.A. (1993). On a functional equation of Aczél and Chung, *Aequationes Math.*, **46**, pp. 99–111.

[24] Baker, J.A. (2001). Distributional methods for functional equations, *Aequationes Math.*, **62**, pp. 136–142.

[25] Baker, J.A. (2005). A general functional equation and its stability, *Proc. Amer. Math. Soc.*, **133**, pp. 1657–1664.

[26] Batko, B. (2008). Stability of an alternative functional equation, *J. Math. Anal. Appl.*, **339**, pp. 303–311.

[27] Batko, B. (2008). On approximation of approximate solutions of Dhombres' equation, *J. Math. Anal. Appl.*, **340**, pp. 424—432.

[28] Benyamini, Y. and Lindenstrauss, J. (2000). *Geometric Nonlinear Functional Analysis*, Colloquium Publications, Vol. 48, American Mathematical Society, Providence, RI.

[29] Bodaghi, A. (2014). Stability of a quartic functional equation, *Sci. World J.*, **2014**, Art. ID 752146, 9 pp.

[30] Boo, D., Oh, S., Park, C. and Park, J. (2003). Generalized Jensen's equations in Banach modules over a C^*-algebra and its unitary group, *Taiwanese J. Math.*, **7**, pp. 641–655.

[31] Borelli, C. (1994). On the Hyers–Ulam stability of Hosszú's functional equation, *Results Math.*, **26**, pp. 221–224.

[32] Borelli, C. and Forti, G.L. (1995). On a general Hyers–Ulam stability result, *Int. J. Math. Math. Sci.*, **18**, pp. 229–236.

[33] Bouikhalene, B. and Elquorachi, E. (2007). Ulam–Gavruta–Rassias stability of the Pexider functional equation, *Int. J. Appl. Math. Stat.*, **7(F07)**, pp. 7–39.

[34] Bourgin, D.G. (1951). Classes of transformations and bordering transformations, *Bull. Amer. Math. Soc.*, **57**, pp. 223–237.

[35] Brzdęk, J. (1994). A note on stability of additive mappings, In *Stability of Mappings of Hyers–Ulam Type*, Th.M. Rassias and J. Tabor (Eds.), Hadronic Press, Palm Harbor, FL, pp. 19–22.

[36] Brzdęk, J. (2009). On the quotient stability of a family of functional equations, *Nonlinear Anal.*, **71**, pp. 4396–4404.

[37] Brzdek, J. (2009). On a method of proving the Hyers–Ulam stability of functional equations on restricted domains, *Aust. J. Math. Anal. Appl.*, **6(1)**, pp. 1–10.

[38] Brzdek, J. and Sikorska, J. (2010). A conditional exponential functional equation and its stability, *Nonlinear Anal.*, **72**, pp. 2929–2934.

[39] Cadariu, L. and Radu, V. (2003). Fixed points and the stability of Jensen's functional equation, *J. Inequal. Pure Appl. Math.*, **4(1)**, Art. 4, p. 15.

[40] Cadariu, L. and Radu, V. (2004). On the stability of the Cauchy functional equation: A fixed point approach, *Grazer Math. Ber.*, **346**, pp. 43–52.

[41] Cao, H.X., Lv, J.R. and Rassias, J.M. (2009). Superstability for generalized module left derivations and generalized module derivations on a Banach module (I), *J. Inequal. Appl.*, **2009**, Art. ID 718020, pp. 1–10.

[42] Cao, H.X., Lv, J.R. and Rassias, J.M. (2009). Superstability for generalized module left derivations and generalized module derivations on a Banach module (II), *J. Pure Appl. Math.*, **10(2)**, pp. 1–8.

[43] Chang, I.S. and Kim, H.M. (2002). On the Hyers–Ulam stability of a quadratic functional equations, *J. Inequal. Appl. Math.*, **33**, pp. 1–12.

[44] Chang, I.S. and Jung, Y.S. (2003). Stability of functional equations deriving from cubic and quadratic functions, *J. Math. Anal. Appl.*, **283**, pp. 491–500.

[45] Hengkrawit, C. and Thanyacharoen, A. (2013). A general solution of a quartic functional equation and its stability, *Int. J. Pure Appl. Math.*, **85(4)**, pp. 691–706.

[46] Cho, Y.J., Saadati, R. and Vahidi, J. (2012). Approximation of homomorphisms and derivations on non-Archimedean Lie C^*-algebras via fix fixed point method, *Discrete Dyn. Nat. Soc.*, **2012**, Art. ID 373904, p. 9.

[47] Cholewa, P.W. (1984). Remarks on the stability of functional equations, *Aequationes Math.*, **27**, pp. 76–86.

[48] Chung, J., Chung, S.Y. and Kim, D. (1994). A characterization for Fourier hyperfunctions, *Publ. Res. Inst. Math. Sci.*, **30**, pp. 203–208.

[49] Chung, J. and Lee, S.Y. (2003). Some functional equations in the spaces of generalized functions, *Aequationes Math.*, **65**, pp. 267–279.

[50] Chung, J. (2005). A distributional version of functional equations and their stabilities, *Nonlinear Anal.*, **62**, pp. 1037–1051.

[51] Chung, J. (2007). Stability of approximately quadratic Schwartz distributions, *Nonlinear Anal*, **67**, pp. 175–186.

[52] Chung, J. (2009). Stability of a generalized quadratic functional equation in Schwartz distributions, *Acta Math. Sin. (Engl. Ser.)*, **25**, pp. 1459–1468.

[53] Chung, J., Kim, D. and Rassias, J.M. (2009). Hyers–Ulam stability on a generalized quadratic functional equation in distributions and hyperfunctions, *J. Math. Phys.*, **50**, Art. ID 113519, pp. 1–14.

[54] Chung, J. (2010). Stability of functional equations on restricted domains in a group and their asymptotic behaviors, *Comput. Math. Appl.*, **60**, pp. 2653–2665.

[55] Chung, J. (2014). Stability of a conditional Cauchy equation on a set of measure zero, *Aequationes Math.*, **87**, pp. 391–400.

[56] Chung, J. and Rassias, J.M. (2014). Quadratic functional equations in a set of Lebesgue measure zero, *J. Math. Anal. Appl.*, **419**, pp. 1065–1075.

[57] Chung, J.K. and Sahoo, P.K. (2003). On the general solution of a quartic functional equation, *Bull. Korean Math. Soc.*, **40(4)**, pp. 565–576.

[58] Chung, S.Y., Kim, D. and Lee, E.G. (2000). Periodic hyperfunctions and Fourier series, *Proc. Amer. Math. Soc.*, **128**, pp. 2421–2430.

[59] Czerwik, S. (1992). On the stability of the quadratic mapping in normed spaces, *Abh. Math. Sem. Univ. Hamburg*, **62**, pp. 59–64.

[60] Czerwik, S. (1994). The stability of the quadratic functional equation, in *Stability of Mappings of Hyers–Ulam Type*, T.M. Rassias and J. Tabor (Eds.), Hadronic Press, Palm Harbor, FL, pp. 81–91.

[61] Czerwik, S. (2002). *Functional Equations and Inequalities in Several Variables*, World Scientific Publishing Co., New Jersey.

[62] Czerwik, S. (2003). *Stability of Functional Equations of Ulam–Hyers–Rassias Type*, Hadronic Press, Palm Harbor, FL.

[63] Davison, T.M.K. (2001). D'Alembert's functional equation and Chebyshev polynomials, *Stud. Mat.*, **4**, pp. 31–38.

[64] Deeba, E., Sahoo, P.K. and Xie, S. (1998). On a class of functional equations in distributions, *J. Math. Anal. Appl.*, **223**, p. 334.

[65] Diaz, J. and Margolis, B. (1968). A fixed point theorem of the alternative for contractions on a generalized complete metric space, *Bull. Amer. Math. Soc.*, **74**, pp. 305–309.

[66] Drljevic, H. (1984). On the stability of the functional quadratic on A-orthogonal vectors, *Publ. Inst. Math. (Beograd) (N.S.)*, **36**, pp. 111–118.

[67] Eghbali, N. and Ganji, M. (2012). Hyers–Ulam–Rassias stability of functional equations in Menger probabilistic normed spaces, *J. Appl. Anal. Comp.*, **2(2)**, pp. 149–159.

[68] Eskandani, G.Z. (2003). On the Hyers–Ulam–Rassias stability of an additive functional equation in quasi-Banach spaces, *J. Math. Anal. Appl.*, **345**, pp. 405–409.

[69] Eskandani, G.Z., Zamani, A.R. and Vaezi, H. (2011). Fuzzy approximation of an additive functional equation, *J. Funct. Spaces Appl.*, **9(2)**, pp. 205–215.

[70] Eskandani, G.Z., Rassias, J.M. and Gavruta, P. (2011). Generalized Hyers–Ulam stability for a general cubic functional equation in quasi-β-normed spaces, *Asian-Eur. J. Math.*, **4**, pp. 413–425.

[71] Eskandani, G.Z. and Rassias, Th.M. (2013). Hyers–Ulam–Rassias stability of derivations in proper JCQ* triples, *Mediterr. J. Math.*, **10**, pp. 1391–1400.

[72] Eskandani, G.Z. and Rassias, J.M. (2014). Approximation of a general cubic functional equation in felbin's type fuzzy normed linear spaces, *Results Math.*, **66**, pp. 113–123.

[73] Fechner, W. (2006). Stability of a functional inequalities associated with the Jordan-von Neumann functional equation, *Aequationes Math.*, **71**, pp. 149–161.

[74] Felbin, C. (1992). Finite dimensional fuzzy normed linear spaces, *Fuzzy Sets Syst.* **48**, pp. 239–248.

[75] Fenyö, I. (1987). On an inequality of P.W. Cholewa, in *General Inequalities*, Vol. 5, Internationale Schriftenreiche zur Numericchen Mathematics, Vol. 80, Birkhäuser, Basel, pp. 277–280.

[76] Filippov, V.T. (1985). n-Lie algebras, *Sib. Mat. Zh.*, **26**, pp. 126–140.

[77] Filippov, V.T. (1998). On n-Lie algebras of Jacobians, *Sib. Mat. Zh.*, **39**, pp. 660–669.

[78] Fochi, M. (2005). An alternative functional equation on restricted domain, *Aequationes Math.*, **70**, pp. 201–212.

[79] Förg-Rob, W. and Schwaiger, J. (1993). On the stability of a system of functional equations characterizing generalized hyperbolic and trigonometric functions, *Aequationes Math.*, **45**, pp. 285–296.

[80] Förg-Rob, W. and Schwaiger, J. (1994). On the stability of some functional equations for generalized hyperbolic functions and for the generalized cosine equation, *Results Math.*, **26**, pp. 274–280.

[81] Forti, G.L. (1995). Hyers–Ulam stability of functional equations in several variables, *Aequationes Math.*, **50**, pp. 143–190.

[82] Forti, G.L. (2004). Comments on the core of the direct method for proving Hyers–Ulam stability of functional equations, *J. Math. Anal. Appl.*, **295**, pp. 127–133.

[83] Gajda, Z. (1991). On the stability of additive mappings, *Int. J. Math. Math. Sci.*, **14**, pp. 431–434.

[84] Găvruță, P. (1994). A generalization of the Hyers–Ulam–Rassias stability of approximately additive mappings, *J. Math. Anal. Appl.*, **184**, pp. 431–436.

[85] Găvruță, P. (1999). An answer to a question of J.M. Rassias concerning the stability of Cauchy functional equation, *Adv. Equ. Inequal, Hadronic Math. Ser.*, pp. 67–71, Hardromic Press, Palm Harbor, Fla. USA.

[86] Gelfand, I.M. and Shilov, G.E. (1968). *Generalized Functions II*, Academic Press, New York.

[87] Ger, R. (1990). Stability of addition formulae for trigonometric mappings, *Zeszyty Nauk. Politech. Slasiej Ser. Mat. Fiz.*, **64**, pp. 75–84.

[88] Ger, R. (1992). On functional inequalities stemming from stability questions, in *General Inequalities*, Vol. 6, W. Walter (Ed.), pp. 227–240, Birkhäuser, Basel/Boston.

[89] Ger, R. and Sikorska, J. (1997). On the Cauchy equation on spheres, *Ann. Math. Sil.*, **11**, pp. 89–99.

[90] Gervirtz, J. (1983). Stability of isometries on Banach spaces, *Proc. Amer. Math. Soc.*, **89**, pp. 633–636.

[91] Ghadir Sadeghi. (2010). A quadratic type functional equation, *Bull. Math. Anal. Appl.*, **2(4)**, pp. 130–136.

[92] Gilányi, A. (2001). Eine zur Parallelogrammgleichung äquivalente Ungleichung, *Aequationes Math.*, **62**, pp. 303–309.

[93] Gilányi, A. (2002). On a problem by K. Nikodem, *Math. Inequal. Appl.*, **5**, pp. 707–710.

[94] Gordji, M.E., Gharetapeh, S.K. and Zolfaghari, S. (2008). Stability of a mixed type quadratic, cubic and quartic functional equation, preprint, arXiv:0812.2939v2 [math.FA].

[95] Gordji, M.E., Kaboli, G.S., Rassias, J.M. and Zolfaghari, S. (2009). Solution and stability of a mixed type additive, quadratic and cubic functional equation, *Adv. Difference Equ.*, **2009**, Art. ID 826130, 17 pp.

[96] Gordji, M.E., Zolfaghari, S., Rassias, J.M. and Savadkouhi, M.B. (2009). Solution and stability of a mixed type cubic and quartic functional equation in quasi-Banach spaces, *Abs. Appl. Anal.*, **2009**, Art. ID 417473, 14 pp.

[97] Gordji, M.E., Gharetapeh, S.K., Park, C. and Zolfaghri, S. (2009). Stability of an additive-cubic-quartic functional equation, *Adv. Difference Equ.*, **2009**, Art. ID 395693, 20 pp.

[98] Gordji, M.E. and Khodaei, H. (2009). On the generalized Hyers–Ulam–Rassias stability of quadratic functional equations, *Abs. Appl. Anal.*, **2009**, Art. ID 923476, 11 pp.

[99] Gordji, M.E., Savadkouhi, M.B. and Park, C. (2009). Quadratic-quartic functional equations in RN-spaces, *J. Inequal Appl.*, **2009**, Art. ID 868423, 14 pp.

[100] Gordji, M.E., Abbaszadeh, S. and Park, C. (2009). On the stability of a generalized quadratic and quartic type functional equation in quasi-Banach spaces, *J. Inequal Appl.*, **2009**, Art. ID 153084, 26 pp.

[101] Gordji, M.E., Gharetapeh, S.K., Park, C. and Zolfaghari, S. (2009). Stability of an additive-cubic-quartic functional equation, *Adv. Difference Equ.*, **2009**, Art.ID 395693, 20 pp.

[102] Gordji, M.E. and Khodaei, H. (2009). Solution and stability of generalized mixed type cubic, quadratic and additive functional equation in quasi-Banach spaces, *Nonlinear Anal.*, **71**, pp. 5629–5643.

[103] Gordji, M.E. (2010). Stability of a functional equation deriving from quartic and additive functions, *Bull. Korean Math. Soc.*, **47(3)**, pp. 491–502.

[104] Gordji, M.E., Khodaei, H. and Khodabakhsh, R. (2010). General quartic-cubic-quadratic functional equation in non-Archimedean normed spaces, *U.P.B. Sci. Bull., Ser. A*, **72(3)**, pp. 69–84.

[105] Gordji, M.E. and Najati, A. (2010). Approximately J^*-homomorphisms: A fixed point approach, *J. Geom. Phys.*, **60**, pp. 809–814.

[106] Gordji, M.E., Cho, Y.J., Ghaemi, M.B. and Majani, H. (2011). Approximately quintic and sextic mappings from r-divisible groups into Serstnev probabilistic Banach spaces: Fixed point method, *Discrete Dyn. Nat. Soc.*, **2011**, Art. ID 572062, 16 pp.

[107] Gordji, M.E. and Kamyar, M. and Rassias, Th. M. (2011). General cubic-quartic functional equation, *Abs. Appl. Anal.*, **2011**, Art. ID 463164, 18 pp.

[108] Gordji, M.E. and Savadkouhi, M.B. (2011). Stability of a mixed type additive, quadratic and cubic functional equation in random normed spaces, *Filomat*, **25(3)**, pp. 43–54.

[109] Gordji, M.E. and Khodabakhsh, R. and Khodaei, H. (2011). On approximate n-ary derivations, *Int. J. Geom. Methods Mod. Phys.*, **8**, pp. 485–500.

[110] Gordji, M.E. and Fazeli, A. (2012). Stability and superstability of homomorphisms on C^*-ternary algebras, *An. Ştiinţ. Univ. Ovidius Constanta Ser. Mat.*, **20**, pp. 173–188.

[111] Gordji, M.E., Ramezani, M., Cho, Y.J. and Baghani, H. (2012). Approximate Lie brackets: A fixed point approach, *J. Inequal. Appl.*, **2012**, p. 125.

[112] Gordji, M.E., Cho, Y.J., Khodaei, H. and Ghanifard, M. (2013). Solutions and stability of generalized mixed type QCA-functional equations in random normed spaces, *An. Ştiinţ. Univ. Al.I. Cuza Iaşi Mat. (N.S.)*, 59, pp. 299–320.

[113] Grabiec, A. (1996). The generalized Hyers–Ulam stability of a class of functional equations, *Publ. Math. Debrecen*, **48**, pp. 217–235.

[114] Gruber, P.M. (1978). Stability of isometries, *Trans. Amer. Math. Soc.*, **245**, pp. 263–277.

[115] Gustavsson, A. (2009). One-loop corrections to Bagger–Lambert theory, *Nucl. Phys. B*, **807**, pp. 315–333.

[116] Hoseinia, H. and Kenary, H.A. (2011). Stability of cubic functional equations in non-Archimedean normed spaces, *Math. Sci.*, **5(4)**, pp. 321–336.

[117] El-Hady, El-S., Förg-Rob, W. and Mahmoud, M. (2015). On a two-variable functional equation arising from databases, *WSEAS Trans. Math.*, **14**, pp. 265–270.

[118] Koh, H. and Kang, D. (2013). Solution and stability of Euler–Lagrange–Rassias quartic functional equations in various quasi-normed spaces, *Abs. Appl. Anal.*, **2013**, Art. ID 908168, 8 pp.

[119] Hoehle, U. (1987). Fuzzy real numbers as Dedekind cuts with respect to a multiple-valued logic, *Fuzzy Sets Syst.*, **24**, pp. 263–278.

[120] Hörmander, L. (1983). *The Analysis of Linear Partial Differential Operators I*, Springer, Berlin.

[121] Hosszu, M. (1964). On the Fréchet functional equation, *Bul. Isnt. Politech. Iasi*, **10(1–2)**, pp. 27–28.

[122] Hyers, D.H. (1941). On the stability of the linear functional equation, *Proc. Natl. Acad. Sci., USA*, **27**, pp. 222–224.

[123] Hyers, D.H. (1983). The stability of homomorphisms and related topics, in *Global Analysis-Analysis on Manifolds, Teubner-Texte zur Mathematik*, Vol. 57, Teubner Verlagsgesellschaftt, pp. 140–153.

[124] Hyers, D.H., Isac, G. and Rassias, Th.M. (1998). *Stability of Functional Equations in Several Variables*, Birkhäuser, Basel.

[125] Jarosz, K. (1985). *Perturbations of Banach Algebras*, Lectures Notes in Mathematics, Vol. 1120, Springer-Verlag, Berlin.

[126] Javadi, S. and Rassias, J.M. (2012). Stability of general cubic mapping in fuzzy normed spaces, *An. Ştiinţ Univ. Ovidius Constanta Ser. Mat.*, **20(1)**, pp. 129–150.

[127] Jensen, J.L.W.V. (1965). Sur les fonctions convexes et les inegalities entre les valeurs myennes, *Acta Math.*, **30**, pp. 179–193.

[128] Jian, W. (2001). Some further generalizations of the Hyers–Ulam–Rassias stability of functional equations, *J. Math. Anal. Appl.*, **263**, pp. 406–423.

[129] Jin, S.S. and Lee, Y.H. (2012). On the stability of the quadratic-additive type functional equation in random normed spaces via fixed point method, *Korean J. Math.*, **20(1)**, pp.19–31.

[130] Johnson, B.E. (1986). Approximately multiplicative functionals, *J. London Math. Soc.*, **34(2)**, pp. 489–510.

[131] Johnson, B.E. (1988). Approximately multiplicative maps between Banach algebras, *J. London Math. Soc.*, **37(2)**, pp. 294–316.

[132] Jordan, P. and Von Neumann, J. (1935). On inner products in linear metric spaces, *Ann. of Math.*, **36**, pp. 719–723.

[133] Jun, K. and Lee, Y. (1999). A generalization of the Hyers–Ulam–Rassias stability of Jensen's equation, *J. Math. Anal. Appl.*, **238**, pp. 305–315.

[134] Jun, K.W., Kim, G.H. and Lee, Y.W. (2000). Stability of generalized gamma and beta functional equations, *Aequationes Math.*, **60**, pp. 15–24.

[135] Jun, K.W. and Kim, H.M. (2001). Remarks on the stability of additive functional equation, *Bull. Korean Math. Soc.*, **38**, pp. 679–687.

[136] Jun, K.W. and Lee, Y.H. (2001). On the Hyers–Ulam–Rassias stability of a Pexiderized quadratic inequality, *Math. Inequal Appl.*, **4**, pp. 93–118.

[137] Jun, K.W. and Kim, H.M. (2002). The generalized Hyers–Ulam–Rassias stability of a cubic functional equation, *J. Math. Anal. Appl.*, **274**, pp. 867–878.

[138] Jun, K.W. and Kim, H.M. (2003). On the Hyers–Ulam–Rassias stability of a general cubic functional equation, *Math. Inequal Appl.*, **6(2)**, pp. 289–302.

[139] Jun, K.W. and Kim, H.M. (2004). On the stability of a general quadratic functional equation and its applications, *J. Chungcheong Math. Soc.*, **17(1)**, pp. 57–75.

[140] Jun, K.W. and Kim, H.M. (2005). Ulam stability problem for quadratic mappings of Euler–Lagrange, *Nonlinear Anal.*, **61**, pp. 1093–1104.

[141] Jun, K.W. and Lee, S.B. (2006). On the generalized Hyers–Ulam–Rassias stability of a cubic functional equation, *J. Chungcheong Math. Soc.*, **19(2)**, pp. 189–196.

[142] Jun, K.W. and Lee, S.B. (2007). On the stability of an *n*-dimensional quadratic equation, *J. Chungcheong Math. Soc.*, **20(1)**, pp. 23–29.

[143] Jun, K.W., Kim, H.M. and Rassias, J.M. (2007). Extended Hyers–Ulam stability for Cauchy–Jensen mappings, *J. Difference Equ. Appl.*, **13**, pp. 1139–1153.

[144] Jung, S.M. (1997). Hyers–Ulam–Rassias stability of functional equations, *Dynamic Systems Appl.*, **6**, pp. 541–566.

[145] Jung, S.M. (1998). Hyers–Ulam–Rassias stability of Jensen's equation and its application, *Proc. Amer. Math. Soc.*, **126(11)**, pp. 3137–3143.

[146] Jung, S.M. (1998). On the Hyers–Ulam stability of the functional equations that have the quadratic property, *J. Math. Anal. Appl.*, **222**, pp. 126–137.

[147] Jung, S.M. (1999). On the Hyers–Ulam–Rassias stability of a quadratic functional equation, *J. Math. Anal. Appl.*, **232**, pp. 384–393.

[148] Jung, S.M. and Kim, B. (1999). On the stability of the quadratic functional equation on bounded domains, *Abh. Math. Sem. Univ. Hamburg*, **69**, pp. 293–308.

[149] Jung, S.M. and Sahoo, P.K. (2002). Stability of a functional equation of Drygas, *Aequationes Math.*, **64**, p. 263.

[150] Jung, S.M., Kim, T.S. and Lee, K.S. (2006). A fixed point approach to the stability of quadratic functional equation, *Bull. Korean Math. Soc.*, **43(3)**, pp. 531–541.

[151] Jung, S.M. and Moslehian, M.S. and Sahoo, P.K. (2010). Stability of a generalized Jensen equation on restricted domains, *J. Math. Inequal.*, **4**, pp. 191–206.

[152] Jung, S.M. (2011). *Hyers–Ulam–Rassias Stability of Functional Equations in Nonlinear Analysis*, Springer, New York.

[153] Kadison, R.V. and Ringrose, J.R. (1986). *Fundamentals of the Theory of Operator Algebras*, Academic Press, New York.

[154] Kaleva, O. and Seikkala, S. (1984). On fuzzy metric spaces, *Fuzzy Sets Syst.*, **12**, pp. 215–229.

[155] Kaleva, O. (1985). The completion of fuzzy metric spaces, *J. Math. Anal. Appl.*, **109**, pp. 194–198.

[156] Kaleva, O. (2008). A comment on the completion of fuzzy metric spaces, *Fuzzy Sets Syst.*, **159(16)**, pp. 2190–2192.

[157] Kannappan, Pl. (1969). Theory of functional equations, *Matsci. Rep.*, **48**.

[158] Kannappan, Pl. (1995). Quadratic functional equation in inner product spaces, *Results Math.*, **27(3–4)**, pp. 368–372.

[159] Kannappan, Pl. (2001). Application of Cauchy's equation in combinatorics and genetics, *Mathware Soft Comput.*, **8**, pp. 61–64.

[160] Kasymov, S.M. (1987). On the theory of n-Lie algebras, *Algebra Logika*, **26**, pp. 277–297.

[161] Kasymov, S.M. (1991). On nil-elements and nil-subsets of n-Lie algebras, *Sib. Math. J*, **32**, pp. 77–80.

[162] Kim, H.M. and Rassias, J.M. (2007). Generalization of Ulam stability problem for Euler–Lagrange quadratic mappings, *J. Math. Anal. Appl.*, **336**, pp. 277–296.

[163] Kim, S.S., Rassias, J.M., Cho, Y.J. and Kim, S.H. (2013). Generalized Hyers–Ulam stability of derivations on Lie C^*-algebras, *J. Adv. Phys.*, **3(1)**, pp. 176–186.

[164] Kim, S.S., Rassias, J.M., Cho, Y.J. and Kim, S.H. (2013). Stability of n-Lie homomorphisms and Jordan n-Lie homomorphisms on n-Lie algebras, *J. Math. Phys.*, **54**, 053501.

[165] Kim, S.S., Rassias, J.M. and Kim, S.H. (2014). Homomorphisms and derivations on C^*-ternary algebras associated with a generalized Cauchy–Jensen type additive functional equation, *J. Adv. Phys.*, **4(3)**, pp. 662–668.

[166] Kim, Y.H. (2002). On the Hyers–Ulam–Rassias stability of an equation of Davison, *Indian J. Pure Appl. Math.*, **33(5)**, pp. 713–726.

[167] Kominek, Z. (1989). On a local stability of the Jensen functional equation, *Demonstratio Math.*, **22**, pp. 499–507.

[168] Kreyszig, E. (1978). *Introduction to Functional Analysis with Applications*, Wiley, New York.

[169] Kuczma, M. (1978). Functional equations on restricted domains, *Aequationes Math.*, **18**, pp. 1–34.

[170] Kuczma, M. (1985). *An Introduction to the Theory of Functional Equations and Inequalities*, Panstwowe Wydawnictwo Naukowe — Uniwersytet Slaski, Warszawa-Krakow-Katowice.

[171] Kurepa, S. (1960). A cosine functional equation in Hilbert spaces, *Canan. J. Math.*, **12**, pp. 45–60.

[172] Lee, H., Kim, S.W., Son, B.J., Lee, D.H. and Kang, S.Y. (2012). Additive-quartic functional equation in non-Archimedean orthogonality spaces, *Korean J. Math.*, **20(1)**, pp. 33–46.

[173] Lee, J., An, J.S. and Park, C. (2008). On the stability of quadratic functional equations, *Abs. Appl. Anal.*, **2008**, Art. ID 628178, 8 pp.

[174] Lee, J.R. and Shin, D.Y. (2004). On the Cauchy–Rassias stability of the Trif functional equation in C^*-algebras, *J. Math. Anal. Appl.*, **296**, pp. 351–363.

[175] Lee, J.R., Jang, S.Y., Park, C and Shin, D.Y. (2010). Fuzzy stability of quadratic functional equations, *Adv. Difference Equ.*, **2010**, Art. ID 412160, 16 pp.

[176] Lee, J.R., Shin, D.Y. and Park, C. (2013). Hyers–Ulam stability of functional equations in matrix normed spaces, *J. Inequal Appl.*, **2013(22)**, 11 pp. DOI:10.1186/1029-242X-2013-22

[177] Lee, S.B., Park, W.G. and Bae, J.H. (2011). On the Ulam–Hyers stability of a quadratic functional equation, *J. Inequal Appl.*, **2011 (79)**, 9 pp. DOI: 10.1186/1029-242X-2011-79

[178] Lee, S.H., Im, S.M. and Hwang, I.S. (2005). Quartic functional equations, *J. Math. Anal. Appl.*, **307**, pp. 387–394.

[179] Lee, S.J. (2011). Quadratic mappings associated with inner product spaces, *Korean J. Math.*, **19(1)**, pp. 77–85.

[180] Lee, Y.H. and Jun, K.W. (1999). A generalization of the Hyers–Ulam–Rassias stability of Jensen's equation, *J. Math. Anal. Appl.*, **238**, pp. 305–315.

[181] Lee, Y.H. and Jun, K.W. (2000). On the stability of approximately additive mappings, *Proc. Amer. Math. Soc.*, **128**, pp. 1361–1369.

[182] Lee, Y.S., Na, J. and Woo, H. (2013). On the stability of a mixed type quadratic-additive functional equation, *Adv. Difference Equ.*, **2013(198)**, 10 pp.

[183] Lee, Y.W. (2000). The stability of derivations on Banach algebras, *Bull. Inst. Math. Acad. Sinica*, **28**, pp. 113–116.

[184] Lee, Y.W. (2002). On the stability of a quadratic Jensen type functional equation, *J. Math. Anal. Appl.*, **270**, pp. 590–601.

[185] Lee, Y.W. (2005). Stability of a generalized quadratic functional equation with Jensen type, *Bull. Korean Math. Soc.*, **42(1)**, pp. 57–73.

[186] Lindenstrauss, J. and Szankowski, A. (1985). *Non-linear Perturbations of Isometries*, Colloquium in Honor of Laurent Schwartz, Vol. I, Palaiseau.

[187] Liguang, W. and Jing, L. (2012). On the stability of a functional equation deriving from additive and quadratic functions, *Adv. Difference Equ.*, **2012(98)**, pp. 1–12.

[188] Losonczi, L. (1996). On the stability of Hosszú's functional equation, *Results Math.*, **29**, pp. 305–310.

[189] Lowen, R. (1996). Fuzzy real numbers, in *Fuzzy Set Theory*, Chap. 5, Kluwer, Dordrecht.

[190] Maligranda, L. (2008). A result of Tosio Aoki about a generalization of Hyers–Ulam stability of additive functions — a question of priority, *Aequationes Math.*, **75**, pp. 289–296.

[191] Matsuzawa, T. (1990). A calculus approach to hyperfunctions III, *Nagoya Math. J.*, **118**, p. 133.

[192] McKiernan, M.A. (1967). On vanishing n-th ordered differences and Hamel bases, *Ann. Polon. Math.*, **19**, pp. 331–336.

[193] Mirmostafaee, A.K. and Moslehian, M.S. (2008). Fuzzy version of Hyers–Ulam–Rassias theorem, *Fuzzy Sets Syst.*, **159(6)**, pp. 720–729.

[194] Mirzavaziri, M. and Moslehian, M.S. (2006). A fixed point approach to stability of a quadratic equation, *Bull. Braz. Math. Soc.*, **37**, pp. 361–376.

[195] Mizomoto, M. and Tanaka, K. (1979). Some properties of fuzzy numbers, in *Adv. Fuzzy Set Theory and Applications*, M.M. Gupto *et al.* (Eds.), North-Holland, New York, pp. 153–164.

[196] Mohamadi, M., Cho, Y.J., Park, C., Vetro, P. and Saadati, R. (2010). Random stability of an additive-quadratic-quartic functional equation, *J. Inequal Appl.*, **2010**, Art. ID 754210, 18 pp.

[197] Moradlou, F., Vaezi, H. and Eskandani, G.Z. (2009). Hyers–Ulam–Rassias stability of a quadratic and additive functional equation in quasi-Banach spaces, *Mediterr. J. Math.*, **6**, pp. 233–248.

[198] Moslehian, M.S. and Rassias, Th.M. (2007). Stability of functional equations in non-Archimedean spaces, *Appl. Anal. Discrete Math.*, **1**, pp. 325–334.

[199] Moslehian, M.S. and Sadeghi, G. (2008). Stability of two types of cubic functional equations in non-Archimedean spaces, *Real Anal. Exchange*, **33(2)**, pp. 375–384.

[200] Movahednia, E. (2013). Fixed point and Hyers–Ulam–Rassias stability of a quadratic functional equation in Banach spaces, *IJRRAS*, **16(1)**, pp. 111–117.

[201] Najati, A. (2007). Hyers–Ulam–Rassias stability of a cubic functional equation, *Bull. Korean Math. Soc.*, **44(4)**, pp. 825–840.

[202] Najati, A. (2007). The generalized Hyers–Ulam–Rassias stability of a cubic functional equation, *Turk. J. Math.*, **31**, pp. 395–408.

[203] Najati, A. and Moradlou, F. (2008). Stability of a quadratic functional equation in quasi-Banach spaces, *Bull. Korean Math. Soc.*, **45(3)**, pp. 587–600.

[204] Najati, A. and Park, C. (2008). On the stability of a cubic functional equation, *Acta Math. Sin. (Engl. Ser.)*, **24(12)**, pp. 1953–1964.

[205] Najati, A. and Eskandani, G.Z. (2008). Stability of a mixed additive and cubic functional equation in quasi-Banach spaces, *J. Math. Anal. Appl.*, **342**, pp. 1318–1331.

[206] Najati, A. and Ranjbari, A. (2008). Stability of homomorphisms for a 3D Cauchy-Jensen type functional equation on C^*-ternary algebras, *J. Math. Anal. Appl.*, **341**, pp. 62–79.

[207] Najati, A. and Jung, S.M. (2010). Approximately quadratic mappings on restricted domains, *J. Inequal Appl.*, **2010**, Art. ID 503458, 10 pp.

[208] Nakmahachalasint, P. (2006). The generalized Hyers–Ulam–Rassias stability of a quadratic functional equation, *Thai J. Math.*, **4(2)**, pp. 321–328.

[209] Nakmahachalasint, P. (2007). On the generalized Ulam–Gavruta–Rassias stability of mixed-type linear and Euler–Lagrange–Rassias functional equations, *Int. J. Math. Math. Sci.*, **2007**, Art. ID 63239, 10 pp.

[210] Nakmahachalasint, P. (2007). Hyers–Ulam–Rassias and Ulam–Gavruta–Rassias stabilities of additive functional equation in several variables, *Int. J. Math. Math. Sci.*, **2007**, Art. ID 13437, 6 pp.

[211] Nakmahachalasint, P. and Towanlong, W. (2008). A quadratic functional equation and its generalized Hyers–Ulam–Rassias stability, *Thai J. Math.* **2008(Special Issue)**, pp. 85–91.

[212] Nambu, Y. (1973). Generalized Hamiltonian dynamics, *Phys. Rev. D* **7**, pp. 2405–2412.

[213] Oxtoby, J.C. (1980). *Measure and Category*, Springer, New York.

[214] Parnami, J.C. and Vasudeva, H.L. (1992). On Jensen's functional equation, *Aequationes Math.*, **43**, pp. 211–218.

[215] Park, C. (2002). On the stability of the linear mapping in Banach modules, *J. Math. Anal. Appl.*, **275**, pp. 711–720.

[216] Park, C. and Park, W. (2002). On the Jensen's equation in Banach modules, *Taiwanese J. Math.*, **6**, pp. 523–532.

[217] Park, C. (2003). Linear functional equations in Banach modules over a C^*-algebra, *Acta Appl. Math.*, **77**, pp. 125–161.

[218] Park, C. (2004). On an approximate automorphism on a C^*-algebra, *Proc. Amer. Math. Soc.*, **132**, pp. 739–745.

[219] Park, C. (2004). Lie $*$-homomorphisms between Lie C^*-algebras and Lie $*$-derivations on Lie C^*-algebras, *J. Math. Anal. Appl.*, **293**, pp. 419–434.

[220] Park, C. (2005). Homomorphisms between Lie JC^*-algebras and Cauchy–Rassias stability of Lie JC^*-algebra derivations, *J. Lie Theory*, **15**, pp. 393–414.

[221] Park, C. (2005). Isomorphisms between unital C^*-algebras, *J. Math. Anal. Appl.*, **307**, pp. 753–762.

[222] Park, C. (2005). Homomorphisms between Poisson JC^*-algebras, *Bull. Braz. Math. Soc.*, **36**, pp. 79–97.

[223] Park, C., Hou, J. and Oh, S. (2005). Homomorphisms between JC^*-algebras and between Lie C^*-algebras, *Acta Math. Sinica*, **21**, pp. 1391–1398.

[224] Park, C. (2006). Isomorphisms between C^*-algebras, *J. Math. Phys.*, **47**, 103512.

[225] Park, C. (2006). Hyers–Ulam–Rassias stability of isomorphisms in C^*-algebras, *J. Chungcheong Math. Soc.*, **19(2)**, pp. 159–175.

[226] Park, C., Jun, K.W. and Lu, G. (2006). On the quadratic mapping in generalized quasi-Banach spaces, *J. Chungcheong Math. Soc.*, **19(3)**, pp. 263–274.

[227] Park, C. (2007). Fixed points and Hyers–Ulam–Rassias stability of Cauchy-Jensen functional equations in Banach spaces, *Fixed Point Theory Appl.*, **2007**, Art. ID 51075, p. 15.

[228] Park, C., Cho, Y. and Han, M. (2007). Functional inequalities associated with Jordan-von Neumann type additive functional equations, *J. Inequal. Appl.*, **2007**, Art. ID 41820, p. 13.

[229] Park, C. and Cui, J. (2007). Generalized stability of C^*-ternary quadratic mappings, *Abs. Appl. Anal.*, **2007**, Art. ID 23282, p. 6.

[230] Park, C. and Najati, A. (2007). Homomorphisms and derivations in C^*-algebras, *Abs. Appl. Anal.*, **2007**, Art. ID 80630, pp. 1–12.

[231] Park, C., An, J.S. and Cui, J. (2007). Isomorphisms and Derivations in Lie C^*-Algebras, *Abs. Appl. Anal.*, **2007**, Art ID 85737, 14 pp.

[232] Park, C. (2008). Generalized Hyers–Ulam–Rassias stability of quadratic functional equations: A fixed point approach, *Fixed Point Theory and Appl.*, **2008**, Art. ID 493751, 9 pp.

[233] Park, C. and An, J.S. (2008). Stability of the Cauchy–Jensen functional equations in C^*-algebras: A fixed point approach, *Fixed Point Theory Appl.*, *2008*, Art. ID 872190, 11 pp.

[234] Park, C. and Rassias, Th.M. (2009). Fixed points and stability of the Cauchy functional equation, *Aust. J. Math. Anal. Appl.*, **6(1)**, Art. ID 14, 9 pp.

[235] Park, C. (2009). Fixed points and stability of Cauchy functional equations in C^*-algebras, *Fixed Point Theory Appl.*, **2009**, Article ID 809232, 14 pp.

[236] Park, C. and Rassias, J.M. (2009). Stability of the Jensen-type functional equations in C^*-algebras: A fixed point approach, *Fixed Point Theory Appl.*, **2009**, Art. ID 360432, 17 pp.

[237] Park, C., Jo, S.W. and Kho, D.Y. (2009). On the stability of AQCQ-functional equation, *J. Chungcheon Math. Soc.*, **22(4)**, pp. 757–770.

[238] Park, C. and Kim, J.H. (2009). The stability of a quadratic functional equation with the fixed point alternative, *Abs. Appl. Anal.*, **2009**, Art. ID 907167, 11 pp.

[239] Park, C. (2009). A fixed point approach to the fuzzy stability of an additive-quadratic-cubic functional equation, *Fixed Point Theory Appl.*, **2009**, Art. ID 918785, 24 pp.

[240] Park, C. (2010). Fuzzy stability of an additive-quadratic-quartic functional equation, *J. Inequal Appl.*, **2010**, Art. ID 253040, 22 pp.

[241] Park, C. (2010). On the stability of an additive functional inequality in normed modules, Chap. 9, in *Functional equations, Difference Inequalities and Ulam stability Notions*, Nova Science Publishers, pp. 107–117.

[242] Park, C. and Shin, D.Y. (2012). Functional equations in paranormed spaces, *Adv. Difference Equ.*, **2012(123)**, 23 pp.

[243] Park, C.G. (2007). Stability of an Euler–Lagrange–Rassias type additive mapping, *Int. J. Appl. Math. Stat.*, **7**, pp. 101–111.

[244] Park, C.G. and Rassias, J.M. (2007). Hyers–Ulam stability of an Euler–Lagrange type additive mapping, *Int. J. Math. Stat.*, **7**, pp. 112–125.

[245] Park, K.H. and Jung, Y.S. (2004). Stability of a cubic functional equation on groups, *Bull. Korean Math. Soc.*, **41(2)**, pp. 347–357.

[246] Park, W.G. and Bae, J.H. (2005). On a bi-quadratic functional equation and its stability, *Nonlinear Anal.*, **62(4)**, pp. 643–654.

[247] Park, W.G., Bae, J.H. and Chung, B.H. (2005). On an additive-quadratic functional equation and its stability, *J. Appl. Math. Comput.*, **18(1)**, pp. 563–572.

[248] Park, W.G. and Bae, J.H. (2008). A functional equation originating from elliptic curves, *Abs. Appl. Anal.*, **2008**, Art. ID 135237, 10 pp.

[249] Petapirak, M. and Nakmahachalasint, P. (2008). A quartic functional equation and its generalized Hyers–Ulam–Rassias stability, *Thai J. Math.* (**Special Issue**), pp. 77–84.

[250] Pietrzyk, A. (2006). Stability of the Euler–Lagrange–Rassias functional equation, *Demonstratio Math.* **39(3)**, pp. 523–530.

[251] Popa, D. (2004). Functional inclusions on square-symmetric groupoids and Hyers–Ulam stability, *Math. Inequal. Appl.*, **7**, pp. 419–428.

[252] Prastaro, A. (1996). *Geometry of PDEs and Mechanics*, World Scientific, River Edge, NJ.

[253] Prastaro, A. (1999). (Co)bordism groups in PDEs, *Acta Appl. Math.*, **59**, p. 111.

[254] Radu, V. (2003). The fixed point alternative and stability of functional equations, *Fixed Point Theory, Cluj-Napoca* **4(1)**, pp. 91–96.

[255] Rassias, J.M. (1982). On approximately of approximately linear mappings by linear mappings, *J. Funct. Anal.*, **46**, pp. 126–130.

[256] Rassias, J.M. (1984). On approximately of approximately linear mappings by linear mappings, *Bull. Sci. Math.*, **108(4)**, pp. 445–446.

[257] Rassias, J.M. (1985). On a new approximation of approximately linear mappings by linear mappings, *Discuss. Math.*, **7**, pp. 193–196.

[258] Rassias, J.M. (1989). Solution of a problem of Ulam, *J. Approx. Theory*, **57**, pp. 268–273.

[259] Rassias, J.M. (1992). Solution of a stability problem of Ulam, *Discuss. Math.*, **12**, pp. 95–103.

[260] Rassias, J.M. (1992). On the stability of the Euler–Lagrange functional equation, *Chinese J. Math.*, **20**, pp. 185–190.

[261] Rassias, J.M. (1994). Complete solution of the multi-dimensional problem of Ulam, *Discuss. Math.*, **14**, pp. 101–107.

[262] Rassias, J.M. (1994). On the stability of the non-linear Euler–Lagrange functional equation in real normed linear spaces, *J. Math. Phys. Sci.*, **28**, pp. 231–235.

[263] Rassias, J.M. (1994). On the stability of a multi-dimensional Cauchy type functional equation, in *Geometry, Analysis and Mechanics*, World Scientific, pp. 365–376.

[264] Rassias, J.M. (1996). On the stability of the general Euler–Lagrange functional equation, *Demonstratio Math.*, **29**, pp. 755–766.

[265] Rassias, J.M. (1998). Solution of the Ulam stability problem for Euler–Lagrange quadratic mappings, *J. Math. Anal. Appl.*, **220**, pp. 613–639.

[266] Rassias, J.M. (1999). On the stability of the multi-dimensional Euler–Lagrange functional equation, *J. Indian Math. Soc.*, **66**, pp. 1–9.

[267] Rassias, J.M. (1999). Solution of the Ulam stability problem for quartic mappings, *Glasnik Mat. Ser. III*, **34(2)**, pp. 243–252.

[268] Rassias, J.M. (2001). Solution of a quadratic stability Hyers–Ulam type problem, *Ricerche Mat.* **1**, pp. 9–17.

[269] Rassias, J.M. (2001). Solution of a Cauchy–Jensen stability Ulam type problem, *Arch. Math. (Brno)*, **37**, pp. 161–177.

[270] Rassias, J.M. (2001). Solution of the Ulam stability problem for cubic mappings, *Glasnik Mat. Ser. III*, **36(56)**, pp. 63–72.

[271] Rassias, J.M. (2002). On the Ulam stability of mixed type mappings on restricted domains, *J. Math. Anal. Appl.*, **281**, pp. 747–762.

[272] Rassias, J.M. and Rassias, M.J. (2003). On the Ulam stability of Jensen and Jensen type mappings on restricted domains, *J. Math. Anal. Appl.*, **281**, pp. 516–524.

[273] Rassias, J.M. (2004). Solution of a quadratic stability Ulam type problem, *Arch. Math. (Brno)*, **40**, pp. 1–16.

[274] Rassias, J.M. (2005). On the Cauchy-Ulam stability of the Jensen equation in C^*-algebras, *Int. J. Pure Appl. Math. Sci.*, **2(1)**, pp. 92–101.

[275] Rassias, J.M. (2005). On the Ulam stability for Euler–Lagrange type quadratic functional equations, *Aust. J. Math. Anal. Appl.*, **2**, pp. 1–10.

[276] Rassias, J.M. and Kim, H.M. (2009). Generalized Hyers–Ulam stability for general additive functional equations in quasi-β-normed spaces, *J. Math. Anal. Appl.*, **356**, pp. 302–309.

[277] Rassias, J.M., Ravi, K., Arunkumar, M. and Senthil Kumar, B.V. (2010). Solution and Ulam stability of a mixed type cubic and additive functional equation, Chap. 13, in *Functional equations, Difference Inequalities and Ulam Stability Notions*, Nova Science Publishers, pp. 149–175.

[278] Rassias, J.M., Arunkumar, M., Ramamoorthi, S. and Hemalatha, S. (2014). Ulam-Hyers stability of a 2–variable AC- mixed type functional equation in quasi-beta normed spaces: direct and fixed point methods, *Malaya J. Mat.*, **2(2)**, pp. 108–128.

[279] Rassias, M.J. and Rassias, J.M. (2005). On the Ulam stability for Euler–Lagrange type quadratic functional equations, *Aust. J. Math. Anal. Appl.*, **2**, pp. 1–10.

[280] Rassias, M.J. (2010). Hyers–Ulam stability of Cauchy type additive functional equations, Chap. 12, in *Functional equations, Difference Inequalities and Ulam stability Notions*, Nova Science Publishers, pp. 143–147.

[281] Ravi, K. and Arunkumar, M. (2007). On the Hyers–Ulam stability of a particular quadratic functional equation, *Appl. Sci. Periodical*, **11(4)**, pp. 250–258.

[282] Ravi, K. and Arunkumar, M. (2007). On the Ulam–Gavruta–Rassias stability of the orthogonally Euler–Lagrange type functional equation, *Int. J. Math. Stat.*, **7(Fe07)**, pp. 143–156.

[283] Ravi, K. and Arunkumar, M. (2007). Stability of a 3-variable quadratic functional equation, in *Proceedings of ICMS07*, Malaysia, pp. 331–342.

[284] Ravi, K., Arunkumar, M. and Rassias, J.M. (2008). Ulam stability for the orthogonally general Euler–Lagrange type functional equation, *Int. J. Math. Stat.*, **3**, pp. 36–46.

[285] Ravi, K., Murali, R. and Arunkumar, M. (2008). The generalized Hyers–Ulam–Rassias stability of a quadratic functional equation, *J. Inequal Pure and Appl. Math.*, **9(1)**, Art. 20, 5 pp.

[286] Ravi, K. (2008). Functional equations, in *Proceedings of the Fifteenth Ramanujan Symposium on Dynamic Equations*, University of Madras, Chennai, pp. 58–67.

[287] Ravi, K., Rassias, J.M., Arunkumar, M. and Kodandan, R. (2009). Stability of a generalized mixed type additive, quadratic, cubic and quartic functional equation, *J. Inequal Pure Appl. Math.*, **10(4)**, Art. ID 114, 29 pp.

[288] Ravi, K., Narasimman, P. and Kishore Kumar, R. (2009). Generalized Hyers–Ulam–Rassias stability and J.M. Rassias stability of a quadratic functional equation, *Int. J. Math. Sci. Engl. Appl.*, **3(2)**, pp. 79–94.

[289] Ravi, K. and Senthil Kumar, B.V. (2010). Solution and stability of 2-variable additive and Jensen's functional equations, *Int. J. Math. Sci. Engl. Appl.*, **4(II)**, pp. 171–185.

[290] Ravi, K., Senthil Kumar, B.V., Thandapani, E. and Pinelas, S. (2010). Solution and generalized Hyers–Ulam stability of a 2-variable cubic functional equation, *Panamerican Math. J.*, **20(4)**, pp. 15–30.

[291] Ravi, K. and Murali, R. (2010). Stability of generalized additive functional equation in C^*-algebras: A fixed point approach, *Int. J. Math. Sci. Engl. Appl.*, **4(I)**, pp. 309–322.

[292] Ravi, K., Rassias, J.M. and Kodandan, R. (2011). Generalized Ulam-Hyers stability of an AQ-functional equation in quasi-beta-normed spaces, *Math. Aeterna*, **1(4)**, pp. 217–236.

[293] Ravi, K., Rassias, J.M. and Narasimman, P. (2011). Stability of a cubic functional equation in fuzzy normed space, *J. Appl. Anal. Comp.*, **1(3)**, pp. 411–425.

[294] Ravi, K., Senthil Kumar, B.V. and Kandasamy, S. (2011). Solution and stability of a mixed type quadratic and additive functional equation in two variables, *Int. J. Math. Comp.*, **13(D11)**, pp. 51–68.

[295] Ravi, K. and Narasimman, P. (2012). Stability of generalized quadratic functional equation in non-Archimedean ℓ-fuzzy normed spaces, *Int. J. Math. Anal.*, **6(24)**, pp. 1193–1203.

[296] Ravi, K. and Sabarinathan, S. (2014). Generalized Hyers–Ulam Stability of a sextic functional equation in paranormed spaces, *Int. J. Math. Trends Tech.*, **9(1)**, pp. 61–69.

[297] Ravi, K. and Edwin Raj, A. (2014). Generalized Hyers–Ulam stability of quadratic functional equation in paranormed spaces, *Indian J. Appl. Res.*, **4(3)**, pp. 285–293.

[298] Rassias, Th.M. (1978). On the stability of the linear mapping in Banach spaces, *Proc. Amer. Math. Soc.*, **72**, pp. 297–300.

[299] Rassias, Th.M. (1990). Problem 16:2, Report of the 27th international symp. on functional equations, *Aequationes Math.*, **39**, pp. 292–293.

[300] Rassias, Th.M. (1991). On a modified Hyers–Ulam sequence, *J. Math. Anal. Appl.*, **158**, pp. 106–113.

[301] Rassias, Th.M. and Šemrl, P. (1992). On the behaviour of mappings which do not satisfy Hyers–Ulam stability, *Proc. Amer. Math. Soc.*, **114**, pp. 989–993.

[302] Rassias, Th.M. and Tabor, J. (1992). What is left of Hyers–Ulam stability? *J. Nat. Geom.*, **1**, pp. 65–69.

[303] Rassias, Th.M. (1997). On a problem of S.M. Ulam and the asymptotic stability of the Cauchy functional equation with applications, in *General Inequalities*, Vol. 7, *International Series of Numerical Mathematics*, Vol. 123, Birkhäuser, Basel, pp. 297–309.

[304] Rassias, Th.M. (1998). On the stability of the quadratic functional equation and its applications, *Studia Univ. Babes-Bolyai*, **43**, pp. 89–124.

[305] Rassias, Th.M. (2000). The problem of S.M. Ulam for approximately multiplicative mappings, *J. Math. Anal. Appl.*, **246**, pp. 352–378.

[306] Rassias, Th.M. (2000). On the stability of functional equations in Banach spaces, *J. Math. Anal. Appl.*, **252**, pp. 264–284.

[307] Rassias, Th.M. (2000). On the stability of functional equations and a problem of Ulam, *Acta Appl. Math.*, **62**, pp. 23–130.

[308] Rassias, Th.M. (2003). *Functional Equations, Inequalities and Applications*, Kluwer Academic Publishers, Dordrecht.

[309] Rätz, J. (1980). On approximately additive mappings, in *General Inequalities*, Vol. 2, E.F. Beckenbach (Ed.), Birkhäuser, Basel, pp. 233–251.

[310] Rätz, J. (2003). On inequalities associated with the Jordan-von Neumann functional equation, *Aequationes Math.*, **66**, pp. 191–200.

[311] Rodabaugh, S.E. (1982). Fuzzy addition in the *L*-fuzzy real line, *Fuzzy Sets Syst.*, **8**, pp. 39–51.

[312] Rolewicz, S. (1984). *Metric Linear Spaces*, Reidel, Dordrecht.

[313] Sadeqi, I. and Salehi, M. (2007). Fuzzy real numbers and their relation to the topological vector space, in *Proceedings of First Joint Congress on Fuzzy and Intelligent Systems*, Ferdowsi University of Mashhad.

[314] Sadeqi, I. and Salehi, M. (2009). Fuzzy compact operators and topological degree theory, *Fuzzy Sets Syst.*, **160(9)**, pp. 1277–1285.

[315] Sadeqi, I. and Moradlou, F. and Salehi, M. (2013). On approximate Cauchy equation in Felbin's type fuzzy normed linear spaces, *Iran. J. Fuzzy Syst.*, **10(3)**, pp. 51–63.

[316] Samanta, T.K., Chandra Kayal, N. and Mondal, P. (2012). The stability of a general quadratic functional equation in fuzzy Banach spaces, *J. Hyperstructures*, **1(2)**, pp. 71–89.

[317] Savadkouhi, M.B., Gordji, M.E., Rassias, J.M. and Ghobadipour, N. (2009). Approximate ternary Jordan derivations on Banach ternary algebras, *J. Math. Phys.*, **50(4)**, p. 042303.

[318] Schwartz, L. (1966). *Théorie des Distributions*, Hermann, Paris.

[319] Šemrl, P. (1994). The functional equation of multiplicative derivation in super-stable on standard operator algebras, *Integral Equations Operator Theory*, **18**, pp. 118–122.

[320] Šemrl, P. (1994). The stability of approximately additive functions, in *Stability of Mappings of Hyers–Ulam Type*, Th.M. Rassias and J. Tabor (Eds.), Hadronic Press, Palm Harbor, pp. 135–140.

[321] Sibaha, M.A. and Bouikhalene, B. and Elquorachi, E. (2007). Ulam–Gavruta–Rassias stability for a linear functional equation, *Int. J. Math. Stat.*, **7(Fe07)**, pp. 157–168.

[322] Sikorska, J. (2009). On two conditional Pexider functional equations and their stabilities, *Nonlinear Anal.*, **70**, pp. 2673–2684.

[323] Skof, F. (1983). Sull'approssimazione delle applicazioni localmente δ-additive, *Atii Accad. Sci. Torino Cl. Sci. Fis. Mat. Natur.*, **117**, pp. 377–389.

[324] Skof, F. (1983). Proprietá locali e approssimazione di operatori, *Rend. Semin. Mat. Fis. Milano*, **53**, pp. 113–129.

[325] Skof, F. (1984). Approssimazione di funzioni δ-quadratiche su dominio restretto, *Atti Accad. Sci. Torino Cl. Sci. Fis. Mat. Natur.*, **118**, pp. 58–70.

[326] Stetkaer, H. (2002). d'Alemberts and Wilsons functional equations on step 2 nilpotent groups, Series No. 8, Aarhus University, Denmark.

[327] Székelyhidi, L. (1986). Note on Hyers' theorem, *C.R. Math. Rep. Acad. Sci. Canada*, **8**, pp. 127–129.

[328] Székelyhidi, L. (1990). The stability of sine and cosine functional equations, *Proc. Amer. Math. Soc.*, **110**, p. 109.

[329] Tabor, J. (1988). On functions behaving like additive functions, *Aequationes Math.*, **35**, pp. 164–185.

[330] Tabor, J. (1990). Quasi-additive functions, *Aequationes Math.*, **39**, pp. 179–197.

[331] Takhtajan, L. (1994). On foundation of the generalized Nambu mechanics, *Commun. Math. Phys.*, **160**, pp. 295–316.

[332] Towanlong, W. and Nakmahachalasint, P. (2008). A quadratic functional equation and its generalized Hyers–Ulam Rassias stability, *Thai J. Math.*, Special Issue (Annual Meeting in Mathematics), pp. 85–91.

[333] Towanlong, W. and Nakmahachalasint, P. (2010). A mixed-type quadratic and cubic functional equation and its stability, *Thai J. Math.*, **Special Issue(Annual Meeting in Mathematics)**, pp. 61–71.

[334] Trif, T. (2000). Hyers–Ulam–Rassias stability of a Jensen type functional equation, *J. Math. Anal. Appl.*, **250**, pp. 579–588.

[335] Trif, T. (2002). On the stability of a functional equation deriving from an inequality of Popoviciu for convex functions, *J. Math. Anal. Appl.*, **272**, pp. 604–616.

[336] Ulam, S.M. (1960). *Problems in Modern Mathematics*, Rend. Chap. VI, Wiley, New York.

[337] Ulam, S.M. (1960). *A Collection of Mathematical Problems*, Interscience, New York.

[338] Ulam, S.M. (1974). *Sets, Numbers and Universes*, MIT Press, Cambridge, MA.

[339] Venkatachala, B.J. (2002). *Functional Equations — A Problem Solving Approach*, First Edition, Prism Books Pvt Ltd.

[340] Wiwatwanich, A. and Nakmahachalasint, P. (2008). On the stability of a cubic functional equation, *Thai J. Math.*, **Special Issue(Annual Meeting in Mathematics)**, pp. 69–76.

[341] Xiao, J. and Zhu, X. (2002). On linearly topological structure and property of fuzzy normed linear space, *Fuzzy Sets Syst.*, **125**, pp. 153–161.

[342] Xiao, J. and Zhu, X. (2004). Topological degree theory and fixed point theorems in fuzzy normed space, *Fuzzy Sets Syst.*, **147**, pp. 437–452.

[343] Xu, T.Z. and Rassias, J.M. and M.J. Rassias and Xu, W.X. (2010). A fixed point approach to the stability of quintic and sextic functional equations in quasi-β-normed spaces, *J. Inequal Appl.*, **2010**, Article ID 423231, 23 pp.

[344] Xu, T.Z. and Rassias, J.M. (2012). On the Hyers–Ulam stability of a general mixed additive and cubic functional equation in n-Banach spaces, *Abs. Appl. Anal.*, **2012**, Art. ID 926390, 23 pp.

[345] Zadeh, L.A. (1965). Fuzzy sets, *Inform. Control*, **8**, pp. 338–353.

[346] Zhao, X., Yang, X. and Pang, C.-T. (2013). Solution and stability of a general mixed type cubic and quartic functional equation, *J. Funct. Spaces Appl.*, **2013**, Art.ID 673810, 8 pp.

[347] Zivari-Kazempour, A. (2014). On the generalized Hyers–Ulam stability for Euler–Lagrange type functional equation, *Int. J. Pure Appl. Math.*, **91(1)**, pp. 49–55.

[348] Zhou, D.X. (1992). On a conjecture of Z. Ditzian, *J. Approx. Theory*, **69**, pp. 167–172.

Index